American Institute of Chemical Engineers

9th Process Plant Safety Symposium

PPSS 2007

Held at the 2007 AIChE Spring National Meeting

April 22-27, 2007
Houston, Texas, USA

Printed from e-media with permission by:

Curran Associates, Inc.
57 Morehouse Lane
Red Hook, NY 12571
www.proceedings.com

ISBN: 978-1-60423-352-0

Some format issues inherent in the e-media version may also appear in this print version.

ISBN: 978-1-60423-352-0

American Institute of Chemical Engineers
Proceedings
Three Park Avenue
New York, NY 10016-5991
Phone: 212-591-8100

www.aiche.org

American Institute of Chemical Engineers

9th Process Plant Safety Symposium
2007

TABLE OF CONTENTS

Session 18 – Safety Culture – Key to Process Safety Performance

Session 24 – Security, Vulnerability Assessments and Mitigation

Session 42 – PSM and Risk Training in the 21st Century – Methods, Tools and Innovations

Session 44 – Safety Instrumented Systems – Identification, Design and Application

Session 64 – Assuring Safety in the Design and Construction of Process Systems

Session 83 – Poster Session on Risk and Process Design

Session 86 – Risk Assessment and Risk Management – New Directions

EVALUATING AND IMPROVING OPERATIONAL DISCIPLINE

James A. Klein
E. I. du Pont de Nemours and Company

Bruce K. Vaughen
DuPont Teijin Films

Prepared for Presentation at the
American Institute of Chemical Engineers
2007 Spring National Meeting
Process Plant Safety Symposium
Houston, TX
April 22 – 26, 2007

April, 2007

UNPUBLISHED

Abstract

DuPont defines operational discipline as a deeply rooted dedication and commitment by each member of the organization to carry out each task, the right way, each time. Operational discipline reflects the strength of an organization's safety culture in making process safety management systems effective and in providing tangible results for preventing injuries and incidents. Achieving high levels of operational discipline can be difficult, though, partly since methods for evaluating and improving performance have not been well characterized. This paper reviews efforts to evaluate operational discipline from both organizational and individual perspectives using self-assessments, audits, and incident investigations and to improve performance using a toolbox of best practices.

Introduction

Managing potentially hazardous materials and processes to prevent serious injuries and incidents is an essential part of safety management programs, requiring the continuous dedication and commitment of everyone in an organization. DuPont's Process Safety Management (PSM) programs have evolved since the company's beginnings [1, 2] and have developed and incorporated an organizational "Operational Discipline" (OD) program [3, 4]. Recent upgrades to our OD program include a next generation approach based on individual OD characteristics in addition to the organizational OD characteristics [5]. With this next generation approach and new emphasis on the individual, sites needed ways to evaluate OD from both organizational and individual perspectives and ways to improve their OD performance. Since we could not find significant guidance in the process safety literature, this paper presents approaches we're now using for evaluating and improving Operational Discipline. Recall, the goal of good Operational Discipline integrated within PSM programs is not only to prevent serious injuries and incidents, but also to contribute to excellent business performance through increased operating excellence, including higher productivity, higher quality, reduced waste, and lower costs.

Evaluating Operational Discipline

This section discusses several methods that are being used to evaluate the strengths of the different Operational Discipline characteristics at a site: OD-Related Metrics, OD Self-Assessments, OD-Related Audits, and Incident Investigations. Each of these methods provides a different "view" of the state of a site's OD health, with different degrees of time and effort to complete. The results of these evaluation methods should help identify the specific OD characteristics that may need the most improvement, helping the site prioritize and focus its resources on the weaker OD characteristics. Once a need is identified, the site can focus OD improvement efforts, using a recommended practice or "tool" from the OD Toolbox to improve the characteristic. The OD Toolbox is described in the next section. See Reference 5 for a more detailed discussion of the DuPont OD program and for definition of the OD characteristics.

OD-Related Metrics

Periodic reviews of Environmental, Health, and Safety (EHS) and operations metrics and activities provide another source of data on site OD performance. Examples include:

- No. of incidents (incidents with OD as a key factor - described below)
- No. of injuries (injuries with OD as a key factor)
- Second Party Audits (PSM OD scores – described below)
- Audits and PHAs completed on schedule
- Overdue findings (incidents, PHAs, audits)
- Overdue PSM Metrics (see example below)
- Overdue test and inspections, procedure revisions
- Environmental deviations
- Excess waste
- Poor product quality
- Customer complaints
- Higher costs
- Lower uptime (see example below)
- Staff turnover rates

Two examples from the OD-Related metrics noted above, Overdue PSM Metrics and Lower Uptime, are described in greater detail below.

A site's "Overdue PSM Metrics" chart could appear like that shown in Figure 1. Overdue metrics may indicate an organizational OD problem in terms of following up on action items from various PSM systems. In particular, a large number of overdue items suggests a problem with Leadership Focus. With the example shown in Figure 1, the recent improved performance may actually indicate improved Leadership Focus, but the number of overdue items is still high, indicating additional improvement is required.

Figure 1

Example OD-Related Metric – One Site's "Overdue PSM Metrics" Chart

	2003	2004	2005	2006
No. Overdue	38	29	32	21
% Overdue	22%	25%	19%	15%

An example of a vessel's utilization rate compared to its actual uptime rate is shown in Figure 2. The utilization rate, a measure of the total time the vessel is used, is consistently higher than the actual uptime rate, a measure of the total time the vessel is being used productively to make quality product versus an expected standard. The question suggested by this data is: "Why is the actual uptime rate so much lower than the utilization rate?" This difference could result from many problems, such as inventory management, scheduling, or equipment problems, but could also indicate fundamental OD problems that are affecting productivity. Which OD characteristic is not being met? Is leadership not providing the schedule in a consistent, disciplined manner? Is the line not running because of maintenance-related equipment downtime? Are the operators following standard operating conditions to ensure quality product? Although other factors may affect the metric results shown in Figure 2, some explanation of these results could be attributed to lack of Operational Discipline.

Figure 2

Example OD-Related Metric –

Comparison Between An Expected Utilization Rate vs. An Actual Uptime Rate

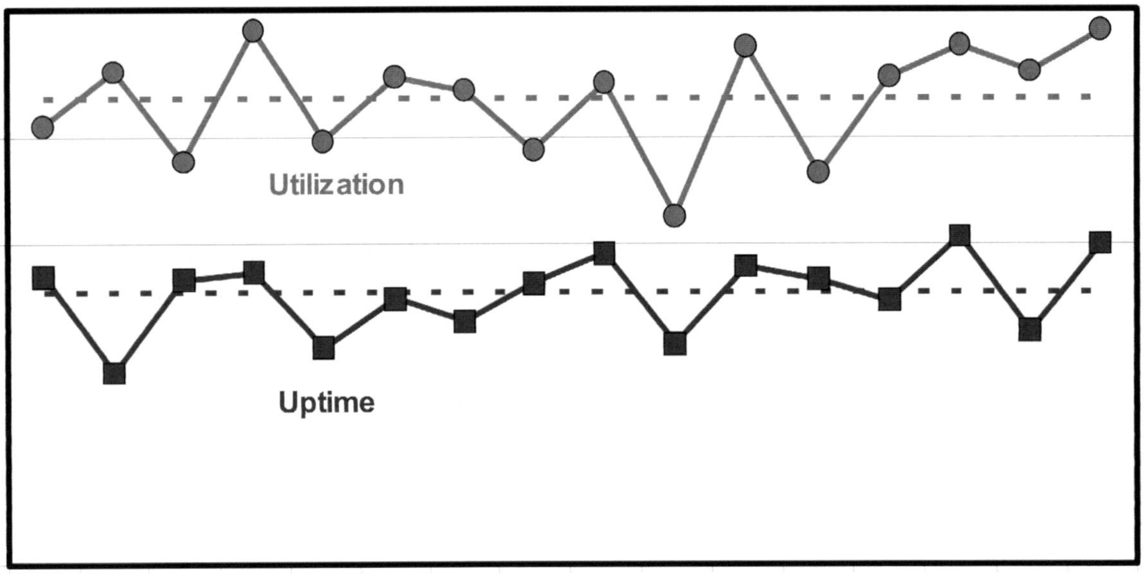

OD Self-Assessments

The Operational Discipline Self-Assessment survey is designed with specific questions for each characteristic and uses SurveyMonkey [6], an internet-based survey tool, to collect the responses. The survey respondents identify themselves from one of six areas:

1) operator, technician, or maintenance (mechanics, electricians, etc.)
2) safety, health, or environmental resource
3) management (including first line supervisors)
4) technical (chemists, engineers, etc.)
5) non-technical (human resources, clerical, site services, etc.)
6) other.

By separating the responses by functions, the responses can be compared between functions (e.g. operators vs. supervisors) to help analyze and interpret the data. The results are analyzed to identify the weaker OD characteristics, if any. Example questions from the Leadership Focus portion of this survey are shown in Figure 3, with the respondent scores ranging from 1 (low, OD is not addressed) to 5 (high, OD is excellent).

Figure 3

Example Questions From a Portion of the OD Self-Assessment Survey

1	**Leadership Focus (Organizational)**	
	Leaders demonstrate an unrelenting passion for SHE and model the behavior they expect from others.	
	1 = not addressed, 2 = significant gaps, 3 = multiple opportunities, 4 = minor opportunities, 5 = excellent	Score
1.1	Does site leadership continually and visibly emphasize SHE and reinforce its importance? Is this priority demonstrated in daily decision-making and by spending time in the field? Is there a "quick fix" or "safety first / do it right" mentality?	
1.2	Does site leadership clearly document and effectively communicate SHE goals and standards? Are specific improvement goals being set? Are relevant metrics and tracking systems used to monitor and communicate progress? Is sufficient training on SHE, <u>including operational discipline</u>, provided to ensure employees can work safely?	

After all the participants have entered their responses, the question's average and its standard deviation are determined, based on survey data as shown in Figure 4. Poor or non-existent operational discipline programs will move the question's average closer to 1; excellent and fully implemented operational discipline programs will move the question's average closer to 5. Good agreement on the response between the respondents will provide a low standard deviation; poor agreement (in other words, the response range is large) will result in a larger standard deviation. Sorting the averages and standard deviations provides a measure of how good or how poor operational discipline is at the site – noting that this is the goal of the self-assessment. Questions in the survey also provide opportunities for suggestions on how to improve OD. Use of the internet survey approach more easily allows feedback to be obtained from a larger portion of the site versus earlier self-assessment approaches, based on small group discussions. Typically, it takes about 30 minutes to complete the survey. An example on how the site uses the results from its self-assessment data analysis is provided in the Operational Discipline Toolbox description section below.

Figure 4

Typical Survey Results

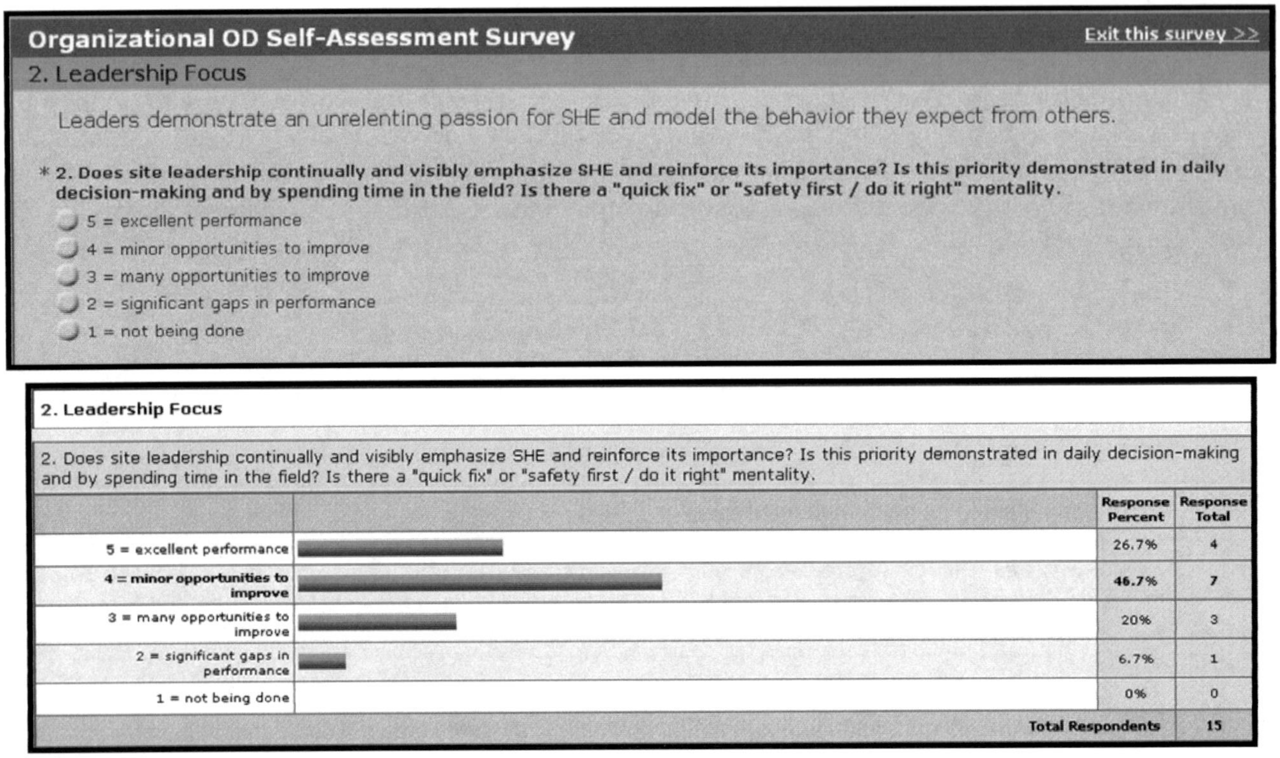

OD-Related Audits

Audits may be specifically directed at the different OD characteristics or they may be incorporated as a part of EHS, operations, or even financial audits. Depending on the type of audit, the time and resource commitment could vary from a couple of hours of two or three people's time to an entire week of the organization's time. Example questions and responses from a simple "Spot OD Audit" and from the corporate "Second Party PSM Audit" are provided below.

"Spot OD Audits" are designed to help evaluate OD performance using minimal resources: one or two people for only an hour or two. Spot audits can be repeated periodically to provide the site with an overall OD-auditing perspective over time. An example question on Excellent Housekeeping is shown in Figure 5. Leadership can develop actions based on the audit noted, such as in this example, locate and publicize Area #1's housekeeping policy, if it exists, or create and publicize a new policy.

Figure 5

Example of an "Excellent Housekeeping Spot Audit"

Housekeeping

Two areas: one routinely patrolled; another one out of the way

	Score	Area #1	Area #2
"0" items needing attention and an area policy on housekeeping	0		
1 - 2 items needing attention and an area policy on housekeeping	1		1
2 - 5 items needing attention and an area policy on housekeeping	2		
2 - 5 items needing attention and no policy on housekeeping	3	3	
Many items needing attention	4		
Average :			2

Good Practices : Nothing stored on tops of lockers (prior finding)
Items needing attention : Could not locate Area #1 policy

DuPont's "Second Party" Process Safety Management (PSM) Audit" protocol has audit teams travel to different sites and audit a site's PSM system. The questions from the PSM audit relating to OD characteristics are identified and are tallied, providing the site with an OD characteristics score. Some scores may show some weaker OD characteristics when compared to the others. As shown in example "results" in Figure 6, Site A may decide that it must focus its energy on its housekeeping efforts, Site B may decide that it simply needs to sustain the systems in place, and Site C may decide that major across-the-board efforts are required to improve OD. These audit results may help "confirm" operational discipline issues that were raised in the site's self-assessment survey noted above.

Figure 6

Example of Possible Operational Discipline Characteristic Scores from PSM Audits

		Site A	Site B	Site C
1	Leadership Focus	90%	95%	76%
2	Employee Involvement	88%	96%	82%
3	Practice Consistent with Procedure	91%	95%	80%
4	Excellent Housekeeping	74%	93%	86%
5	Knowledge	90%	94%	79%
6	Commitment	89%	95%	81%
7	Awareness	92%	97%	89%

Incident Investigations

Incident investigations are used to determine "key factors" or "root causes" to help ensure that incident learnings are captured and that recommendations are implemented to prevent recurrence. When these human, equipment, or system key factors are identified, "band-aid" solutions are avoided. Key factors often relate to human error or OD issues, either in not following procedures or in other poorly-operating systems. An example of possible data based on the percentage of incidents with OD as a key factor for one site's metrics is shown in Figure 7. Although this type of data may be influenced by several factors, this example shows that following improvement one year, the trend has steadily gotten worse, possibly indicating increased OD problems at the site. Identification of which OD characteristics primarily contributed to the data by incident investigation teams would help guide the site's actions when choosing its OD improvement efforts.

Figure 7

Example from Incident Investigations: "Incidents Noted With OD As A Key Factor"

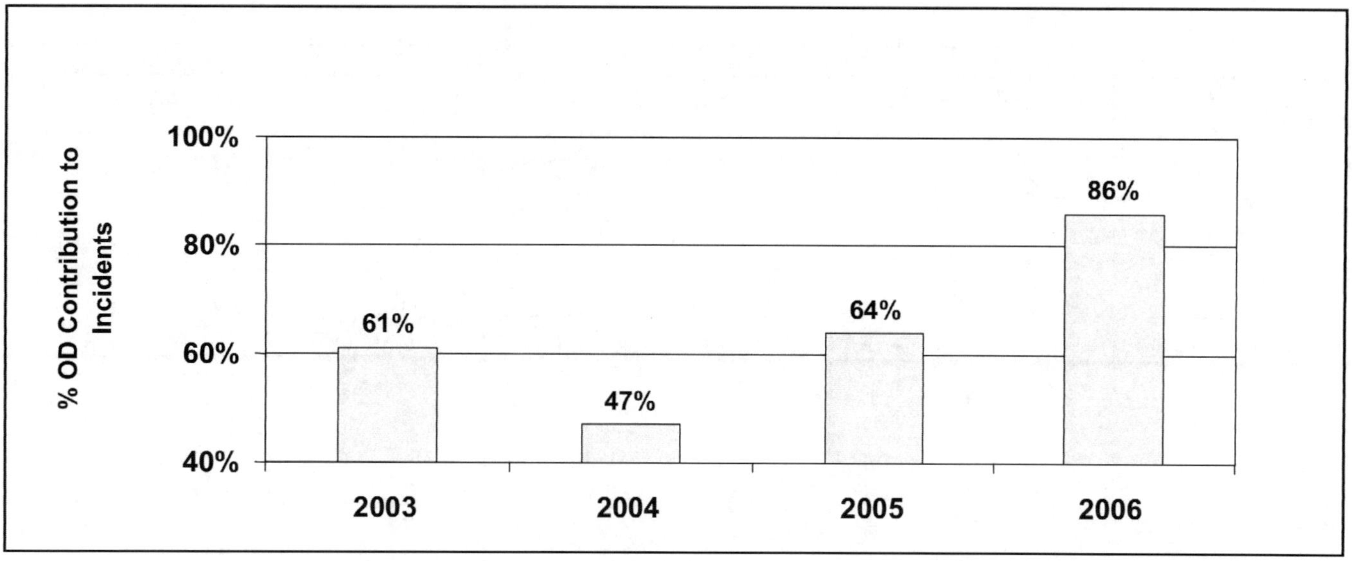

Evaluating Operational Discipline Summary

Each of these evaluation methods, the OD-Related Metrics, the Self-Assessments, the OD-Related Audits, or the Incident Investigations, will provide the site with a view of the weaknesses and the strengths of their OD performance. Using multiple evaluation methods may provide additional insights or trends on the OD characteristics. Where there are weaknesses, there are opportunities for improvement. Recall that OD reflects the strength of an organization's safety culture: OD helps make PSM systems effective and provides tangible results for preventing injuries and incidents.

Improving Operational Discipline

This section presents ways sites can improve weak OD characteristics using various recommended practices. The toolbox image is used: similar to a car mechanic who chooses different tools from his toolbox depending on the type of maintenance. The site chooses a different "OD-improvement tool" (recommended practice) from the "OD Toolbox" database depending on which characteristic the site has selected for improvement. Since most sites have differences, what works at one site may not work well at another. The toolbox approach helps to provide flexibility to sites in choosing how to improve OD. The OD Toolbox database is organized using the "roadmap" shown schematically in Figure 8 [5]. After SHE system data is analyzed to determine the current status of OD (e.g., using the OD evaluation methods), the site prioritizes which organizational and / or individual OD improvement efforts it will pursue. The site then chooses "tools" selected from the OD Toolbox database.

Figure 8

Operational Discipline Toolbox Roadmap

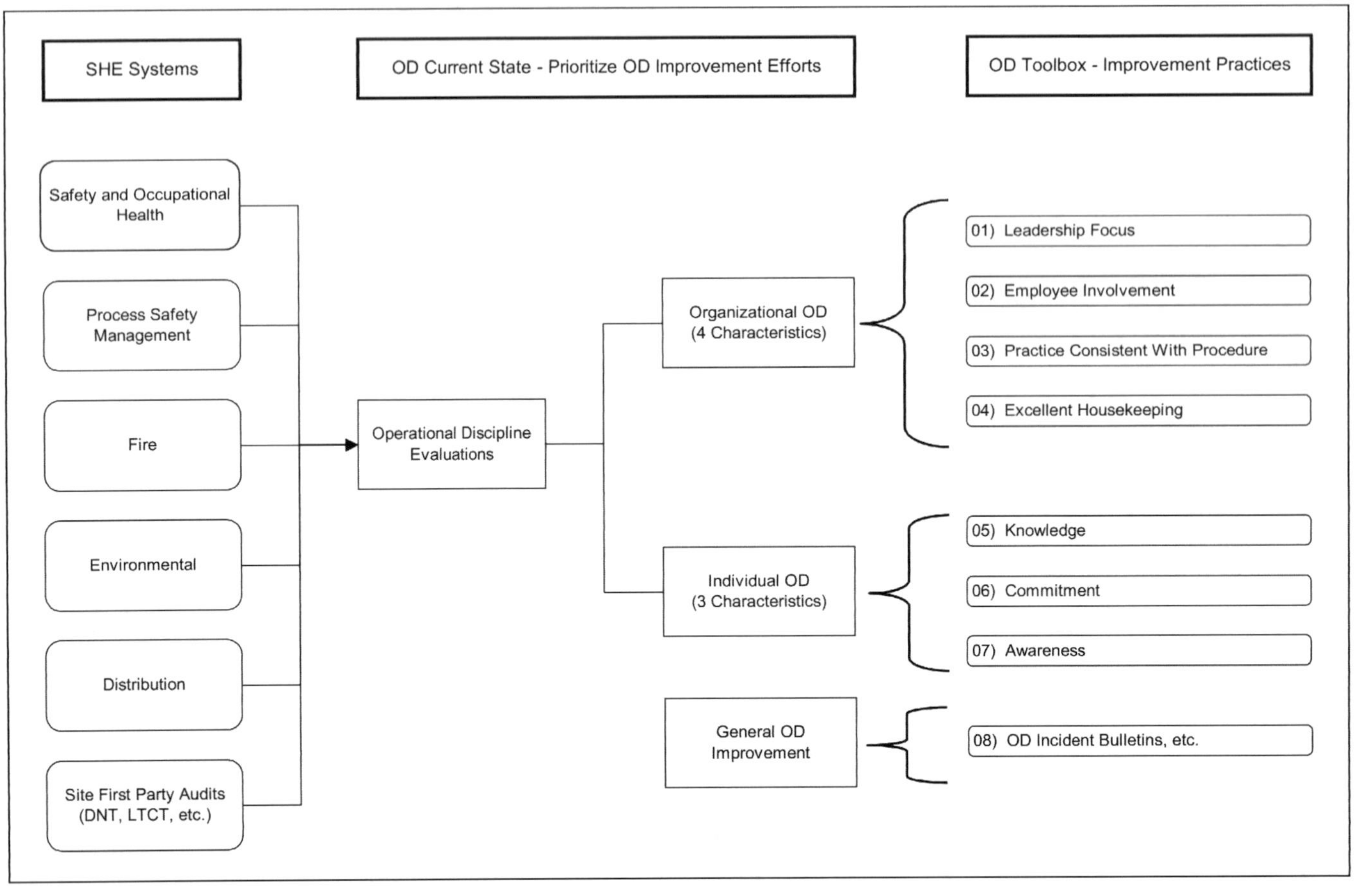

The site's OD improvement process is best described using an example that evaluates the results of the OD Self-Assessment survey example described above. The steps are:

1) the site analyzes the results of their survey
2) the site may combine the survey results with the results of other OD characteristic evaluation methods (metrics, audits, incident investigations, etc.)
3) the site selects and prioritizes the OD characteristics that need improvement
4) the site uses the "OD Toolbox" to locates recommended practices that can be best applied at their site.

Since some recommended practices may require more effort than others, the site must determine how many resources it has available to support the recommended practice, as well, before choosing between different approaches.

Figure 9 lists the OD Characteristics and shows example results from a site's OD Self-Assessment survey. Sorting these results by the average or the standard deviation highlights the characteristics that have the best (or worst) OD scores or the most agreement (disagreement). The results from the example below indicate that the Individual OD characteristics have the best OD (all three have the highest scores) and that there is general agreement between personnel who completed the survey (all three have the lowest standard deviation). Although the organizational characteristic Leadership Focus has the lowest average of 3.76, this is close to 4, the "minor OD improvements" level (not bad!). This characteristic, though, may still be targeted for improvement, depending on site needs.

Figure 9

Example OD Self-Assessment Survey Results – Listed by OD Characteristic

		Score	Std Dev
1	Leadership Focus (Organizational)	3.76	0.78
2	Employee Involvement (Organizational)	3.87	0.85
3	Practice Consistent With Procedures (Organizational)	4.22	0.85
4	Excellent Housekeeping (Organizational)	3.91	0.87
5	Knowledge (Individual)	4.44	0.61
6	Commitment (Individual)	4.53	0.59
7	Awareness (Individual)	4.33	0.69

Sorting the Self-Assessment survey questions by their averages (shown in Figure 10) and also by their standard deviation (if desired) reveals the specific questions that have "issues" worth additional discussion. In this example, of the five lowest scoring questions, four have "leadership" issues, two are directly from the Leadership Focus characteristic, and all are from the organization OD characteristics. These results specifically identify the "issue," and confirm the results in Figure 9, suggesting that the Leadership Focus OD Characteristic is a potential area for improvement. Do note that the scores are still relatively high, indicating only minor improvement is needed in most areas. These results should be compared to other metrics for confirmation, as well, since the perceptions of the survey participants may be different than the site's actual operating experience.

Figure 10

Example OD Self-Assessment Survey Results – Sorted by Individual Question

The Five Lowest Average Scoring Questions

		Score	Std Dev
1.5	Does site leadership periodically review staffing needs and turnover? Are site activities adequately staffed with qualified personnel? Is there an effective process for managing change in key operational, technical, and SHE roles? Are employee stress levels reasonable or strained?	3.33	0.62
4.3	Does site leadership recognize/reward excellent housekeeping?	3.47	0.92
1.7	Does site leadership have an effective reward/recognition program to acknowledge and celebrate individual and team accomplishments?	3.53	0.83
2.4	Is there an effective process for obtaining and evaluating feedback and suggestions from employees?	3.60	0.83
3.4	Does site leadership periodically schedule and/or require use of management of change systems to review and authorize equipment or procedure changes prior to use? Are site employees involved in the review and revision of site procedures? Is there a way for employees to suggest when new procedures are necessary or existing procedures should be modified?	3.80	1.21

Based on these results, the site in this example could choose to focus their resources on the Leadership Focus OD Characteristic. As is shown in the "OD Toolbox Roadmap" in Figure 8, the site then searches the database and selects the characteristic from the menu. This database is set up with a list of the available tools, a short description of what the tool addresses, and a contact. An example "screen shot" of tools from which the site can select for the Leadership Focus OD Characteristic in the database is shown in Figure 11. Similar Toolbox selections have been created for the other characteristics listed in the Roadmap shown in Figure 8. Depending on the available resources and the specific OD characteristic improvement items that the site needs to address, it is intended that this database will contain an appropriate tool or tools to support the site's OD improvement efforts.

Figure 11

Example Selection Options from the Operational Discipline Toolbox Database

OD Characteristic: Leadership Focus

Reference Number (Click on link)	Tool	Description	Contact
01.01	Bradley Model Assessment	Improve leadership "behavioral safety" using the Bradley Model: preventing injuries by progressing organizations from dependent through independent to interdependent behavior - individuals to team	John Q. Bradley
01.02	Personal Safety Awareness Control	Improve safety with personal control of Attitude, Awareness, and Actions (AAA); Help for leaders in leading changes in "Personal Influence" drivers ("Outside ---> In" to "Inside ---> Out")	John Q. Aware
01.03	etc.		

Improving Operational Discipline Summary

The OD Toolbox database is evergreen, where more recommended practices are added as more sites across the corporation share their best practices. As shown in the example, the results from the OD-Self Assessment guide the site to its lowest scoring characteristics and provide the site with direction in selecting its improvement tools from the OD Toolbox.

This paper shows how OD-Related Metrics, OD Self-Assessments, OD-Related Audits, and Incident Investigations can be used to evaluate OD, from both the organizational and the individual perspectives. This paper also shows how OD performance can be improved by prioritizing the evaluation results and then selecting an appropriate OD improvement practice from an OD Toolbox of recommended practices. At the time of this paper, several pilot "Next Generation Operational Discipline Leadership" training workshops have been held, which have helped tailor this approach to taking operational discipline to a new level. It is believed that the goal of good Operational Discipline integrated within PSM programs will not only prevent serious injuries and incidents, but will also contribute to excellent business performance through increased operating excellence, including higher productivity, higher quality, reduced waste, and lower costs.

Literature Cited

1 Burk, Arthur F., *Achieving Excellence in Process Safety & Risk Management*, Managing Safety & Health in the Workplace, Chicago, IL, June 13-14, 1994

2 Klein, James A., C. Curtis Clements, and David E. Cummings, *Two Hundred Years of PSM at DuPont*, Risk Management: The Path Forward, AIChE, 2005 (20th Annual CCPS International Conference, April 11-13, 2005)

3 DuPont, *Achieving Operating Excellence Through Operational Discipline*, seminar manual, 1997

4 Klein, James A., *Operational Discipline in the Workplace*, Process Safety Progress, Vol. 24, No. 4, p. 228-35, December, 2005

5 Klein, James A., Bruce K. Vaughen, *The Next Generation Approach to Operational Discipline at DuPont*, Advancing Process Safety Through Design and Operations, AIChE, 2007 (22nd Annual CCPS International Conference, April 22-26, 2007)

6 http://www.surveymonkey.com

Safety Climate from the Safety Experts: In their Own Words

Stephanie C. Payne, Mindy E. Bergman, Jaime B. Henning, and Carolyn J. Stufft
Department of Psychology, Texas A&M University

Prepared for Presentation at
American Institute of Chemical Engineers
2007 Spring National Meeting
Process Plant Safety Symposium
Houston, TX
April 22-27, 2007

Abstract

Researchers define safety climate as shared employee perceptions of the policies, practices, and procedures concerning safety (Zohar, 2000) and have demonstrated how these perceptions relate to various safety-related outcomes (e.g., Barling, Loughlin, & Kelloway, 2002; Clarke, 2006; Hofmann & Stetzer 1996). In an effort to further understand the meaning of safety climate, we surveyed 62 safety experts about their experiences and asked them to respond to a series of questions in their own words. Some of the questions included: (a) How do you know when you are in a safe or unsafe work environment?; (b) What features or characteristics of a workplace (general observations about the workplace) indicate it is a safe or unsafe place to work?; (c) What do managers do to ensure a good safety climate is maintained (does not go away)?; (d) What do managers do that jeopardizes the safety climate?; (e) In what ways do contractors have a positive and/or negative influence on the safety climate of an organization?; and, (f) In a merger, acquisition, or divestiture, what are some effective and ineffective ways that you have observed organizations communicate their safety climate to the employees? We also asked them to share critical incidents involving how a company effectively or ineffectively dealt with a near miss as well managed safety climate across different locations, businesses, and/or nations. We summarize these responses and highlight consistencies and inconsistencies across respondents as well as in comparison to the research literature.

Safety Climate from the Safety Experts: In their Own Words

Safety climate refers to shared employee perceptions of the policies, practices, and procedures concerning safety. Consistent with this definition, Zohar (2000) offered a more parsimonious definition of safety climate as *the relative priority of safety in an organization*. Organizations often have competing operational goals, and one goal that often competes with safety is productivity. Thus, the focal priority of interest in safety climate is the relative priority given to safety versus productivity.

A greater understanding of what safety climate involves is necessary, given that research has shown safety climate relates to various safety-related outcomes, such as safety-related behavior and compliance, injuries, and accidents (e.g., Barling, Loughlin, & Kelloway, 2002; Clarke, 2006; Hofmann & Stetzer 1996). While a variety of research studies have focused on safety-related outcomes, less research has examined the development and maintenance of safety climate over time, as well as across locations, businesses, and/or nations. This can be further complicated by organizational structure and events, such as the presence of contractors and organizational mergers, acquisitions, and/or divestitures. The purpose of this study is to gather information about these topics from individuals who deal with safety issues on a daily basis.

Organizational culture and climate researchers have described the many components of organizations that contribute to its overall culture and its component climates. Rousseau (1990) describes several layers of culture that vary in their direct observability and the extent to which they are surface examples vs. deeply held and ingrained cognitions. Artifacts include material manifestations of the organization, including employee handbooks, posted signs, logos, and mottos. Structures are patterns of behavior that can be observed by organizational outsiders and serve to address basic organizational activities, such as communication. Behavioral norms are the beliefs that employees have about acceptable behavior; O'Reilly and Chatman (1996) have suggested that behavioral norms guide individuals to continue to act in concert with organizational goals when no one is watching. Values are the explicit beliefs about what is important to the organization and its members, because of their membership. Finally, at the core of this model are the fundamental assumptions, the unspoken (even unspeakable), unconscious beliefs upon which all other components rest.

Notably, although fundamental assumptions are what are at the core of safety climate, these are unspoken, unspeakable, and unconscious beliefs that are not easily accessible even to the persons that hold them; these assumptions are so fundamental to our views of the organization that they essentially are the view of the world that we operate from, and therefore we have difficulty even knowing that they are there, because they are always there. They often go unrecognized unless they are challenged, such as when an organization that believes it has high safety standards experiences a fatal accident.

Therefore, to change and improve safety climate, we must focus on the other levels—artifacts, structures, norms, and values—that can be observed and even quantified. Each of these informs employees about the organization's safety climate. For example, employees may evaluate the priority given to safety based on the statements and actions of supervisors and managers (Neal & Griffin, 2004; Zohar, 2003). This can include actions which promote safe behavior and punish unsafe behavior, as well as actions viewed as being consistent with the organizational safety policies, practices, and procedures in place. However, employees might also attend to the organization's physical environment to understand the organization's safety climate. Further, the artifacts, structures, behavioral norms, and values of an organization can stand in stark contrast to the mottos of safety climate improvement programs. That is, although organizations might talk the talk, artifacts, structures, norms, and values walk the walk.

The specific interest here is that organizational climates are not static; they adapt and change over time as individuals act and interact in organizations (Schneider, 1987). Yet artifacts, structures, norms, values, and assumptions change slowly. Programs persist even after their progenitors have left for other organizations. Individuals are hired into jobs and they behave much in the same way as their predecessors by fulfilling the same tasks and duties. Thus, we are interested in what artifacts, structures, norms, and values contribute to the maintenance of good safety climate.

With this in mind, we sought the following information from safety experts:

(1) features or characteristics of safe/unsafe work environments
(2) managerial behaviors that maintain/jeopardize safety climate
(3) effective/ineffective ways to deal with a near miss
(4) contractor influence on safety climate
(5) managing safety climate across locations, businesses, and/or nations
(6) communicating safety climate during a merger, acquisition, or divestiture

Method

Sample

Attendees of a process safety symposium were recruited to participate in a survey study on safety climate. Sixty-two usable surveys were returned from 47 males and 12 females (3 did not provide this information). The majority of participants (74%) were Caucasian, and the average age was 49 years. Thirty-nine percent had an advanced degree and an additional 57% percent had a college degree. Most of the respondents (74%) were employed in the chemical process industry, 14.5% were consultants, 5% were academics, 3% worked for the government, and 3% did not indicate their professional affiliation. On average, the respondents had been working for their current employer for eight years, had 22 years of experience in industry, and had eight years of experience in consulting.

Procedure

Surveys were distributed to registrants after they checked in at the registration table at the conference. In addition to the demographic questions, the surveys contained six open-ended safety-related questions. For each safety question, we asked respondents to indicate positive examples and negative examples. Upon completion of the survey, respondents had the opportunity to include their name in a raffle drawing for free registration to the next annual symposium.

Responses to each question were reviewed and organized into general themes. Important themes were identified by the extent to which concepts were suggested by several respondents. The repetition of answers from multiple sources with different work experiences within a domain of knowledge (i.e., process safety and chemical engineering) suggests that these themes are important across the many organizations, stakeholders, and industry positions in the chemical processing industry.

Results

The following results are organized by the questions asked in the safety survey, followed by the themes that our analysis uncovered.

(1) How do you know when you are in a safe or unsafe work environment? What features or characteristics of a workplace indicate it is a safe or unsafe place to work? (Note: We are not necessarily looking for technical answers, such as the kinds of chemicals being processed, but more general observations about workplaces.)"

This question was designed to tap into the artifacts that are present in organizational sites. Responses were classified into six categories or themes. The first theme centered on *cleanliness, lighting, and noise level in the environment*. Safe environments were depicted as having good housekeeping as evidenced by clean and organized spaces. In contrast, unsafe environments were depicted as dirty, untidy, and cluttered. Further, safe environments were described as well-lit, whereas unsafe environments were described as having poor lighting and a noisy control room with alarms going off.

The second theme was oriented around *maintenance and age of the equipment*. Safe environments were described as including well-maintained and modern equipment, as well as no evidence of leaks or spills. Unsafe environments included poorly maintained and outdated equipment, rusty/corroded pipes, evidence of spills/leaks and erosion, and parts lying all over the place. A particularly illustrative comment summarized this category by noting that safe environments involve "investment in infrastructure that does not make money."

The third theme focused on *restricted access and safety-related signs*. Safe environments have barriers or restrictions to prevent entry or movement without authorized access, well-labeled chemicals and tanks, and safety/hazard signs or posters, as well as records and awards, whereas unsafe environments either do not have safety precautions or signs or they have old ones.

Additionally, the respondents focused on having *personal protective equipment* (PPE) readily available with instructions on how to use it. Safe environments require visitors to wear PPE, and personnel always wear it. On the contrary, in unsafe environments PPE is in poor condition, and there is "spotty" or no usage by personnel.

Management presence was the fifth theme. Supervisors make periodic visits and checks and maintain open communication with their personnel in safe environments. Unsafe environments have no presence of management, or there are "labor/management issues" and inconsistent practices.

Finally, the sixth theme described the *personnel* or workers. Safe environments were described as having professional, well-trained, and calm personnel who always follow procedures. They were also described as having positive attitudes, a sense of pride and ownership in their work, and being "empowered to do what is right." Personnel in unsafe environments seem preoccupied, lack an awareness of what is going on, and are often rushing around. With regard to attitudes, these personnel are complacent, have "not my job attitudes" and act as though "safety is done because 'I have to' not because 'I want to.'"

(2) What do managers do to ensure a good safety climate/culture is maintained (does not go away)? Also, what do managers do that jeopardizes the safety climate?

This question was designed to understand the structure that managers promote regarding safety in the organization and how this affects the behavioral norms and values that employees perceive. We cued the respondents to consider how managers maintain safety climate and/or jeopardize it. We believed that this was important because the absence of a climate maintenance behavior may not be the same as the presence of a climate jeopardizing behavior; that is, being "passively not good" might not be the same as being "actively bad." Six themes were also produced in the responses to the second question. The first theme can be summarized as *prioritizing safety over productivity*, consistent with the parsimonious definition of safety climate (Zohar, 2000). In safe environments, managers talk about safety before profit, whereas in unsafe environments production is stressed over safety, and managers tend to focus too much on production numbers.

The second theme oriented on *being proactive about safety rather than reactive*. Illustrative comments came more from the list of *jeopardizing* behaviors to include "knee jerk reactions after an incident; new procedures within 24 hours that cause more problems than they solve, " and "having a no blame policy; it is never their [management's] fault. They always pass responsibility to someone else."

The third theme involved *promoting two-way communication with employees about safety*. Maintenance behaviors include being vocal about safety, such as starting every meeting with a safety update and finding new ways to present safety updates daily, as well as fostering incident and near miss reporting. Jeopardizing behaviors were described as a lack of communication between management and employees to the extreme of not listening, colloquially described as having a "just do it mantra," "brain not engaged, body only working from shoulders down." Other related behaviors include a general lack of employee involvement in which "management works 'in a box' without input from personnel."

The fourth theme in the responses to the second question was *financial investment in safety* and came primarily out of the *maintenance* behavior answers. These included investing money and personnel in safety and ensuring that personnel have adequate training. Responses in this theme were generally about the extent to which management proactively spends money to ensure that employees are prepared cognitively, behaviorally, and materially to work in safe ways.

The fifth theme also came primarily out of the *maintenance* behavior answers and can be summarized as *modeling good safety behaviors*. In order to maintain a good safety climate, managers need to "walk the talk" and be a good example for their employees (e.g., wear PPE). This is further evidence that managers are powerful role models in the organization.

The final theme consists of *maintaining a presence and exercising authority*. Maintenance behaviors included management by walking around (MBWA), rewarding employees for safe behaviors by including safety in annual evaluations and raises, and consistently holding employees accountable for their actions. However, a disclaimer was expressed about using discipline: "disciplinary treatment of willful safety violations but not honest mistakes." Jeopardizing behaviors, on the other hand, include no MBWA, no consequences for unsafe behaviors, and inconsistencies in policies and practices (rewarding and punishing).

An overarching theme across these responses could be summarized as behavioral manifestations of a *commitment to safety*. In other words, employees assume a commitment to safety based on the way that managers act, and these are the behaviors that convey that commitment. Additionally, employees know, based on managers' actions, when managers are not committed or do not care about safety. Note that throughout this section there were distinct differences in the answer to the "maintenance" and "jeopardizing" questions. Although it is certainly the case that if managers do not actively try to maintain a good safety climate, they will not improve their safety climate and may even be likely to lose it. Thus eventually, a good safety climate can be actively destroyed.

(3) In your experiences, what are some good and not so good examples of how a company handles a near miss? This might include communications about the near miss within the company, how the incident and its seriousness are described in reports, the treatment of the worker(s) involved, etc.

This question also examined the structures of the organization as they related to safety climate. This question focuses on the broader organizational processes of investigation and communication, rather than the day-to-day supervision provided by managers. The answers to the third question also seemed to have an overarching theme; one for the good handling and one for the poor handing of near misses. Good handling of a near miss involves treating it as a serious event, or as one respondent suggested: "*treat it as a near hit* rather than a near miss." This involves *identifying* the near miss (which assumes employees have had the training needed to help them to be able to detect it), (2) *reporting and/or recording* the near miss (maintaining statistics and trending these numbers), (3) *fully investigating* it as if it were an incident in order to find the root cause, which means that managers must "support the time and effort to allow adequate investigation," (4) *communicate* to all employees at all locations about the near miss and treat it as a learning opportunity, and finally, (5) *stroke/reward* the employee who reported the near miss and make it easy to do. Some extremes of the latter two behaviors include "commemorate the event so that it stays fresh for a long time," "have a 'near miss of the month' award," and tie "performance stats into how well near misses are reported and evaluated."

Poor handling of near misses involves not doing many of the behaviors listed above and attributing "the event to bad luck," or thinking "it was 'good luck' that something worse didn't happen." Additional examples of poor handling of near misses included punishing or blaming the people involved and not providing "support to people that have been physically or psychologically affected." Beyond not reporting and not conducting an investigation, poor handling also involves no or limited communication (e.g., only to the affected unit) and a general "no harm, no foul" reaction. Whereas reporting and recording appear to be good ways to treat near misses, the focus should not be on the number reported, and requiring employees to report a specified number does not necessarily yield helpful information. Instead, the focus must be on developing an understanding of what should be reported, when, where, and why.

(4) In what ways do contractors have a positive and/or negative influence on the safety climate/culture of an organization?

Although contractors are not a unique feature of chemical processing organizations, the presence of contractors for building, operating, and maintaining equipment and sites appears to be somewhat commonplace in the chemical processing industry. This is an organizational structure factor that might be of importance because contractors are individuals who do not have the same relationship with the organization as core employees. Contractors may work on-site for a short term. Even if they work long term, they may not be as well integrated into the organization as are core employees. Additionally, they have a dual employment role (i.e., their hiring or "home" firm and their work site) that could have conflicting safety climates.

One respondent provided a strong response to this question: "I have never seen a contractor have a positive effect on safety in 30 years!" But most respondents were able to articulate some positive features that contractors contribute. Responses to this question fell into two primary categories: (1) serving as an additional resource, and (2) adopting the company safety climate. Contractors contribute positively to the safety climate when they *serve as an additional physical and cognitive resource*. They "expand the number of hands/eyes to find problems;" "add workers for seasonal or shutdown work;" and can bring in new safety ideas, higher standards, and unique experiences/expertise. Contractors also have a positive influence on the safety climate when they *adopt the company's safety culture,* abide by the company's safety practices, act as good examples, and "take a partnership role in site safety; [don't] hesitate to point out safety shortcomings of site and [recommend] measures for improvement."

Negative experiences with contractors involved primarily the opposite behaviors, through which they actually hinder safety (e.g., hiding injuries or near misses). Further, when contractors do not adopt the company's safety culture and do not abide by the company's safety practices, they jeopardize the safety climate. Some respondents identified some broader reasons for contractors to have a negative impact on the safety climate, including "contractor management encourages driving near miss reporting underground," and "contractors usually have incentives to take shortcuts." They further point out that language barriers prohibit effective communication with contractors, because oftentimes contractors are less-educated, not trained about safety, and come from an organization that may have lower safety standards. Some contractors were also described as having a "'get it done' mentality," which emphasizes productivity over safety.

(5) What are some good and not so good examples of effectively managing safety climate/culture across different locations? different businesses? different nations?

This question was designed to tap into all four levels of climate: artifacts, structure, norms, and values. Transporting safety culture to various sites might require very different artifacts (for example, different kinds of signs), structures (for example, the ways that rules are communicated), and norms (for example, different safety behavior in sites that have different inherent processing risks) in some instances, but similar values in all instances. Responses to this question fell into four categories: (1) core company safety culture, (2) customized climate for each location, (3) management presence and monitoring, and (4) communication.

First, the company must maintain a *core company safety culture* in which safety procedures are consistently communicated and applied across locations. "Insist on a common company safety culture and communicate it globally." Make sure everyone understands it: "Orientations, signs, etc. [are] painted and communicated in multiple languages for all to understand." One respondent summarized this succinctly: "My company applies safety rules across all boundaries…We all speak the same safety language." This theme reflects the importance of having a core set of values that are communicated and shared across the many organizational sites.

Second, allow each location to *develop a customized climate* that maintains the core safety culture (e.g., "allowing an individual location to develop and implement additional/customized risk management safeguard measures"). In general this recognizes that off-the-shelf programs or one-size-fits-all approaches are not generally effective. It further recognizes cultural differences in the use and effectiveness of discipline. This is an example of how values may be the same across sites, but that artifacts, structures, and norms may differ depending on the specific sites' demands and needs.

Third, *management maintains a presence and does periodic monitoring*. Consistent with the whole premise of a regional manager, have regional managers ensure that policies are endorsed across locations. Even further, one respondent suggested that companies "rotate key people among sites to permeate safety culture through operations." Thus, the organization hierarchy must be carefully designed to ensure that proper oversight of the safety climate is possible.

Finally, once again, *communication* is a central theme to these answers. In particular, sharing across locations is advocated (e.g., "global best practice safety sharing conferences [internal to the company])." In general, knowledge needs to be transferred, and lessons learned (e.g., from near miss incidents) need to be shared. Finally, one respondent wisely pointed out that a global safety culture is not fostered when managers "encourage or allow competitiveness in performance numbers between locations in the same company." Such competition emphasizes productivity (e.g., performance numbers) over safety.

(6) In a merger, acquisition, or divestiture, what are some effective and ineffective ways that you have observed organizations communicate their safety climate/culture to the employees?

Organizational events such as merger, acquisitions, and divestitures are equivalent to major changes in the personnel in an organization. In some cases, many key personnel are not radically changed—top management may be jettisoned from an acquired organization, or high talent may be retained from a divested subunit—but in all, the people of the organization change dramatically in such an event. The particular individuals, as well as the specific event of mergers/acquisitions/divestitures, may speak volumes about organizational values in general and safety in particular. This question drives to the heart of the meaning of climate: it is made by the people in the place. If every person were replaced in an organization in a single moment, the climate would change radically as well.

Respondents' answers to the sixth question could be classified into two primary categories: (1) audit/study of the two cultures and (2) communication of the new culture. Respondents had specific recommendations for both of these actions. With regard to the *audit/study of the two cultures*, one respondent recommended "initially have an audit/survey/study to evaluate the level of safety program and culture" on both sides and "establish pre-merger team to develop uniform and same policies/procedures." On the negative side, beyond not doing an evaluation/audit of safety practices, respondents indicated not identifying procedures that worked well in the first place and adopting "we know best since we bought you" attitudes, as well as "running down or criticizing the safety climate/culture of the former employer" as ineffective merger management behaviors.

With regard to *communicating the new culture*, respondents listed: "management sets clear expectations that the intention of the merger is to take the best of both," "once uniform policies are established, insist that training be done on them and implemented without exception," create a "video of high level executives discussing specific safety topics," "empower new management to promote the safety culture," and "get buy-in from employees on safety program." On the negative side, respondents warned against assuming the new organization will have high standards and that "written policy alone will establish safety culture." They further suggested the need to discuss the new mission/vision with the employees, explaining and getting buy-in from the employees on the safety rules and policies rather than forcing the rules on the employees, and the need to "silence the rumors." Finally, respondents pointed out that change takes time and that it is ineffective to "expect wholesale changes immediately."

Discussion

Research suggests that safety climate is related to several safety-related outcomes. Therefore, a greater comprehension of safety climate may be useful in the prediction and understanding of safety-related outcomes. The purpose of this study was to gather information from safety experts on the following key topics:

(1) features or characteristics of safe/unsafe work environments
(2) managerial behaviors that maintain/jeopardize safety climate
(3) effective/ineffective ways to deal with a near miss
(4) contractor influence on safety climate
(5) managing safety climate across locations, business, and/or nations
(6) communicating safety climate during a merger, acquisition, or divestiture

Consistent with the literature on safety climate, management commitment to safety is a critical underlying factor that impacts all important and effective safety-related behaviors. A number of characteristics were identified and associated with safe and unsafe work environments. Communication, both verbal and nonverbal, is key to creating and maintaining safety climate in organizations. Safe work environments are communicated through safety signs and barriers to restricted areas. Two-way communication encourages on open dialogue about safety between managers and employees, in which both sides are able to discuss safety issues and share their knowledge and questions regarding safety within their organization. When handling near miss situations, supervisors must communicate with employees regarding the near miss incident, what can be learned from it, and ways to prevent future incidents. Contractors can positively influence safety through communication of new safety ideas and standards and by sharing their particular expertise. Safety climate is effectively managed across locations by communicating knowledge gained and lessons learned from near misses and by sharing safety policies through communication across sites. Finally, communication is central to introducing employees from mergers and acquisitions into an organization's safety climate/culture. Communication of the new climate is vital to helping new employees fit into the organization, just as communication of safety with veteran employees is vital to sustaining the safety climate of an organization. Effective communication regarding safety issues can foster effective organizational safety climates, while ineffective communication or a lack of communication about safety can lead to unsafe work environments.

We also saw evidence that safety climate permeates through the various levels of culture. Importantly, safety climate can be observed through the artifacts and structures that even outsiders can observe, making it possible for organizational outsiders such as human resource consultants to be able to aid organizations in their safety climate interventions. Additionally, safety climate has a number of behavioral norms associated with it, as well as deep-rooted values.

Interestingly, our work speaks to the importance of solidifying values and yet tailoring artifacts, structures, and norms to specific site needs. Different sites vary in inherent risk, education and source of its workforce, operational tempo and requirements, and national values and culture, among other factors. Our experts suggest that tailoring the climate to the specific worksite is possible and even necessary to maintaining an overall strong safety record. Yet it is also clear that this must be done so that the overall organizational values are not compromised.

However, it is important to note that there can be disconnects across all level of an organization's culture. Merely changing the artifacts and structures of an organization may not lead to the kinds of changes that the organization truly needs to improve safety—either across or within sites. Hanging signs and walking around will not ensure that individuals do what they need to do when no one else is watching. These kinds of changes also do not make people believe that safety is important to the organization and therefore it is something that they should strive to achieve. Truly changing a safety climate requires changes at all levels of the culture. Especially good changes will alter the fundamental assumptions that individuals hold, as the artifacts, structures, norms, and values trickle down to alter the worldview that organizational members have.

References

Barling, J., Loughlin, C., & Kelloway, E. K. (2002). Development and test of a model linking safety-specific transformational leadership and occupational safety. *Journal of Applied Psychology, 87,* 488-496.

Clarke, S. (2006). The relationship between safety climate and safety performance: A meta-analytic review. *Journal of Occupational Health Psychology, 11,* 315-327.

Hofmann, D. A., & Stetzer, A. (1996). A cross-level investigation of factors influencing unsafe behaviors and accidents. *Personnel Psychology, 49,* 307-339.

Neal, A., & Griffin, M. A. (2004). Safety climate and safety at work. In J. Barling & M. R. Frone (Eds.), *The psychology of work place safety* (pp. 15-34). Washington, DC: American Psychological Association.

O'Reilly, C. A., III, & Chatman, J. A. (1996). Culture as social control: Corporations, cults, and commitment. *Research in Organizational Behavior, 18,* 157-200.

Rousseau, D. M. (1990). Normative beliefs in fund-raising organizations: Linking culture to organizational performance and individual responses. *Group & Organization Studies, 15,* 448-460.

Schneider, B. (1987). The people make the place. *Personnel Psychology, 40, 437-453.*

Zohar, D. (2000). A group-level model of safety climate: Testing the effect of group climate on microaccidents in manufacturing jobs. *Journal of Applied Psychology, 85*, 587-596.

Zohar, D. (2003). Safety climate: Conceptual and measurement issues. *Handbook of Occupational Health Psychology*, 123-142.

PROCESS SAFETY CONCEPT IN INDIAN CHEMICAL INDUSTRIES: EVOLUTION AND SOCIO-ECONOMIC-LEGAL CHANGES IN THE AFTERMATH OF THE BHOPAL DISASTER

Dr. R. K. Elangovan
Inspector of Factories, Government of Tamilnadu, Arkonam, Tamilnadu, India
E mail : elanrk@yahoo.co.in ; elanrk @ hotmail.com

Abstract

After Independence, the Indian government planned for green revolution and this brought in lot of chemicals in the form of fertilizers, pesticides etc. and for producing them world major chemical Industries were putting up their plants in India to meet the need of the agricultural activity. The expansion of chemical industries worldwide had also brought complex chemical plants to India either due to less stringent environmental and safety regulations or the need to operate them in developing countries, to ensure proximity to the consumer.

The process safety concept in India was at embryo stage, when these chemical multinationals were establishing factories in India. Neither the government nor the society knew anything about process safety except the details provided by the companies themselves to them, even though at a miniscule level. The Socio – Economic - Legal system was totally not aware of the concept of process safety and its effects on society and environment, until the Bhopal gas leak that struck in 1984 at Bhopal, India. Later on, many changes have been brought in even though a lot more needs to be done on this.

This paper discusses on evolution of process safety concept in India and the Socio-Economic-Legal changes in the aftermath of Bhopal gas leak disaster as well as the required changes as of now, to derive the advantages of process safety concept to ensure the safety of human beings and protection of the environment.

Key words

Process safety, Socio - Economic - Legal System and Chemical Industries.

Introduction

After Independence, the Indian government planned for green revolution and this brought in lot of chemicals in the form of fertilizers, pesticides etc. and for producing them world major chemical Industries were putting up their plants in India to meet the need of the agricultural activity. The expansion of chemical industries worldwide had also brought complex chemical plants to India due to the need to operate them in developing countries, to ensure proximity to the consumer. The process safety concept in India was at embryo stage, when these chemical multinationals were establishing factories in India. The Socio – Economic - Legal system was totally not aware of the concept of process safety and its effects on society and environment, until the Bhopal gas leak that struck in 1984 at Bhopal, India. This paper discusses on evolution of process safety concept in India and the Socio-Economic-Legal changes in the aftermath of Bhopal gas leak disaster as well as the required changes.

Conceptual Frame Work

Process Safety

Process safety is the activity of ensuring safety in processes involving a highly hazardous chemicals including using, storing, manufacturing, handling or moving such chemicals at site, or any combination of these activities(OSHA 3132).

Process Safety Management

The process safety management is the application of management principles, methods, and practices to prevent, and control accidental releases of process chemicals or energy (DOE-HDBK-1101-2004).

PSM Elements

The various elements of PSM are given as follows(DOE-HDBK-1101-2004):

- Employees Participation
- Process Safety Information
- Process Hazard Analysis
- Operating Procedures
- Training
- Sub-Contractors Safety
- Pre-start up Safety Review
- Mechanical Integrity
- Non-routine Work Authorization
- Management Charge
- Incident Investigation
- Emergency Planning and Response
- Compliance Audit
- Trade Secrets

If the PSM elements were adequately implemented in the chemical industry, the process safety would have ensured the safe operation of plants

Background for the Union Carbide plant at Bhopal

The discovery of Dichloro-Diphenyl-Trichloroethane (DDT), an insecticide formulated by Mr. Herman Muller and subsequent findings on its use revealed contamination of the earth and thereby the ban of it in developed countries led to the research on finding new safe pesticides[3]. In 1940, Union Carbide India Ltd (UCIL) started production of batteries in Calcutta and set up two factories in Madras and Hyderabad, India. In 1957 Union Carbide developed a new pesticide, *sevin,* which used deadly MIC in its production. The expanding global market of Union Carbide and the green revolution in India between 1966 and 1971 brought Union Carbide to sell pesticides in India. In 1966, the Indian Government permitted UCIL to import one thousand two hundred tons of *sevin*, and in 1969 work began on construction of *sevin* formulation unit in Bhopal. In 1970, UCIL applied for production license for carbaryl, the active ingredient in *sevin*. In 1979, MIC production unit was added to Bhopal facility according to the design and operating procedures specified by UCC and UCIL, Bhopal, was licensed to manufacture 5,250 tonnes of MIC based pesticides per year. In 1981, production of *sevin* in Bhopal plant touched around 2700 tones. In 1983, UCIL suffered heavy financial losses and to cut costs, many safety measures were withdrawn. In 1984, November, UCC directed UCIL to close down the Bhopal plant and prepare it for sale. As there were no buyers in India, plans were made to dismantle the plant and ship it to another country. The disaster occurred on December 2-3, 1984, which released MIC into atmosphere, and at the time of accident MIC division was dysfunctional for over 9 months but it stored 60 Tones of MIC in the factory premises(Lapierre D and Moro J and Mukherjee S).

Bhopal Gas Leak Disaster

Methyl iso cyanate (MIC) and other gases leakage from Union Carbide's plant in Bhopal, Madhya Pradesh, India, in 1984 December 2-3, is the largest industrial disaster ever experienced in the world, in which between 3000 and 10,000 people died within the first weeks; between 1, 00,000 and 2, 00,000 may have permanent injuries, and over 5, 00,000 persons were exposed to the released toxic gases (Eckerman I).

The probable direct cause for the gas leak was the entry of large amount of water used for water washing of lines along with iron rust particles and residue of salt components in to the tank due to non usage of slip plate causing a runaway reaction, thereby releasing the MIC and other gases into the atmosphere (Eckerman I).

Method

Process Safety Management (PSM) in India

Before the Disaster

The PSM concepts were new to the country like the chemical industries. The PSM was the right of the industries and they decided and implemented them at their wish, as there were neither legal statues nor awareness present in the community. It may not be exaggerated if it is said that PSM was not known to anyone connected with the chemical industries including government, society, and political system and even to the employees. PSM concepts were kept secret like trade secret, as the management of chemical industries was not under duty to disclose them to the government or public.

The practicing and adhering to PSM concepts was at the hands of the management and hence the causes for the Bhopal disaster were too many which means there had been multiple failures. If there had been PSM in the Bhopal Plant, most of the causes or even all the causes might have been removed and even if an accident occurred, the possibly of it turning to be a disaster might have been remote thereby restricting the losses and damages.

After the Disaster

In the United States of America, there is a specific Rule for PSM, 29 CFR 1910.119, the OSHA's "Rule for Process Safety Management of Highly Hazardous Chemicals". In India, there is not any specific law governing PSM concepts in highly hazardous chemical industries. But it covers some of the PSM elements either in a direct or indirect way.

Even though some of the elements are provided under Indian Statutes in an explicit manner, many of them are not covered under completely. The lesson learnt from Bhopal disaster needs determined work in making PSM an integral part of highly hazardous chemical by providing a specific law.

The Causes of the Bhopal Disaster

There had been lot of research on the Bhopal gas leak disaster and putting them all would be out of focus for this paper. The primary causes for this accident are basically process safety failures, which clearly depict on non-availability of process Safety concepts in the plant prior to the accident as well as during the release. The general causes of the disaster to sensitize on non-availability of process safety concepts are summarized as follows(Chouhan T R et al, Eckerman I, Lapierre D and Moro J, Lees. P and Mukherjee S):

Prior to the release

- Non-existence of Inherent safety – Failure to use two-step process without MIC and Storage of large quantity of MIC in large tanks.
- Manual monitoring instead of automated controls.
- No "Knock down" tank
- Usage of carbon steels in adjacent pipelines instead of stainless steel.
- Location of the plant in the densely populated area.
- Use of water (a highly reactive chemical with MIC) to wash the pipeline.
- Non-implementation of safety audit recommendations.
- The tank was not pressurized.
- Tank temperature and other parameters were not logged.
- Water curtain designed to absorb the gas up to 15m where as the actual release up to 33m was realized.
- 30m high flare towers was disconnected and the pilot flame extinguished.
- The vent gas scrubber had seen turned off to a stand by position and was made non-operational.
- Non-availability of skilled workers.
- Reduction of safety management staff.
- Insufficient maintenance of the plant.
- Lack of emergency planning within and outside the plant.

During the Release

- MIC had to be stored under 15°C but it was stored at ambient conditions. The factory was equipped with 30 Tones refrigeration system but the refrigeration system had been closed down a few months earlier as a cost cutting measure.
- Spare tank had to be kept empty in order to divert any excess MIC into it. But in panic, nobody thought of opening the valve in order to allow the MIC in to the empty tank.
- High temperature alarm did not function.
- Evacuation tank was not empty.
- No emergency management within and outside the plant.

Social Aspects of Process Safety

Before the Disaster

The society at large was not aware of hazards of chemical industries and the chemical industries are viewed as safe as their homes. Every one of the societies was of the opinion that these industries were there to bring fortune to them and to alleviate their poverty. The political system also did not have any idea of process safety and, seriousness of accidents and consequences of such accidents on society. The trade unions also did not have any opinion on hazards with chemical industries. But a small group of trade unions working around the chemical factories later gained some knowledge by the experiences of workers working in chemical industries to a smaller extent. The government departments too were not aware of the process safety concepts, which were completely new to them. By and large the society was completely unaware of the process safety concepts and failures in chemical industries.

After the Disaster

The society is yet to recover from the shock of Bhopal gas leak disaster. It has created negative feeling about chemical industries and in spite of safety measures in these industries the public is under fear and they go against erecting new plants in and around their habitation. Many multinational companies could not put their factories and they are forced to shift from the selected locations only due to public pressure. The trade unions are also to an extent aware of process safety in chemical industries and the legal protections and lateral amendments had also improved their say in process safety areas. The government machinery have some additional tools like amended statutes and new statutes, with which some of the process safety concepts are taken care even though not to a larger extent. Political system also developed awareness on the process safety concepts to an extent. But the complete understanding of the process safety concept by the society is still a dream.

Economical Aspects

Before the Disaster

The economy was very particular on the growth of chemical industries in India since it could support green revolution by improved productivity and yield by the use of fertilizers and pesticides. The chemical industries also took economical angle, i.e. cost cutting as a measure towards removing, reducing and compromising safety systems and protective measures. The Bhopal Plant managers were permitted to remove safety devices as a cost cutting measure. The compensation for losses and damages were not provided under the law.

After the Disaster

The various measures introduced in the legal system has provided for siting and layout against economical considerations, need to declare the characteristics of the hazardous materials, emergency preparedness plans and, compensation for damages and losses. The factory management cannot remove safety systems for economical reasons, since such measures would invite legal actions as well as they are also aware that ensuring safety is cheaper than operating the plant unsafely. An accident, which can result in disaster, can make the industry extinct by looking into the cost that it has to bear in terms of compensation towards losses and damages, and legal expenses.

Legal Aspects of Process Safety

Before the Disaster

The Factories Act 1948 was the only Act in force to deal with safety in factories. The first Indian Factories Act was passed in 1881, which gave limited protection to children, fencing of machinery and reporting of accidents. The Factories Act 1948 contained provisions for safety, health and welfare of workers. This Act initially contained provisions on eye-protection, control of explosives and inflammable dust pertaining to safety. The Factories Act 1948 was then amended in 1976, which included provisions regarding appointment of safety officers and inquiry into accidents. The Factories Act 1948 did not contain by and large any specific provisions on process safety prior to the Bhopal disaster(Srivastava K D).

After the Disaster

After the Bhopal Disaster, various major amendments were made to the Factories Act 1948 pertaining to process safety. The following provisions were made in the Factories Act 1948 by Factories (Amendment) Act 1987 pertain to process safet (Mohanty S K and Srivastava K D):

- Safe guards to be adopted against use and handling of hazardous substances by the occupiers of factories.
- Laying down of emergency standards and measures.
- Siting of hazardous industries.
- Compulsory disclosure of information by the occupier.
- Workers participation in safety management.
- Permissible limits of exposure of chemical and toxic substances.
- Right of workers to warn about imminent danger.

Apart from the above amendments to the existing Factories Act 1948, following laws were introduced as a measure to ensure process safety in India:

- The Manufacture, Storage and Import of Hazardous Chemicals (MSIHC) Rules 1989.
- The Chemical Accidents (Emergency Planning, Preparedness and Response) Rules 1996.
- The Public Liability Insurance Act 1991 and The Public Liability Insurance Rules 1991.
- The Disaster Management Act 2005.

The MSIHC Rules 1989, provides for notifications of major accidents, approval and notification of sites, safety reports and safety audit reports, preparation of on-site and off-site emergency plans, information to persons liable to be affected by a major accident and, collection and development, and dissemination of information about hazardous chemicals, which are applicable to the specified industries utilizing listed chemicals under the Act equal or above the defined threshold quantity.

The Chemical Accident (Emergency Planning Preparedness and Response) Rules 1996, provides for defining chemical accidents, crisis groups at central, state, district and local levels and their functions and information to public.

The Disaster Management Act 2005 deals with creation of national, state and district Disaster Management Authority (DMA), as measures by the government for disaster management and creation of National Institute of Disaster Management (NIDM) for managing both natural and industrial disasters.

The Public Liability Insurance Act 1991 and The Public Liability Insurance Rules 1991 provides for public liability insurance for the purpose of providing immediate relief to the persons affected by accident occurring while handling hazardous substances and for matters connected to it. This Act provides duty on the owner to take out insurance policies and imposes liability to give relief, claim and award of relief, establishment of environment relief fund, power to make applications to courts for restraining the owner from handling hazardous substances and provides for dealing with offences by the government departments.

Conclusion

The process safety in India still in its infant stage and a lot of work needs to be done in this respect to make all PSM elements enforceable by providing a specific law towards PSM for highly hazardous chemical units in India. The socio – economic - legal changes in process safety aftermath of Bhopal disaster need to be further transformed into a refined way to ensure process safety as a basic principle of erection, operation and maintenance of highly hazardous chemical industries. The actions in bringing PSM into force by letter and spirit would make man made technological disasters a thing of the past.

References

1. Chouhan T R et al (2004), *Bhopal; The inside story*, second updated edition, Other India Press, India and The Apex Press, Newyork, United States of America, pp. 28-44 .

2. Eckerman I (2005), *The Bhopal saga,* universities Press (India) Private Limited, Hyderabad, India, pp. 21- 78 .

3. Lapierre D and Moro J (2001), *It was Five Past Midnight in Bhopal,* Full Circle, Publishing, New Delhi – 110003, India, pp. 19, 50 - 164.

4. Lees. P (2003), *Loss Prevention in the Process Industries,* Volume 3, Second edition, Butterworth-Heinemann, Oxford, pp. A5/1-A5/11.

5. Mohanty S K (2002), *Environment & Pollution Law Manual,* Third Edition, Universal Law Publishing Co. Pvt. Ltd., New Delhi – 110003, India, pp.156-229.

6. Mukherjee S (2002), *Bhopal gas tragedy,* Tulika Publishers, Chennai, India, pp. 6- 17.

7. Skelton B (1997), *Process Safety Anlaysis*, Institution of Chemical Engineers, Warwickshire, UK, pp. 23-41.

8. Srivastava K D (2001), *Factories Act 1948,* Fifth Edition, Eastern Book Company, Lucknow, India, pp. 2-68, 456-470.

9. OSHA 3132 (2000) Process Safety Management, U.S. Department Labour, Occupational Safety and Health Administration, pp. 1- 41

10. DOE-HDBK-1101-2004 (2004), Process Safety Management for Highly Hazardous Chemicals, U.S. Department of Energy, Washington, D.C. 20585, pp. xii, 1- 5.

Development of a Miniature Calorimeter for Identification and Detection of Energetic Compounds

Yen-Shan Liu [a], Victor M. Ugaz [b], William J. Rogers [b,c], M. Sam Mannan [b,c]

[a] *Department of Chemical Engineering, University of Louisiana at Lafayette, Lafayette, LA 70504-4130*

[b] *Department of Chemical Engineering, Texas A&M University, College Station, TX 77843-3122*

[c] *Mary Kay O'Connor Process Safety Center, Texas A&M University, College Station, TX 77843-3122*

ABSTRACT

The development of a versatile system capable of providing rapid, portable, and inexpensive detection of energetic compounds is needed critically to satisfy a wide range of applications in the fields of homeland security, forensic analysis, emergency response, and industrial hazards analysis. The hand-held nanocalorimeter will serve as a first-of-its-kind screening tools for energetic compounds directly in the settings where they are needed with high efficiency, reduced cost, and simplicity with ease of use. Calorimetric techniques can provide direct measurements of energy release potential and therefore this system do not depend on prior knowledge of familiar compounds.

INTRODUCTION

Thermal analysis or calorimetry is a widely used technique for obtaining both qualitative and quantitative information about thermal transitions associated with a particular material or process. Because all chemical reactions and most physical changes are associated with intake or release of heat, a quantification of thermal energy provides a simple and universal method for characterizing materials and processes. One drawback of conventional macroscale differential scanning calorimetry (DSC) technology is the large thermal mass of the calorimetric cell itself and its associated hardware. Consequently, the required energy input may be large compared with the energy associated with the thermal transitions to be measured. This can pose problems in cases where only minute sample quantities are available for testing. The ability to perform calorimetric analyses over much faster time scales using reduced sample amounts is critical to meeting future needs. The advantages of miniaturizing this technology include reduced costs (including hardware and reagents), faster screening (suitability for high-throughput operation), portability, and ease of use. The ability to perform thermal analysis in an inexpensive and portable format also has the potential to greatly expand the use of these techniques in the areas of quality control and on-line measurement of chemical processes [1]. Stricter customs procedures are currently being implemented in an effort to improve national security. These efforts, however, require the development of new and effective screening tools to identify trace amounts of energetic and hazardous materials in a rapid and cost effective manner. These miniaturized systems must incorporate a very small thermal mass and be able to use reagent quantities in the nanogram/nanoliter range. In order for these systems to operate effectively, rapid and uniform heating and cooling must be achieved while maintaining a high level of temperature homogeneity.

Recent developments in the area of nanoscale analysis of energetic materials have motivated the development of calorimetric instrumentation for testing of nanoscale samples (ng, nL). These miniaturized systems are also inherently safer because the reactor size is very small, so hazardous materials cannot cause serious explosions or generate excessive heat. Lightweight lab-on-a-chip devices will allow detection of chemicals to be completed on site in seconds rather than hours. This feature will assist process engineers, armed forces, police officers, and other authorities during emergency response. When dealing with energetic compounds, the nanocalorimetry system will allow a rapid characterization to be performed so that an appropriate level of initial response can be deployed while subsequent detailed analyses are performed to confirm the fingerprint result, as shown in Figure 1.

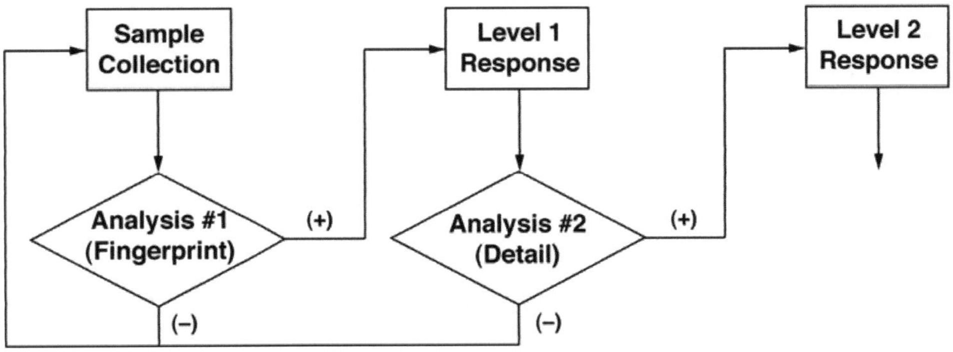

Figure 1. Overview of the Energetic Material Fingerprinting System

After sample collection and concentration, the miniaturized calorimeter and data analysis capabilities will be used to provide a rapid chemical fingerprint of the compounds in question. If signatures consistent with energetic materials are identified, an initial level of response can be quickly deployed while subsequent detailed analyses (e.g., gas chromatography/Fourier transform infrared (FTIR) spectrometer, gas chromatography /mass spectroscopy) are performed to confirm the fingerprint result.

CALORIMETER DESIGN AND FABRICATION

Although thermal analysis is an established technique for material characterization and miniaturized calorimeter systems have been reported in the literature [2-4], none of them are suitable for field measurements. The basic design of the calorimetry cell consists of (i) a thermal control module incorporating arrays of micro-fabricated heaters and temperature sensors, as well as necessary electronic interconnections, and (ii) a sample encapsulation module incorporating etched enclosures designed to accommodate either solid or liquid samples. These two components are fabricated separately and assembled to form the completed calorimetric cell, as shown in Figure 2.

A number of thermal isolation strategies are available involving etching the substrate from the backside of the wafer to create a barrier between the reference and sample locations in order to minimize lateral heat flow. A simplified approach to provide an enhanced degree of thermal isolation involves a dicing saw to machine trenches in the backside of the wafer. Preliminary 3-D heat transfer simulations were performed using a finite element method to indicate that greatly enhanced thermal isolation can be achieved using these strategies relative to an unmodified wafer. Also, a series of additional simulation studies were performed to explore the optimal design to achieve the thermal isolation necessary for our device applications.

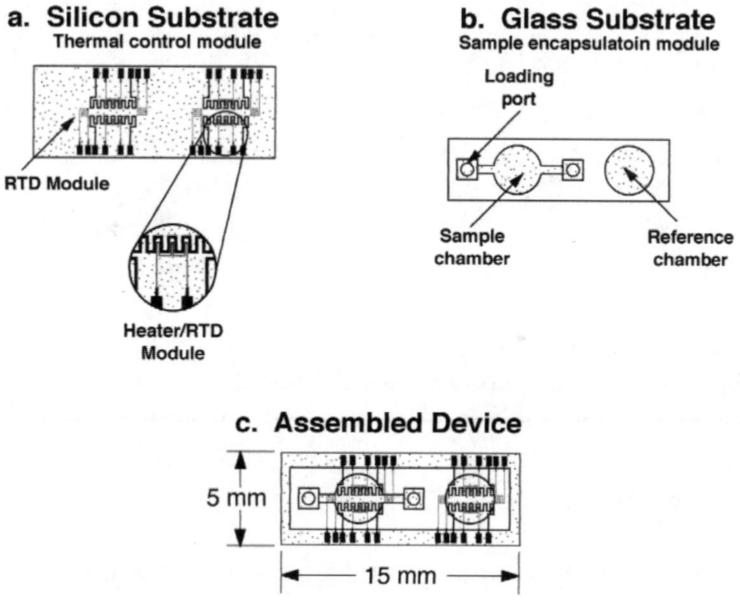

Figure 2. Schematic Illustration of the Microfabricated Calorimetry Cell

The device consists of separate thermal control and sample encapsulation components, which are assembled to form the complete device. The sample enclosure configuration depicted here is suitable for use with liquid samples. At this size scale, over 200 devices can be fabricated on a single 6-inch diameter wafer.

In addition to silicon, transparent substrates can be used for fabrication of the thermal control components in order to provide the unique capability to achieve simultaneous optical and thermal characterization of the sample. The heater and temperature sensor array can be fabricated on the surface of glass or z-cut quartz wafers using essentially the same processes described for silicon substrates. To allow maximum light transmission through the substrate, the thermal control elements can either be positioned around the perimeter of the sample chamber, or be fabricated using polysilicon or indium tin oxide (ITO) thin films that allow light transmission with minimal attenuation. The use of glass or quartz substrates is also potentially advantageous from the standpoint of providing reduced thermal conductivity, thereby offering improved thermal isolation between the sample and reference cells. Such devices have the potential to provide simultaneous optical, x-ray, spectroscopic (e.g., FTIR), or light scattering measurements during the course of a calorimetric experiment so that an enhanced degree of material characterization and fingerprinting can be achieved. It is possible that a sufficient level of thermal isolation may not be achievable in certain specialized reaction systems. Even if this is the case, our device design can still be used by separating the sample and reference chamber portions of the device into two physically distinct dies.

The heaters and resistance temperature detectors (RTDs) are located directly beneath the sample in order to provide a highly accurate reading of the temperature inside the sample and reference chambers. Coordinated experiments and simulations were also performed to determine the performance of heating and sensing elements fabricated using different metals (e.g., Pt, Al, Au, Ti, Cr). Temperature control is achieved using resistive heaters patterned on the thermal control module. Because the resistance of the heaters is known, the electrical output current can be monitored to determine the power supplied to the heater. Similarly, temperature is measured by monitoring the change in resistance of similar elements as a function of temperature. For platinum heating elements, resistances in the vicinity of 2 kΩ provide temperature sensitivity to within 0.2 °C depends on precision of voltage drop measurements. A 4-lead arrangement is used to isolate the resistance change in the vicinity of the sample from that associated with the connector leads. The readings of several temperature sensors can be used to account for spatial inhomogeneities in the temperature field.

The sample enclosure design and loading process largely depend on the type of sample to be studied. For liquid samples, the enclosure consists of a fluidic channel intersecting a chamber centered over the heaters and temperature control circuitry. Access holes are machined into the glass substrate, allowing individual samples to be loaded using a syringe. An externally pumped fluidic manifold can be constructed to allow automated loading and unloading for processing of multiple samples. After loading the sample into the channel, the access ports are sealed using epoxy, through the application of mechanical pressure to cover the holes or through the use of sealing valves in a fluidic manifold. Solid samples can be inserted directly into an etched chamber, after which the chamber is covered and sealed. Additional mechanical pressure can be applied using a loading fixture to maintain sample confinement during the run.

A key question is the extent to which the calorimetric cell can be scaled down. The potential loss of sensitivity due to a reduced sample quantity must be balanced by allowance for a sufficient number of devices to be fabricated on a single wafer in order to realize the enormous cost benefits from photolithographic fabrication techniques. Because these micromachined calorimetry cells occupy a small surface footprint, they can be closely spaced to maximize the number of devices produced on a wafer. This level of cost savings also is expected to allow disposable single-use calorimetry cells, thereby reducing complexities and potential contamination issues associated with unloading and reloading multiple samples.

DATA ACQUISITION AND ANALYSIS

The control algorithm and associated data acquisition system for the nanocalorimeter employs a simplified proportional control circuit whereby the same heating power ($P_{avg.}$) is supplied to both sample and reference chips, in order to change their mean temperature according to a preset scanning rate. When a reaction (or any thermal event) occurs in the sample, a temperature difference is created between the two chambers. This temperature differential is fed to a proportional controller that compensates for the reaction heat-flow rate by increasing or decreasing the heating power by an increment ΔP. The compensating heating power is proportional to the measured temperature difference ($\Delta T = T_S - T_R$) according to the relationship $\Delta P = K_1 \Delta T$, where K_1 is a constant. A heat flow-rate, $\varphi = K_2 \Delta T$, is then assigned to the measured temperature difference, where K_2 is determined by calibration.

The control algorithm was implemented via a modular distributed I/O system and controlled by a Windows-based PC. The modular distributed I/O system consists of a terminal base incorporating analog voltage outputs with a range of zero to five volts as well as a resistance temperature detector (RTD) module. This unit enables the device to impose a user controlled constant heat rate on the sample. The PC connected to the distributed I/O system can independently address and control the various voltage outputs of the modules using a graphical programming language. This modular design will allow the development and testing of a variety of operation and analysis modes to suit the needs of a range of samples and applications. This system is also highly amenable to portable operation using a laptop computer or personal digital assistant (PDA) to perform instrument control and data acquisitions outside the laboratory environment.

Calibration of the temperature sensors was performed by placing each device in an oven and monitoring the RTD resistance at a series of temperatures. A fit to these data provided a calibration curve for the temperature response, which was linear within the range of interest for these applications. On-chip temperature control was achieved by resistive heating in response to a potential applied to the heater elements using feedback control from the RTD sensors. The data acquisition system was used to monitor and control the thermal output, as well as measure the changes in power applied to the heaters in order to identify thermal transitions.

To test the concept of the simplified nanocalorimeter design, the thermal transition associated with the boiling point of acetone ($T_{b.p.}$ = 56.2 °C) was studied. A 350 nL volume of acetone was loaded into the sample channel of the prototype device shown in Fig. 2c, after which access holes were sealed with epoxy to fully encapsulate the sample. The nanocalorimeter chip was then heated at a rate of 1 °C/min while the power supplied to the heaters was monitored. Using this protocol, a peak was observed consistent with the liquid-vapor transition, as shown in Figure 3.

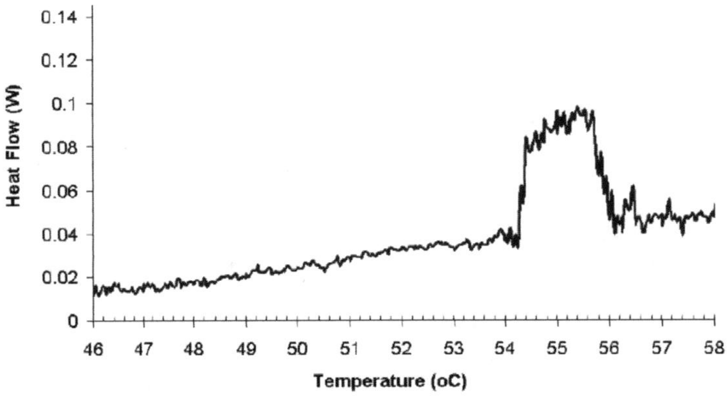

Figure 3. Thermal Analysis Data from the Nanocalorimeter Prototype

Plot of heat flow as a function of temperature during a 1°C/min heating scan of a 350 nL acetone sample using the prototype nanocalorimeter device shown in Figure 2c. The observed curve centered near 55.2 °C is roughly consistent with literature values for the boiling point (56.2 °C).

The measured phase transition temperature for acetone is about 1 °C lower than the literature value because of thermal lag from the sample to the RTDs and high surface to volume ratio of the sample chamber. Calibrations can be performed in applications where more accurate temperatures are required for phase transitions or other measured phenomena due to thermal lag.

Redesign and Improvements

The preliminary results obtained in the prototype device illustrated proof of concept for calorimetry, and the nanocalorimetry cell was then redesigned and fabricated. Key improvements incorporated in this nanocalorimeter cell design, shown in Figure 4 a, include:

- A reduction in thermal mass. With simplicity and ease of fabrication in mind, a reduction in the device surface area footprint is about 11% of the prototype size.
- A simplified single-mask layout. By moving the temperature sensors to the outer perimeter of the sample chamber, the sample is not in direct contact with the sensing elements. Consequently, passivation layer deposition and the corresponding second lithography step to etch the contacts are not necessary. This reduces time and costs associated with fabrication. The photomask of the new chip layout was drawn using L-edit Pro, a CAD software, after a series of designs and metals were compared (Figure 4). Redesign of the thermal control elements to use aluminum instead of platinum metal can further reduce fabrication costs.

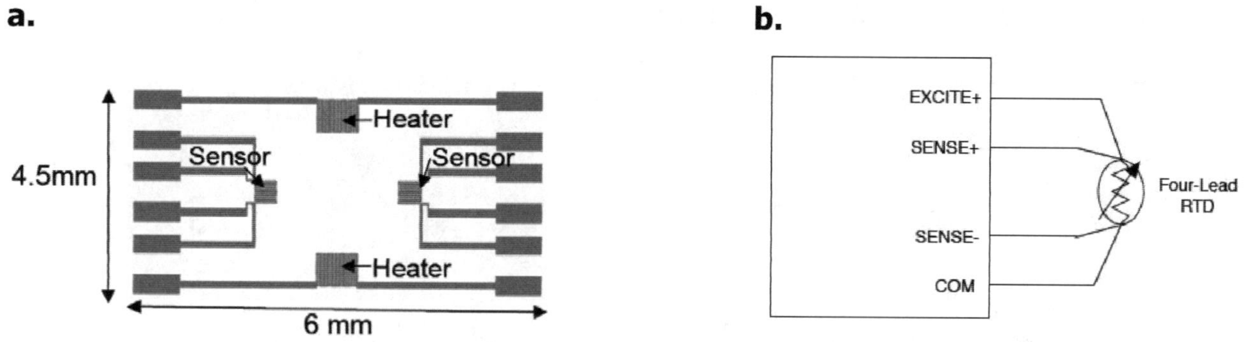

Figure 4. Key Elements of the Redesigned Nanocalorimeter Chip
(a) Heater and temperature sensor layout. The sample chamber will be centered on the chip, with temperature control elements located at the perimeter. (b) Circuit schematic of 4-lead RTD arrangement.

- An improved temperature sensing circuit. A four-wire RTD element (Figure 4 b) is employed in the redesigned device in order to provide a more accurate measurement of temperature localized at the sample chamber. This is achieved by adding two additional wires and operating the circuit with the lead resistance canceled so that only resistance changes at the sensing element are detected.
- Improved thermal isolation. Because of the small device size, sample and reference cells can be physically separated and placed in an enclosed environment to eliminate thermal crosstalk and heat isolation issues.

EXPERIMENTAL RESULTS

The boiling point of pentane was measured using the new nanocalorimeter cell, as shown in Figure 5. For the redesigned device, the measured input energy is about 0.680 J, or about 4.8 times the required input energy for the liquid-vapor transition of pentane (~0.140 J). In the original design, the measured input heat during the vapor-liquid transition of acetone was about 4.50 J, or about 32.4 times the required input energy for this phase change (~0.139 J). Thus, the reduced input energy reflects the reduced thermal mass in the redesigned device. In addition, the measured phase transition temperature, 38.7 °C, is shifted from literature values for the boiling point (36.2 °C) because of several possible factors such as thermal lag from the sample to the RTD sensors and heaters, and high surface to volume ratio of the sample chamber. Calibrations can be performed in applications where more accurate temperatures are required for phase transitions or other measured phenomena due to thermal lag. When the depth of a thin film is less than 100 nm, interfacial forces can have a controlling effect on phase change processes. The dynamics of a fluid nanodomain is controlled by the sum of excess intermolecular interactions, which manifest as the disjoining pressure. Examples of intermolecular/surface forces are apolar van der Waals, polar acid-base, electrostatic, structural, vapor thrust, and curvature-effect.

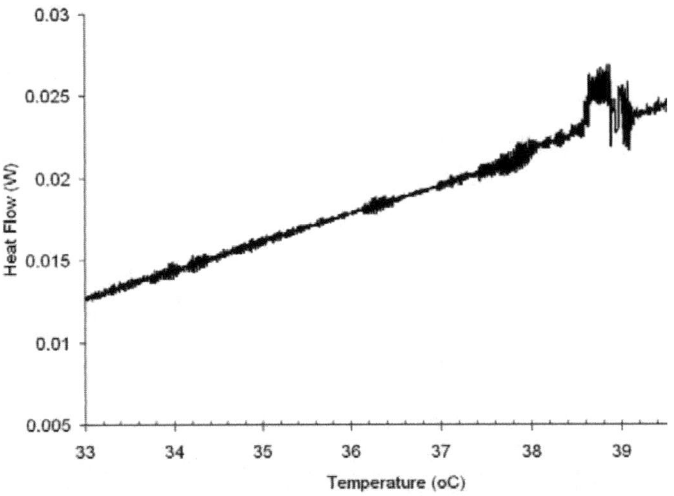

Figure 5. Thermal Analysis Data from the Redesigned Chip

Plot of heat flow as a function of temperature during a 1°C/min heating scan of a 600 nL pentane sample using the redesigned device. The observed curve centered near 38.7 °C is shifted from literature values for the boiling point (36.2 °C).

OPERATION

A calorimetric experiment captures thermal transitions (phase change, reactions) that occur within the system. These transitions can be exothermic or endothermic as indicated by a positive or negative deflection in the measured power. The accuracy and sensitivity of the calorimetry system will be measured through the use of well-characterized calibration standards, as shown in Table 1 [5].

Table 1. Typical calorimetric calibration standards [5]

Material (transition)	T_{trans} (°C)	ΔH_{trans} (J/g)
Gallium (s → l)	29.8	79.9
Benzoic acid (s → l)	123	148
Indium (s → l)	156.6	28.6
Tin (s → l)	231.9	60.4

The onset temperatures, reaction enthalpies, and other reaction parameters will be measured in the nanocalorimeter with sample sizes of ~ 1μg and compared with the same parameters measured in an Automatic Pressure Tracking Adiabatic Calorimeter (APTAC) or an Accelerated Rate Calorimeter (ARC) that use a larger sample size of ~ 2 - 50 g. An extensive library of experimental data for over 1,000 chemicals, mostly measured using ARC and APTAC, are provided by industrial partners (Table 2). The decomposition parameters for the selected compounds can also be compared with DSC measurements on the same compounds reported by Grewer [6], Ando [7] , and Eastman Kodak Co. [8]. Further refinements to the device design and control algorithm will compensate for systematic deviations and optimize performance.

Table 2. Library of energetic compounds to be used for instrument testing

Sr. no.	Chemical	Mass gm.	Onset temperature °C	Max. temperature °C	Heat of reaction cal/g
1	Di-ter-butyl peroxide in toluene	0.71	115	184	49
2	Hydrogen peroxide	8	130	240	700
3	Ammonium nitrate	0.5	201	291	-
4	Styrene	2.2	100	201	172
5	Nitro-styrene	0.95	170	264	297
6	Acrylonitrile	0.79	-	-	389
7	1,3 Butadiene	12	158 ± 10	445 ± 19	260
8	50 wt.% hydroxylamine	2	133	216	420
9	Hydroxylamine nitrate	4.2	171	196	350
10	Hydroxylamine sulfate	4.5	151	185	-

Pressure inside the sample cell

Highly energetic compounds that can decompose or react rapidly with the production of a large volume of gas. If a decomposition reaction occurs, non-condensable gases such as nitrogen are usually formed. The volume of those non-condensable gases can be 100 to 1000 times of the original sample volume depending on the reaction conditions. With the increasing vapor volume, the pressure inside the sample cell will also increase significantly. The working pressure of Pyrex borosilicate glass can be up to 800 psi depending on its diameter and wall thickness (i.e., about 50 times the atmospheric pressure). Thus, there is a possibility that the nanocalorimeter cell may fail during tests of rapid decomposition or combustion reactions of energetic compounds. Nevertheless, the nanocalorimeter chip is designed to be disposable with a relatively low cost. In addition, with nanogram amounts of samples, consequences of rapid runaway reactions will be minimized.

CONCLUSIONS

The hand-held nanocalorimeter will serve as a new generation of identification and detection tools for energetic compounds with high efficiency, reduced cost, and simplicity with ease of use. Initial work has led to successful fabrication of a chip capable of sampling nano-sized solid or liquid compounds. Control algorithms incorporating the DSC principle have also been written using a graphical programming language. Future work will focus on modifying the chip design and control algorithm to improve accuracy and sensitivity, developing trace analysis software to link it to a database of energetic compound information, and adapting different fabrication procedure for high temperature operation and large scale production.

References

1. Cheng, Z.D., Li, C.Y., Calhoun, B.H., Zhu, L., & Zhou, W.W. (2000), "Thermal Analysis: The Next Two Decades," Thermochimica Acta, 355, pp. 59-68.

2. Mathonat, C., Bourdin, G., Parazza, C., Daudon, J.L., & Parlouer, P.L. (1998), "IC - Calorimetry a New Field for Thermal Analysis," 26th North American Thermal Analysis Society (NATAS), Cleveland, OH.

3. Parlouer, P.L. (2000), "Portable Thermal Analyzer a New Calorimetric Concept," 28th North American Thermal Analysis Society (NATAS), Orlando, FL.

4. Olson, E.A., Efremov, M.Y. , Zhang, M. , Zhang, Z., & Allen, L.H. (2003) "The Design and Operation of a MEMS Differential Scanning Nanocalorimeter for High-speed Heat Capacity Measurements in Ultrathin Films," Journal of Microelectromechanical Systems, 12, pp. 355-364.

5. Höhne, G.W.H., Hemminger, W., & Flammersheim, H.-J. (2003). Differential Scanning Calorimetry: an Introduction for Practitioners, second and enlarged ed., Springer-Verlag Berlin and Heidelberg, New York.

6. Grewer, T. (1994). Thermal hazards of chemical reactions. Amsterdam, Netherlands: Elsevier Science.

7. Ando, T., Fujimato, Y., & Morisaki, S. (1991). "Analysis of Differential Scanning Calorimetric Data for Reactive Chemicals," Journal of Hazardous Materials, 28 (3), 251.

8. Chervin, S., & Bodman G.T. (2001). "Method of estimating decomposition characteristics of energetic chemicals," 29th North American Thermal Analysis Society (NATAS), September 24-26, St Louis, MO.

9th Process Plant Safety Symposium
AIChE 2007 Spring National Meeting
Houston, Texas
April 22-26, 2007

Chemical Process Security Performance Standards

David A. Moore, PE, CSP
H. M. Leith
Michael J. Hazzan, PE
Brad Fuller
AcuTech Consulting Group
2001 North Beauregard Street
Alexandria, VA 22311
www.acutech-consulting.com

Abstract

Section 550 of the Homeland Security Appropriations Act of 2007 ("Section 550"), enacted on October 4, 2006, provided the Department of Homeland Security(DHS) with authority to promulgate "interim final regulations" for the security of certain high risk chemical facilities in the United States.[1]

Key relevant provisions of Section 550 included the need for DHS to establish risk-based performance standards for security of high risk chemical facilities, to require vulnerability assessments (VA), and for industry to develop and implement site security plans (SSP) that address these requirements.

Pursuant to the outcome of the SVA all chemical sites will need to prepare a Site Security Plan. The SSP and VA will need to explain how the site met the risk-based performance standards set by DHS. This is a new realm in the world of chemical industry security, so a key aspect of meeting the regulatory requirements will be interpreting these standards and determining the best path to achieving them.

This paper explains the requirements and gives advice on how to achieve them and how to document them in the SVA and SSP. The timing of submission of this paper precluded full knowledge of the language of the interim final regulation, since it is due to be released not later than April 4, 2007.As such the reader is cautioned to refer to the language of the Interim Final Rule and any related requirements or guidance published by DHS

Introduction

Until October 4, 2006, when the President signed the 'Department of Homeland Security Appropriations Act of 2007' (the Act), the Department of Homeland Security (DHS) lacked the authority to regulate the security of the chemical industry[2]. This gives the Department of Homeland Security no later than 6 months after the date of enactment of the Act to issue interim final regulations to identify "high risk" chemical facilities, and to require security vulnerability assessments and site security plans for these high risk facilities based on DHS established risk-based performance standards. DHS intends to implement an appropriate regulatory program under Section 550 of the Act by the deadline imposed on them by Congress (April 4, 2007), focusing its resources first on those chemical facilities in our nation that present the highest levels of security risk.

The legislative language is very brief - about 740 words – so DHS has extensive flexibility to define the program requirements. Some of the relevant Section 550 language to this paper includes the following requirements:

1. "Such regulations shall permit each such facility, in developing and implementing site security plans, to select layered security measures that, in combination, appropriately address the vulnerability assessment and the risk-based performance standards for security for the facility."
2. "Provided further, that the Secretary may not disapprove a site security plan submitted under this section based on the presence or absence of a particular security measure, but the Secretary may disapprove a site security plan if the plan fails to satisfy the risk-based performance standards established by this section."

Likely Impacts

This Act is a very significant development for the chemical sector because up to this point there was no Federal regulation that defined security expectations for chemical facilities except the U.S. Coast Guard Maritime Security Regulations[3]. The Maritime Transportation Security Act of 2002 (MTSA) did not explicitly define risk-based performance standards or tiers. While the MTSA regulations are performance-based for the most part, the USCG does have the authority to specify security measures, whereas DHS is not given this authority by the Act.

At this time the advanced notice of rulemaking, published in December, 2006, provides insights to the expected language of the Interim Final Rule[4]. Following the promulgation of the 'Chemical Facility Anti-Terrorism Regulations' in April, 2007, there will be published requirements three key regulatory components:

1. Risk-Based Performance Standards (RBPS)
2. Security Vulnerability Assessment (SVA)
3. Site Security Plans (SSP)

The general sequence of the regulatory scheme is illustrated in Figure 1.

Figure 1
Strategic Approach
Chemical Facility Anti-Terrorism Regulations

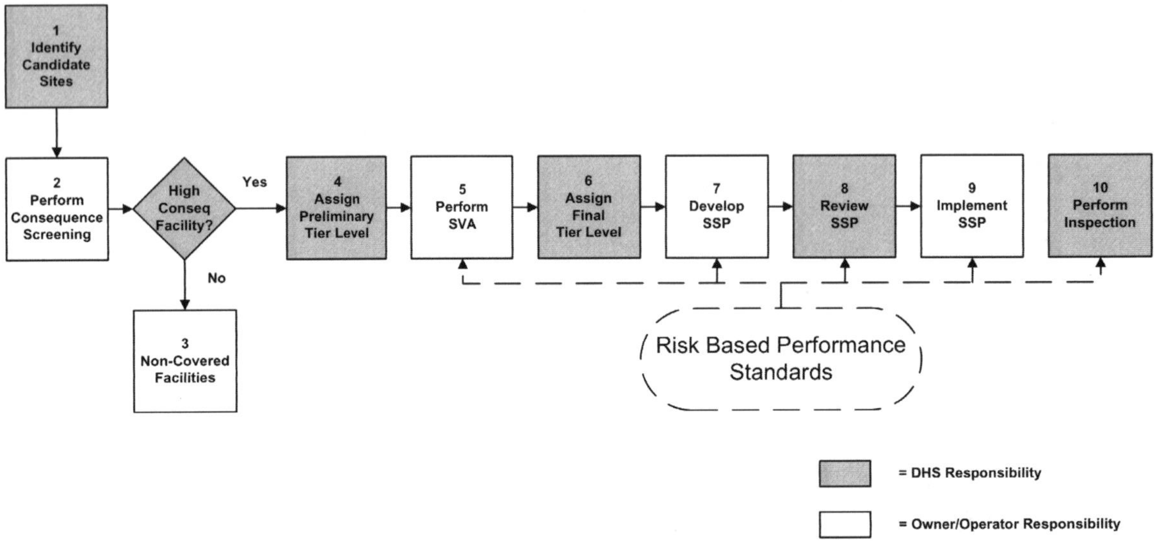

The program proposed by this notice would be implemented in phases, and DHS would address chemical facilities with the most significant risk profiles as early in the program as possible. For each phase, the program contains several basic steps:

- Chemical facilities manufacturing, storing, or using certain chemicals will first complete a "Top-screen" accessible through a secure Department web application (Chemical Security Assessment Tool, or CSAT). The Department intends to use this consequence-based screening to determine if a chemical facility "present[s] a high level of security risk" and should be covered by this program.
- If the Department determines that a chemical facility qualifies as "preliminary high risk," the Department would require the facility to prepare and submit a Security Vulnerability Assessment (SVA) and a Site Security Plan (SSP), and would provide technical assistance to the facility as appropriate. The SVA and SSP are also intended to be completed via secure Department web applications (CSAT).

- Following a facility's submission of these materials, the Department would review the submissions for compliance with the risk-based performance standards. The Department would follow up with a site inspection and audit.
- If the facility's SVA or SSP is found deficient or if other problems arise, the facility could seek further technical assistance from the Department, and could consult, object, or appeal depending on the stage of the process. If the SVA and/or SSP are ultimately disapproved, the covered facility would be required to revise its security plan and resubmit the materials to meet the Department's performance standards, or face the penalties and other remedies set forth in the statute.
- If the covered facility's submissions are approved, the SSP would be fully implemented and if the facility is otherwise in compliance, the Department would issue a Letter of Approval to document the determination. The Department would also then notify the facility of its continuing obligations—based on its level of risk—to maintain and periodically update its SVA and SSP.

CSAT Process

DHS intends to require a two-part process to perform both a preliminary 'consequence' analysis and a more thorough vulnerability assessment on certain of the 'high risk' chemical facilities. The first segment of CSAT is a questionnaire known as the 'Top-Screen'. For chemical facilities, the Top-Screen solicits answers to a series of questions intended to assess different impacts, and levels of impact, that could result from a terrorist incident at the facility. For chemicals with a potential for offsite impacts the Top-Screen relies in part on similar data from the EPA's Risk Management chemical safety program ("RMP," discussed below). For example:

- Does the facility operate any RMP Program 2 or 3 processes?
- Does the facility manufacture, use, or store any of the covered chemicals?
- What is the quantity of each covered chemical onsite?
- What is the potential extent of offsite impact if the chemical(s) were released?

In addition to release of chemicals from the facility, the Top-Screen also includes questions regarding the manufacture and storage of explosives materials, and seeks information on chemical substances and precursors addressed by the Chemical Weapons Convention. See 22 U.S.C. 6701, as well as other toxic compressed gases that could be stolen or diverted from the facility. Finally, the Top-Screen asks several questions designed to determine the economic criticality of the facility and its products to specific critical infrastructures, and the nation as a whole.

The second segment of the CSAT process is to conduct a facility Security Vulnerability Assessment, which will also be conducted and submitted to DHS via a secure website.

This second segment will only be requested to certain high risk facilities determined by DHS based on the Top-Screen results, and other information. The SVA has three fundamental steps, each with detailed instructions:

1. Identify the critical assets on the facility based on the results of the Top-Screen and as directed by DHS.
2. Apply specified potential attack scenarios, as applicable, to each identified critical asset to refine the consequence estimates from the Top-Screen.
3. Apply specified potential attack scenarios, as applicable, to each critical asset in light of the security measures in place to evaluate the gaps between the existing security measures and the risk based performance standards.

Limitations of DHS to Specify Security Measures

The limitation on DHS to define any particular security measure is particularly significant. It allows industry the flexibility of defining security measures that achieve a certain level of performance while requiring DHS to define a risk-based performance standards process that adequately achieves a given security risk level under those constraints.

This is a performance-based and risk-based regulation, fortunately, since it is inconceivable that a regulation could be developed that prescribes a security regime that would be appropriate for all of the possible applications across the industry. This is especially true in the case of the broadly defined 'chemical sector' over which DHS has authority.

The Act restrains DHS from requiring any specific measures and, in particular, is silent on the application of inherently safer technology (IST) strategies. Earlier proposed bills had mentioned IST directly, but in this legislation it is not addressed. The omission is consistent with testimony given before the Senate Environment and Public Works Committee in 2006 when the Senate was considering the feasibility of including inherently safer provisions in Congressional legislation[5], as well as other references on the topic.[6]

Risk-Based Performance Standards per Tier Level

Section 550 provides the Department with authority to regulate those chemical facilities "that, in the discretion of the Secretary, present high levels of security risk." It is expected that DHS will use consequence-based tier levels that will result in the application of a specific set of Risk Based Performance Standards (RBPS) at each level. The facilities will have the flexibility to select layered security measures that, in combination, appropriately address the gaps identified in the SVA and RBPS requirements.

The results of the SVA and SSP will allow DHS to determine if the specific security measures reported by a covered facility meet the requirements for the specified tier level. The security measures in the facility's SSP should address:

- Strategies of prevent, detect, delay, and respond.
- Security gaps identified in the SVA.
- The RBPS associated the appropriate tier level of the facility and its critical assets.
- Objectives for intentional events relevant to the site which include the prevention of:
  1. Intentional release of a regulated chemical substance from a covered facility with the intent to cause harm to the public or employees.
  2. Theft or diversion of a regulated chemical substance.
  3. Contamination of a reactive regulated chemical substance stored in a transportation container.
  4. Damage or disruption to economically critical chemical process with the intent to degrade the production capacity or availability.

Site Security Plans

Under Section 550, the Department must also require that "high risk" chemical facilities develop and implement "Site Security Plans." The statute specifies that the Department "shall permit each facility, in developing and implementing Site Security Plans, to select layered security measures that, in combination, appropriately address the Vulnerability Assessment [for the facility] and the risk-based performance standards for security for the facility."

A SSP must address both the "Security Vulnerability Assessment" for the covered facility and the applicable "risk-based performance standards." To address the Vulnerability Assessment, the plan must identify and describe the function of the measures the covered facility will employ to close the gaps between the existing security measures and the RBPS for their assigned tier.

As published in the ANR, a covered facility's SSP "must identify how the layered security measures selected by the covered facility meet the Department's risk-based performance standards." The Interim Final Rule could specify different RBPS requirements, but the Advanced Notice published an indication of the types of performance standards that may be required:

- Secure and monitor the perimeter of the facility;

- Secure and monitor restricted areas or potentially critical targets within the facility;
- Control access to the facility and to restricted areas within the facility by screening and/or inspecting individuals, deliveries, and vehicles as they enter; including,
  1. Measures to deter the unauthorized introduction of dangerous substances and devices that may facilitate an attack or actions having serious negative consequences for the population surrounding the facility; and
  2. Measures implementing a regularly updated identification system that checks the identification of facility personnel and other persons seeking access to the facility and that discourages abuse through established disciplinary measures;
- Deter vehicles from penetrating the facility perimeter, gaining unauthorized access to restricted areas or otherwise presenting a hazard to potentially critical targets;
- Secure and monitor the shipping and receipt of hazardous materials from the facility;
- Deter theft or diversion of potentially dangerous chemicals;
- Deter insider sabotage;
- Deter cyber sabotage, including by preventing unauthorized onsite or remote access to critical process controls, Supervisory Control And Data Acquisition (SCADA) systems, and other sensitive computerized systems;
- Develop and exercise an emergency plan to respond to security incidents internally and with assistance of local law enforcement and first responders;
- Maintain effective monitoring, communications and warning systems, including,
  1. Measures designed to ensure that security systems and equipment are in good working order and inspected, tested, calibrated, and otherwise maintained.
  2. Measures designed to regularly test security systems, note deficiencies, correct for detected deficiencies, and record results so that they are available for inspection by the Department.
  3. Measures to allow the facility to promptly identify and respond to security system and equipment failures or malfunctions.
- Ensure proper security training, exercises, and drills of facility personnel;
- Perform appropriate background checks on and ensure appropriate credentials for facility personnel, and as appropriate, for unescorted visitors with access to restricted areas or potentially critical targets;
- Escalate the level of protective measures for periods of elevated threat;

- Address specific threats, vulnerabilities, or risks identified by the Assistant Secretary for the particular facility at issue;
- Report significant security incidents to the Department;
- Identify, investigate, report, and maintain records of significant security incidents and suspicious activities in or near the site;
- Establish official(s) and an organization responsible for security and for compliance with these standards;
- Maintain appropriate records; and
- Address any additional performance standards the Assistant Secretary may specify."

The types and robustness of measures necessary to satisfy these requirements will of course depend on the risk-based tier of the covered facility and critical assets. Covered facilities will also have a continuing obligation, which will vary based on their risk-based tier, to maintain and periodically update their SSP. Further, the statute specifically prohibits the Department from rejecting a SSP, simply because it does not incorporate a specific type of security measure: "[T]he Secretary may not disapprove a Site Security Plan submitted under this section based on the presence or absence of a particular security measure."

Example of RBPS

To illustrate the application of RBPS by tier, assume a facility is found through the Top-Screen and SVA to be in one of the higher tier levels, and therefore demands a higher security posture. Referring to the RBPS previously mentioned in the ANR, DHS may clarify the general performance, but not specific requirements, of the desired level of security given the potential consequences.

One of the RBPSs the ANR refers to is 'deter vehicles from penetrating the facility perimeter, gaining unauthorized access to restricted areas or otherwise presenting a hazard to potentially critical targets', Figure 2 shows a protective scheme that may be commensurate with the higher level of security desired of a higher tier facility.

This could include barriers surrounding the restricted areas of the facility including retractable active vehicle barricades and a robust fence (aircraft cable reinforced chain linked fence) as shown on Figure 3.

As shown in Figure 4, another RBPS may concern the issue of site monitoring. This could be worded such as the ANR language 'secure and monitor the perimeter of the facility' or to 'secure and monitor restricted areas or potentially critical targets within the facility'.

The risk-based performance standards will allow a covered facility flexibility, and the ability to select a set of security measures that, when applied in sufficient layers, meet the intent of the RBPS.

If no specificity is allowed, then the RBPS must be detailed enough to be measurable, but not specific to any given security measure, while providing a credible level of security against a specified threat.

Approval of Vulnerability Assessments and Site Security Plans

Section 550 states that "the Secretary shall review and approve each vulnerability assessment and site security plan required under this section." See Pub. L. No. 109-295, sec. 550(a). To implement this provision of the statute, and consistent with the implementation plan discussed herein, the Department will require all covered facilities to submit Security Vulnerability Assessments and Site Security Plans to the Department. The Department will review and approve or disapprove each assessments in accordance with proposed § 27.215. If the Department approves the assessment, the Department will issue a letter to the covered facility so stating.

After a review of the Site Security Plan, the Department will preliminarily approve or disapprove it. In the case of a preliminary approval, the Department will issue a Letter of Authorization to the covered facility. After preliminarily approving a Site Security Plan, the Department will inspect each facility in order to determine compliance with the requirements of this part. After the inspection, if the Department concludes that the Site Security Plan addresses the vulnerabilities identified in the Security Vulnerability Assessment, satisfies the risk-based performance standards, and has been satisfactorily implemented, the Department will issue a Letter of Approval to the covered facility. The Secretary may disapprove a site security plan if the plan fails to satisfy the risk-based performance standards established by this section. If the Department concludes that the Site Security Plan has not been satisfactorily implemented, the Department will consult with the covered facility as provided in proposed 27.240(b) and schedule a second inspection.

Conclusion

The Department of Homeland Security has proposed a regulatory framework for enhancing the security of the chemical sector with respect to acts of terrorism based upon the following principles:

- Not all chemical facilities present the same level of risk (consequence, vulnerability, and threat).
- Emphasis should be focused on those that, if attacked, could endanger the greatest number of lives, have the greatest economic impact, or present other

significant consequences, but also where vulnerabilities and threats create high risk.

- The concept is to base facility security on enforceable performance standards tied to the types and severity of potential risks posed by terrorists, but allowing facilities the flexibility to select among appropriate security measures that will most effectively address those risks.
- In creating this regulatory framework, DHS intends to recognize the progress many responsible companies have made to date and will build on that progress.

This regulatory philosophy is both progressive and realistic, as there is no single security solution for the wide variety of applications that constitute the 'chemical sector'. Once implemented, this initiative will significantly improve the overall security posture of high risk chemical facilities across the nation.

Figure 2
Restricted Area and Access Portals

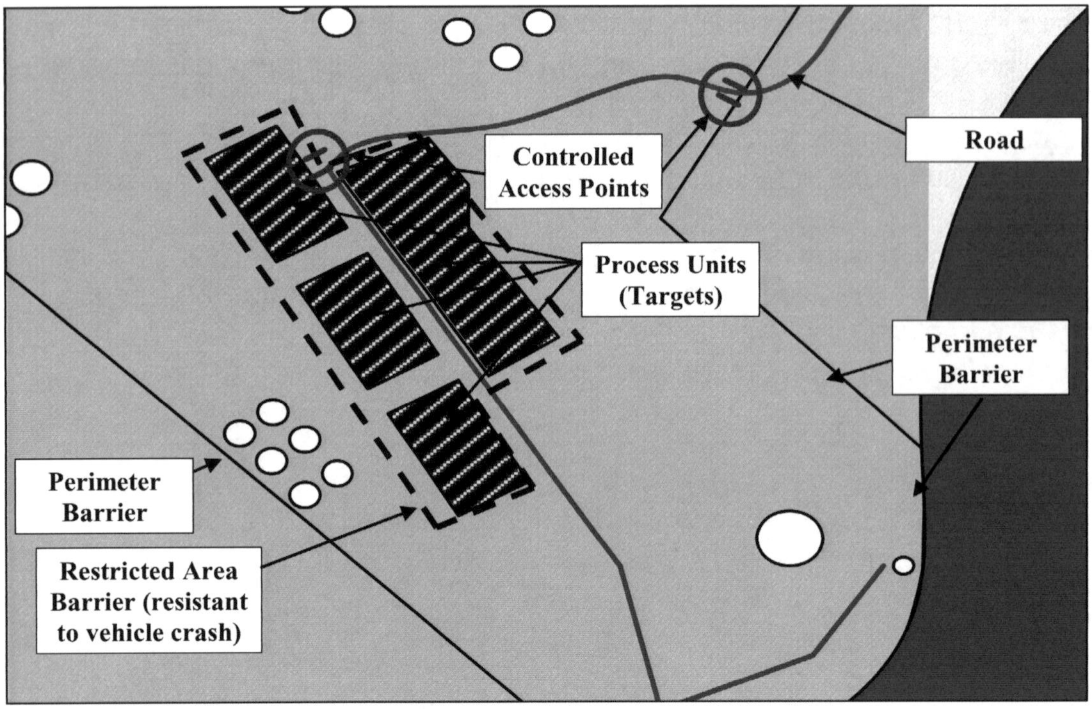

Figure 3
Barriers for Vehicle Access

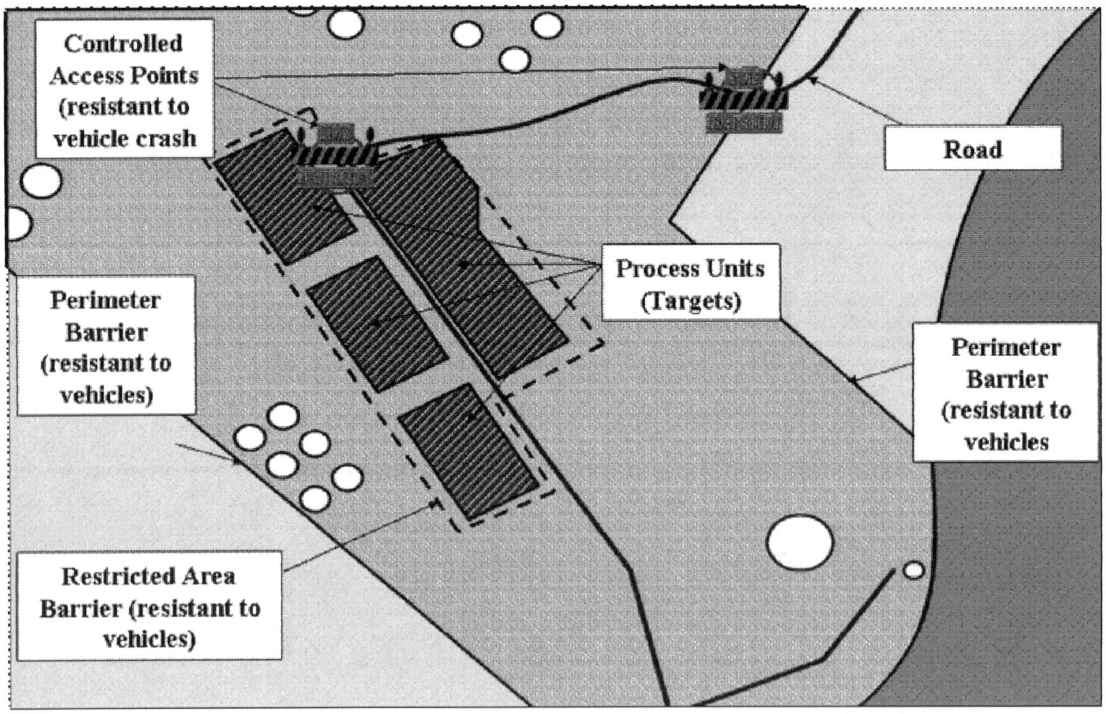

Figure 4
Site Monitoring

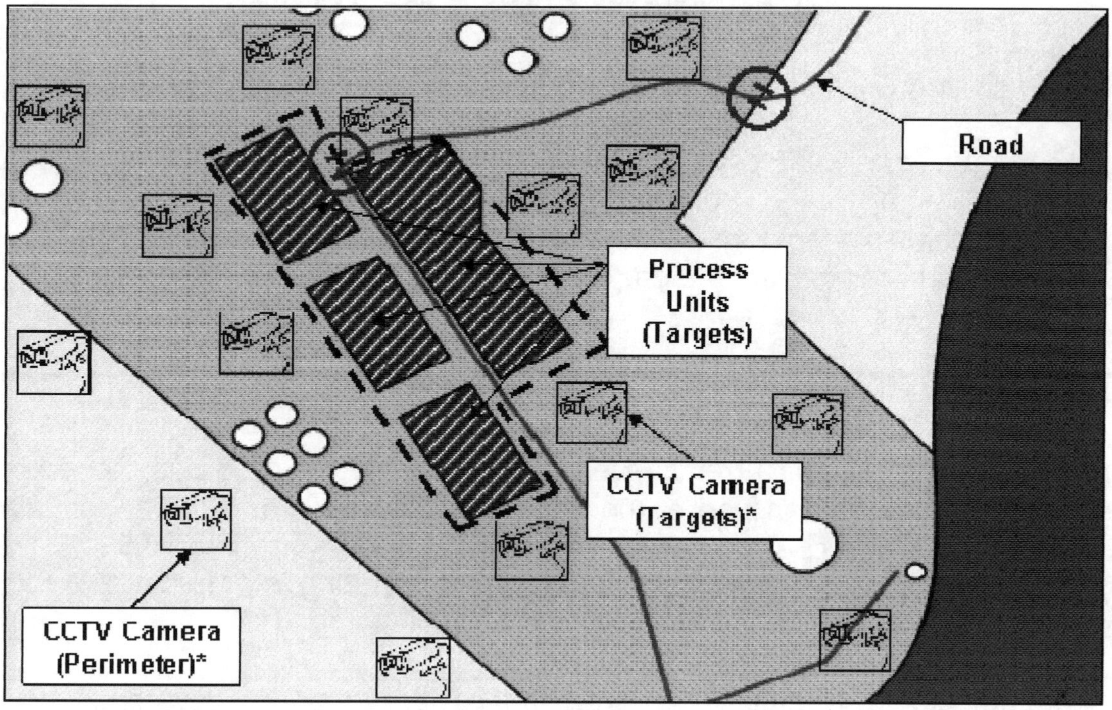

*** CCTV or perimeter intrusion detection system**

---

[1] For the complete docket on the Chemical Facility Anti-Terrorism Standards see the Federal eRulemaking Portal: http://www.regulations.gov.

[2] See Pub. L. No. 109-295, sec. 550

[3] Pub. L. No. 107-295, Nov 25, 2002

[4] Federal Register: December 28, 2006, Volume 71, Number 249, Proposed Rules, Pages 78275-78332.

[5] Testimonies of Moore, David A., and Hendershot, Dennis C., before the Senate Committee On Environment And Public Works, Full Committee Hearing on Inherently Safer Technology in the Context of Chemical Site Security , Wednesday, June 21, 2006.

[6] Mannan, Sam, White Paper – "Challenges in Implementing Inherent Safety Principles in New and Existing Chemical Processes", Texas A&M University, 2002.

OPTIMUM ROUTE SELECTION FOR HAZARDOUS MATERIALS TRANSPORTATION INCORPORATING SAFETY AND COST-EFFECTIVENESS CONSIDERATIONS

Anna Qiao, Energy Solution, Det Norske Veritas (DNV) U.S.A, Inc., 16340 Park Ten Pl, Suit 100, Houston, TX 77084
Nir Keren, Agriculture and Biosystems Engineering, Iowa State University, 102 I Ed II Building, Ames, IA 50011
Dr. M. Sam Mannan, Mary Kay O'Connor Process Safety Center, TAMU, 246 Jack E. Brown Engineering Bldg., 3122, College Station, TX 77843

INTRODUCTION

Hazardous materials (HazMat) traffic levels in the U. S. now exceed 800,000 shipments per day and result in the transport of more than 3.1 billion tons of HazMat annually [1]. According to the statistics from the Department of Transportation, 156,483 HazMat transportation incidents occurred from 1995 to 2004, resulting in a total of 226 fatalities and 3,218 injuries [2].

The large amount of HazMat transportation activities and the associated potential hazards has received increased public awareness on the potential risks of HazMat transportation. Transportation Risk Analysis (TRA) is required by the Code of Federal Regulations for the shipment of certain HazMat. Based on the TRA results, decision makers can select appropriate risk reduction alternatives by demonstrating the effectiveness of various alternatives. It is helpful in making judgments about the tolerability of existing or increased movement levels.

TRA is also helpful in choosing among alternate routes by providing information on the relative risks associated with each route. On the other hand, the vulnerability of HazMat transportation could be reduced by the judicious selection of route. Policy makers are taking steps to improve the safety of transportation systems, including by optimizing the transportation route. This paper presents optimum routing models based on both safety concern and cost analysis, which can be employed by decision makers from both industry and government to solve HazMat logistical problems.

BACKGROUND

Security and Cost Effectiveness Consideration

Risk Assessment on HazMat Transportation

Risk assessment is a sequential process, beginning with an understanding of the level of involvement, the frequency and type of incident, and the consequence for a given incident. The way those components are defined and determined depends on the data available and the purpose of the risk assessment. In the past two decades, attention has been focused on risk analysis of HazMat within transportation networks, and the techniques of quantitative risk analysis initially developed for fixed plants have been extended to transportation risk analysis.

The most common risk measurements used to perform a quantitative risk analysis for a major incident hazard plant, i. e., individual risk and societal risk, are also employed in transportation risk analysis. Individual risk reflects the frequency of a specific health effect at a geographical location. Societal risk represents the overall risk associated with an activity to a particular population.

In 1995, Erkut and Verter proposed that a path between a given origin-destination pair can be represented by a set of road segments, where the road characteristics are uniform within each segment [3]. In 1999, Leonelli et al. proposed a detailed procedure to evaluate individual and societal risk for a transportation network [4].

However, the numerical methodology was difficult to employ directly by decision or policy makers, because only the methodology was proposed without input data, or because the methodology was too complicated to obtain available data as input. For example, for most of the methodologies, incident frequency and conditional release probability data were assumed to be available, but the acquisition of the required data required considerable effort. The methodology should be developed according to the availability of input data, otherwise the meaning of the methodology is in doubt. In this paper, each risk component, including frequency, release scenario, consequence, and resulted fatality, is assessed based on the available data/database. The individual risk in the transportation network is then assessed by a numerical methodology, step by step. The results from TRA will serve as the baseline for the further transportation route selection.

Cost Assessment on HazMat Transportation

Both direct and indirect costs could be involved in the transportation of HazMat. The direct costs include:

- The cost for transporting the material, depending mainly on the amount of the shipment and the length of the route
- The costs for late-delivery of the materials, which is the penalty paid for the late-delivery.

If an accident occurs in the transportation process, additional cost could be accrued, including:

- The cost for the loss of the transported material in the accident
- The cost for compensation, which may involve the personal injuries and environmental damage

If only the direct costs are considered in the transportation, it would be the route with the shortest distance that would always be selected. However, the potential risk associated with different route may not only aggravate both the transporter and society, but also increase the transportation cost indirectly. Leonelli et al. proposed to sum the indirect risk related cost to the direct transporting cost, and to select the route with the minimum total cost [4]. The direct cost can be quantified and relatively easily evaluated, but it is hard to quantify the indirect cost like the compensations to the personal injury or fatality. The trade-off between cost and risk has been debating issue for many years.

In this paper, only the direct cost is accounted for the cost evaluation, the indirect cost is addressed in the form of risk assessment. By this way, the risk factor would not be counted twice. A routing methodology is developed to avoid the tradeoffs between the risk and cost. Since it is relative straightforward to evaluate the direct transportation cost, it would not be explained further in this study.

Optimum Routing Selection Methodology

When considering the simplest deterministic case, Abkowitz [5] and many other authors proposed to solve the routing problems by employing the classic shortest path algorithm. It involves finding the route from an origin to a destination which minimizes a single measure. The measure must be additive across links. The risk draws much concern as well as cost to be used as the measure. List and Mirchandani [6] analyzed tradeoff between risk & cost. Zografos et. al. [7] and List and Mirchandani [6] proposed bi-objective models to address the effects of both cost and risk in routing decision.

In some cases the bi-objective model causes problems to decision makers, i.e., combining both the transportation cost and risk is hard for the decision makers to determine weights that would make such a combination meaningful; and it is also hard to choose the weights that would differentiate equitably between the risk and the transport cost for the entire network.

In this study, two optimum models are presented by taking account of decision makers' perception and avoiding the tradeoffs between cost and risk. The optimum models could be developed and solved in optimization software such as LINGO.

TRA METHODOLOGY

Preliminary Network Analysis

A transportation route can be represented as a risk source consisting of links and nodes over which risk is summed, since a release can occur at each of its points. Each link in the network must have uniform properties at all its points, which can be accomplished by adding nodes to the network.

A transportation route can be considered to be composed of links and nodes. When performing the transportation risk analysis each link must be characterized by certain properties. For a generic link l of a set of links N_{link}, the uniform properties include: the transportation network type that the link belongs to, i.e., road network, rail, inland waterway, or pipeline, the physical conditions including road type and road conditions, vehicle type, and yearly shipment.

After characterization of the links, the risk source has been defined. The next step is to define the possible consequence and the impact area, i.e., the information on environmental condition and population distribution is needed to characterize the transportation network. Then the risk associated with transportation activities in the network can be quantified by assessing both accident frequencies and potential consequences.

Frequency Assessment

The frequency for transportation accident is assessed by incorporating effects of numerous parameters describing the nature of the truck configuration, operation, environment, and road conditions, and it is the function of *Lane # (x_1), Weather (x_2), Surface condition (x_3), Alignment (x_4), Month of year (x_5), Day of week (x_6), Hour of day (x_7), Lane Divided/undivided (y_1), Container capacity (y_3), Container type (y_4), Driver experience (y_5), and Driver status (y_6).* For each link, the *Lane # (x_1), Surface condition (x_3), Alignment (x_4), and Lane Divided/undivided (y_1)* is the same during the entire year; the *Container capacity (y_3)* and *Container type (y_4)* are defined prior to the calculation of transportation risk; and the effects from driver status and driver experience are not considered in this paper, so for each lane the transportation frequency is the function of *Weather (x_2), Month of year (x_5), Day of week (x_6),* and *Hour of day (x_7).* Thus the incident frequency is expressed as $P_{incident}(w, m, d, h)$, where w, m, d, and h are weather, month, day, and hour, respectively. Detailed assessment on HazMat accident frequency assessment is presented in referent [8].

Consequence Assessment

The estimated consequences of an incident involving a shipment of HazMat depends on a variety of factors including the amount released, toxicity of the chemical, health effects from the released materials, population and environmental characteristics adjacent to the incident site, and the weather conditions at the time of the incident. The interplay among these factors produces the risk spectrum for transporting HazMat.

The most popularly used consequence assessment package in the oil and gas industry, PHAST, is employed to assess the transportation incident consequence. Given a release at any point risk source of the transportation route, the potential hazards can be modeled using PHAST, including toxic vapor cloud, flammable vapor cloud, unconfined vapor cloud explosion, pool fire, torch fire, flare, BLEVE, confined explosion, and others.

Risk Indicator

Both individual risk and societal risk are employed as the risk indicators in this study. Individual risk considers the risk to an individual who may be in the effect zone of an incident or set of incidents The individual risk at a certain point of the transportation network is calculated by integrating the consequences associated with all links in the network. Societal risk reflects the frequency of health effects (usually fatalities) in a specific population, occurring as the result of exposure to a specific HazMat. The societal risk is often expressed in terms of frequency distribution of multiple fatalities (f-N curve). Both the individual risk and societal risk are evaluated by risk assessment package SAFETI.

These risk indicators are important in evaluating the potential risk associated with the transportation activity in the network, and more important, they provide the input to the transportation routing models, and make it possible to optimize the transportation activity based on both cost consideration and safety concern.

ROUTING METHODOLOGY

Fuzzy Logic Optimum Routing Methodology

Fuzzy Sets on Risk and Cost

The human perception in the decision making process has been addressed in many existing mathematical models indirectly. However, few of these models have an effective method of measuring the public's perceptions and incorporating these perceptions into the strict mathematical framework required by optimization techniques. In this study, fuzzy logic is employed to provide one way of directly integrating human perceptions.

Figure 1 and Figure 2 represent the linear fuzzy sets used to represent the public's acceptance on the transportation of certain HazMat transportation, with respect to the cost and risk associated with that activity. If an activity has zero risk, its acceptability with respect to risk, μ_r, is unity. As risk exceeds the value A_r, the acceptability decreases to zero at A_r' linearly. A similar fuzzy membership function is used for the acceptability with respect to cost. The mathematical expressions on the membership functions on these two variables, risk and cost, are as follows:

$$\mu_r \le \frac{M_r - (Risk - A_r)}{M_r} \tag{1}$$

$$\mu_c \le \frac{M_c - (Cost - A_c)}{M_c} \tag{2}$$

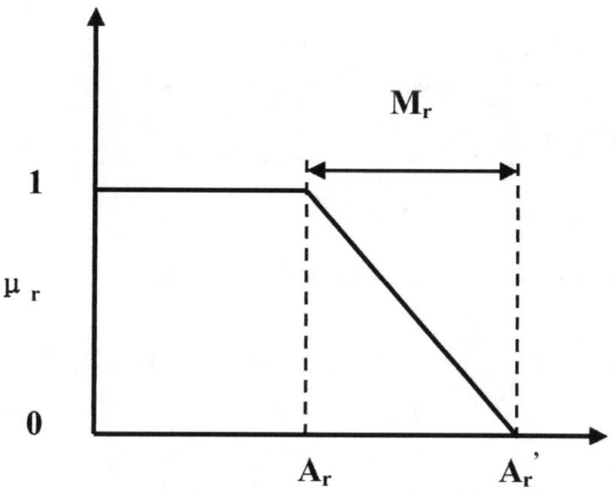

Figure 1. Linear Risk Fuzzy Membership Function.

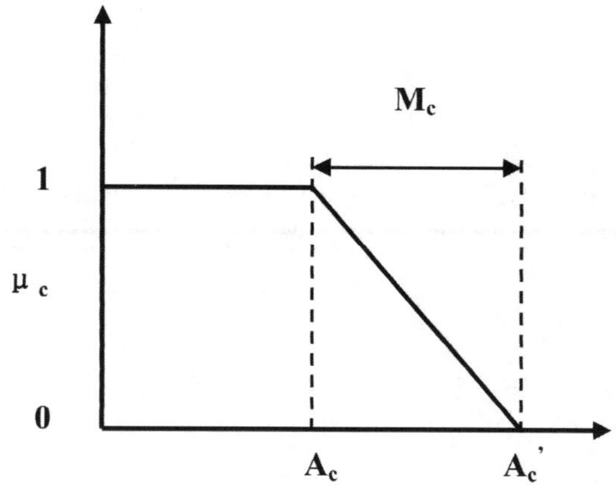

Figure 2. Linear Cost Fuzzy Membership Function.

Fuzzy Logic Optimum Model Setup

For model setup, perceptions on risk and cost are incorporated into the model using fuzzy sets. A single, non-controversial, non-fuzzy objective function is formulated while all fuzzy sets are represented as constraints. In this manner, the algorithm can be used to arrive at a solution that optimizes the objective function subject to the fuzzy constraints as well as the physical constraints governing the transport and HazMat.

The network needs to be defined a little bit in routing process different than in the TRA process. The characterization process is shown in the following part.

- Network G={N, L}
- Link (i, j): from node i to node j
  - l (i, j): length of link (i, j)
  - C (i, j): link cost; R (i, j): link risk
  - T (i, j): number of trucks traveling on link (i, j)
  - Cap (i, j): capacity of link (i, j)
- Node (i)
  - b(i): truck's supply or demand
  - b(i)>0 for supply nodes, <0 for demand nodes, =0 for other nodes

All the above parameters are known or could be determined directly by users except the capacity of link (i, j). It is determined by combing the risk acceptance and the risk associated with the link. The Dutch government has general risk criteria for fixed facility, which is the individual risk less than 10^{-6}. This criterion is implemented in transportation decisions in our study. Since transportation risk is additive, the individual risk caused by each link, R(i, j) should be less than 10^{-6} to ensure that the overall risk associated with the whole transportation route can meet this criterion. Thus the capacity of link is determined as:

$$Cap(i,j)=\frac{10^{-6}}{R(i,j)} \tag{3}$$

The objective is to maximize the overall acceptability of the routing solution with respect to cost and risk. Mathematically, this objective is expressed as

$$\max = \mu_r + \mu_c \tag{4}$$

This objective function is subject to several mathematical constraints that are necessary to physically define the HazMat transportation problems:

- Mass balance for node
- Link capacity constraint for each link
- Fuzzy membership function for total cost and risk

The first constraint ensures that all the HazMat originating at the origin is transported though the transportation route and delivered to the destination, and the truck's supply and demand in the intermediate nodes are 0. The constraint is expressed as:

$$\sum_{(i,j)\in A} T(i,j) - \sum_{(i,j)\in A} T(j,i) = b(i) \qquad (5)$$

$$b(0) = Number\ (Transportation\ Truck)$$

$$b(i) = 0, \qquad (1 \le i \le N-1)$$

$$b(N) = -Transportation\ Truck$$

The second constraint guarantees the truck transport capacity for each link. The number of trucks passing through each link can not exceed the capacity of that link. By this way the risk associated with this segment is guaranteed not beyond the risk acceptance level.

$$0 \le T(i,j) \le Cap(i,j) \qquad (6)$$

The acceptability of the policy includes acceptability with respect to both cost and risk. The overall cost reflects the total cost of transporting HazMat. Mathematically, the cost is defined as:

$$Cost = \sum_{(i,j)\in A} T(i,j) \times C(i,j) \qquad (7)$$

The acceptance level to cost is defined in Equation (1).

The overall policy risk also reflects the total risk associated with the HazMat transportation. Assuming independence, this objective is mathematically defined as:

$$Risk = \sum_{(i,j)\in A} T(i,j) \times R(i,j) \qquad (8)$$

The acceptance level to cost is defined in Equation (2).

Solutions to this model formulation completely specify the routing strategy corresponding to the maximum acceptance. The model solutions are governed by the physical constraints defining the routing process and driven by overall risk and cost considerations. The solution could be obtained from optimization software LINGO once all data are input to the model.

Users – Determiner Optimum Model

Fuzzy logic optimum model presented in Section 4.1 is a good routing approach when users already have the perception towards the risk and cost associated with HazMat transportation and the acceptance levels of risk and cost can be changed to some degree.

However, if the decision makers already have a concept about the exact data of maximum acceptable risk, the model presented above needs to be modified to meet users' demand. In this section, the optimum route under given risk tolerance is selected based on an alternative optimum model.

Since the maximum acceptable risk has been defined by users, the objective function in the model is only the cost associated with the transportation activity:

$$\min = Cost = \sum_{(i,j)\in A} T(i,j)\times C(i,j) \tag{9}$$

The constraints to the objective function include:

- Risk constraint set by decision maker
- Mass balance for each node i
- Link capacity constraint for link (i, j)

The first constraint is defined by decision makers regarding the acceptance of risk:

$$Risk = \sum_{(i,j)\in A} T(i,j)\times R(i,j) \le R_{constraint} \tag{10}$$

Here $R_{constraint}$ is given by decision maker.

The constraints on mass balance and link capacity are the same as those described in Section 4.1.

After the setup of the model, the optimum route can be obtained by using linear optimization software LINGO to get the solution to the input model. With the change of risk acceptance level, different routes can be utilized to ensure the minimum cost. The decision makers can then choose the optimum routes based on their own requirements for acceptable risk level.

CONCLUSIONS

Both cost and Risk are considered in optimizing the route of HazMat transportation. The TRA is a process resulting from the interaction between the transportation network, the vehicle or traveling risk source, and the impact area. It provides the baseline for the optimum route selection.

This paper outlines two solution methods for the HazMat routing problem. Fuzzy concept is employed in the linear optimum models, which takes account of decision makers' perception. The tradeoffs between the cost and risk are avoided in this way. The optimum models could be developed and solved in optimization software such as LINGO. The proposed routing methods could be easily employed in supporting the strategic, tactical, and operational decisions in HazMat transportation system.

References:

1. Office of Hazardous Materials Safety (OHMS) (Accessed 2005), "Hazardous materials transportation guides". http://ntl.bts.gov/DOCS/hmtg.html.

2. US Department of Transportation (US DOT) (Accessed 2005), "Incidents, deaths, and injuries by mode and incident year". http://hazmat.dot.gov/pubs/inc/data/tenyr.pdf.

3. Erkut, Erhan and Verter, Vedat, "A Framework for Hazardous Materials Transport Risk assessment", Risk Analysis, 15 (5), pp. 589-601.

4. Leonellli, Paolo, Bonvicini, Sarah, and Spadoni, Gigliola (1999), "New Detailed Numerical Prodedures for Calculating Risk Measures in Hazardous Materials Rransportation", Journal of Prevention in the Process Industries, 12, pp. 507-515.

5. Abkowitz, M. P. and Cheng, P.D-M (1988), "Developing a Risk/Cost Framework for Routing Truck Movements of Hazardous Materials", Accident Analysis and Prevention, 20 (1), pp. 39-51.

6. List, George and Mirchandani, Pitu (1991), "An Integrated Network/Planar Multiobjective Model for Routing and Siting for Hazardous Materials and Wastes", Transportation Science, 25 (2), pp. 146-156.

7. Zografos, Konstantinos G. and Androutsopoulos, Konstantinos N. (2004), "A Heuristic Algorithm for Solving Hazardous Materials Distribution Problems", European Journal of Operational Research, 152, pp. 507-519.

8. Qiao, Yuanhua, Mannan, Sam M., and Gentile, Michela (2004), "Fuzzy Logic Methodology for Accident Frequency Assessment in Hazardous Materials Transportation", CCPS Annual International Conference & Workshop, June, Orlando.

Create and Deliver Effective Training Programs Using Instructional Systems Design and E-Learning, Traditional and Blended Learning

Philip M. Myers, Advantage Risk Solutions, Inc.
P.O. Box 510, Sunbury, OH 43074 USA
(740) 965-6304 pmyers@ARiskSolution.com

Brent Watson and Molly Watson, VisionWeaver Media
37 ½ East Granville Street, Sunbury, OH 43074 USA
(740) 965-9723 bwatson@VisionWeaverMedia.com

ABSTRACT

In an ever changing business environment, enterprise-wide and plant site-based training initiatives take on added importance. There are training requirements to address Process Safety Management (PSM), risk management, security, and EHS needs and regulations as well as training programs to address related internal company and business focused directives and initiatives. These ever-changing topics and increasing requirements present ample challenges to companies attempting to provide quality training to their workforce for human performance improvement.

Much has been written regarding the promise of e-learning technologies in meeting these needs. However, technology itself will not solve training challenges. Training programs built using a formal, structured approach such as Instructional Systems Design (ISD) ensures that a highly effective, quality training program is built and consistently delivered. Instructional Systems Design utilizes the following sequential steps: analysis, design, development, implementation, and evaluation. Use of Instructional Systems Design assures that the proper target audience and specific instructional needs are identified, instructional objectives are established, an appropriate instructional strategy is selected, and the evaluation strategy is devised and implemented.

Care must be taken to assess and implement the appropriate delivery mechanisms based on each company's unique organizational needs and operating environment. A coordinated and thorough analysis should be undertaken within organizations to lessen potential missteps in implementing training initiatives, through use of audience and technical infrastructure assessments. Integrated, collaborative learning environments allow incorporation of newly developed and existing training materials while providing a framework for innovative approaches to the training effort. Secure access and content delivery, as well as employee tracking are requirements for adequate protection. New technologies allow engaging and efficient mechanisms for dissemination of training materials, but must be assessed in relation to an overall training plan. When properly applied, Learning Management Systems, Web 2.0, social media and multimedia content provide efficient, potent tools for workplace training.

A summary of related issues for Instructional Systems Design of training programs using the ADDIE model, as well as a review of new technologies is presented herein.

INTRODUCTION

In the areas of Process Safety Management (PSM), risk management, security, and in Environment, Health and Safety (EHS) human performance is critical. Businesses and society rely upon persons involved in these activities to protect and provide competitive advantage to the business while also safeguarding the workforce, public, and environment. Unfortunately, accidents continue to happen periodically, some of them quite significant. In 2005, three significant industry impacting incidents occurred – the Graniteville, South Carolina train derailment involving the release from chlorine tank cars; the BP refinery explosion in Texas City, Texas; and the Buncefield Oil Depot fire in the United Kingdom. In each of these accidents, as is the case for most events, training is considered to be a contributing factor. Many incident investigations lead to recommendations advocating training or refresher training to prevent future occurrences. A number of actions involved in Management of Change (MOC) programs involve additional training. In everyday operations at processing plants many Process Hazards Analysis (PHA) recommendations include the need for improved training.

Of course there are many other operational, compliance and business applications where training is essential to business performance improvement. There is a continual push to improve performance—and therefore a need for effective training programs to affect both behavior and performance. "As business leaders focus on top-line growth, the value of learning is becoming better understood and more relevant. Those leaders who understand how to drive business results in an increasing competitive, global environment recognize that a better-trained workforce improves performance, and investing in employee learning and development is critical to achieving success." [1] A number of sources and corporate CEOs have indicated a substantial return on corporate training investments – sometimes factors of two or three or even as high as thirty. As in all aspects of business, training programs need to meet business and compliance needs cost-effectively. There are often significant restrictions on training in terms of both timing and economic constraints. So in spite of the critical need for highly effective training, and the clear benefits, often training can get "cut short". Sometimes it is due to a lack of knowledge regarding how to go about creating effective training, what can be achieved in the process, and the business return to the organization.

These constraints may be more often encountered when the training is viewed by the sponsors as solely a compliance activity.

One (now outdated) training "model" historically used in the process safety and EHS fields was to have the company expert in a given technical area, with years and years of experience, pull together his ideas and "throw together" some slides for a course to convey this information to others. This also may be through memos, procedures, meetings or classroom style lectures and training. However, while the company Subject Matter Expert (SME) certainly can play an important role in the design or development of training material, the SME is not necessarily a good trainer. Another mistake sometimes made by technical experts is that they view training similarly to writing a book on the subject. So while all of that expertise is available to the company, it may not be delivered in a cohesive manner that his well suited to learning. The older model was of classical "teaching" – focusing solely on the teacher and the student - that is an expert conveying knowledge to students intended to absorb the content.

Current models for learning typically utilize a "systems" view. In a systems view, all aspects that can impact learning are addressed. Further, in creating effective training materials the approach taken is to use current adult learning pedagogies and to divide the content into discrete learning objectives. Specific skills and behaviors can then be improved through master-able objectives together with exercises and practice for reinforcement. To accomplish all of this, a systematic approach to training and human performance improvement needs to be applied. There is clearly a need for a structured training approach to ensure consistent, quality training that is complete, accurate, and delivered effectively to meet specifically identified business needs. Use of a structured approach is good for meeting new training requirements and for significant improvement of existing courses and programs. Within the areas of process safety, risk management, security and EHS there are a wide range of training programs in use – from quality training assembled using a systematic approach to those put together informally. There is also a wide range of available media and technologies for use in training, from more traditional ones to promising new technologies recently developed and currently developing. Instructional Systems Design (ISD) is a valuable tool that can be used to address business needs for increasing human performance through a structured, systematic approach and integrating multimedia and new technologies.

Formal Approach – Instructional Systems Design and the ADDIE Model

Instructional Systems Design, or ISD, uses a holistic, systematic approach to training. It draws from the technology of education and instruction as an applied science. It is rooted in cognitive and behavioral psychology. Theorists such as B.F. Skinner studied operant learning where the training focus is on observable behaviors while Benjamin

Bloom developed his taxonomy. Bloom's taxonomy is still used in education and instruction today. Bloom's taxonomy covers three domains of learning – cognitive (what we know), psychomotor (physically what we do) and affective (how we feel and our attitudes). The cognitive portion breaks down into knowledge, comprehension, application, analysis, synthesis, and evaluation – the last three of these are higher level thinking. Knowledge includes recall and remembering previously learned material. Comprehension addresses the ability to translate it and grasp the meaning of the material. Application then involves the capacity to generalize and use learned material in new and concrete situations. Analysis involves the aptitude to break down information and discover – breaking down material into parts so that it can be easily understood. Synthesis then is the faculty for being able to compose or put together material to form a new whole. Finally, evaluation is being able to judge the value of material learned for a specific purpose.

In the 1970s, with the advent of computers, an information processing approach was taken in training. This was followed by cognitive load theory in the late 1980s – recognizing that the format of materials does affect learners. In the 1990s the focus turned to media options involving CD-ROM based training and the use of intranets. Cognitive load theory is now gaining international acceptance. The current emphasis has moved from a focus on training courses in an "installment" view to human performance improvement and human performance technology.[2]

High quality, effective training programs currently utilize a systems approach. That is, training is viewed as a system that involves the trainer, learners, material, the learning environment and all that is required to create and deliver a training program. Instructional Systems Design (ISD) is a scientifically derived process "intended to optimize gains in knowledge and performance from precisely engineered (and designed) instruction." [3] ISD ensures that the:

- Proper target audience is identified
- The instructional need is identified
- Instructional objectives are established
- Instructional strategies are selected
- Evaluation strategy is devised and implemented

There are numerous formal ISD models, though the primary ones include the ADDIE model,[4] and the model by Dick, Carey and Carey ("The Systematic Design of Instruction") [5] and the model by Kemp. In fact, nearly all ISD models utilize the ADDIE model as its basis, with variations. The ADDIE model is typically applied to both products and programs; the Dick, Carey and Carey model to programs and curriculum; and the Kemp model to large scale projects. The ADDIE model is the focus of this paper as it is the most common ISD model, it serves as the fundamental basis for nearly all other ISD models (including those named here), and provides a valuable tool for training specialists in creating and delivering effective programs. ADDIE is an acronym that stands for:

- **A**nalysis – learner characteristics, environment, and infrastructure
- **D**esign – specific objectives, instructional strategies
- **D**evelopment – create instructional materials
- **I**mplementation – deliver the training program
- **E**valuation – verify that the objectives (learning, performance, business) have all been met. In addition, evaluation takes place at each stage (formative evaluations) and at the "end" (summative evaluations).

The ADDIE ISD model provides a rigorous, organized approach or process from beginning to end for training programs. In its simplest form, the goal of ISD (and the goal in using the ADDIE model) is to systematically improve human performance. The ADDIE model, used by a training specialist, saves time while producing a high quality product as compared to projects where structure is missing. ADDIE is a step-by-step process that is used to plan and create training programs. Figure 1 shows a simple schematic of the ADDIE process. As there are numerous variations in ISD models, there may be slightly different placement of specifics (e.g. objectives and level or degree of specificity, pilot programs). However these minor variations are not important. What is important is use of a systematic, structured approach. The ADDIE model came about as a result of the need to train fairly technical, complex tasks to large numbers of people both quickly and effectively. The ADDIE process covers the entire training process from asking "What do our personnel need to learn?", to "Did they learn what is necessary and has performance improved?" to "What can be revised to improve training?"

Many persons involved in training know of the needs, but don't really have knowledge of the formal training process. In fact, most of those persons that could benefit greatly from the ADDIE process are not aware of it. ADDIE can provide a structured guide or roadmap for the entire training project.

ANALYSIS

Analysis is the first step of the ADDIE process once a training need is identified. This step is neglected most by persons who do not know of or do not understand the ISD process, and may view it as a waste of time and money. However, over the course of most training projects it saves time and money – the right tools are applied to the identified need with a desired outcome. In the analysis step, we ask the "Who?", "What?", "Where?", "Why?" and "By whom?" types of questions. We analyze the

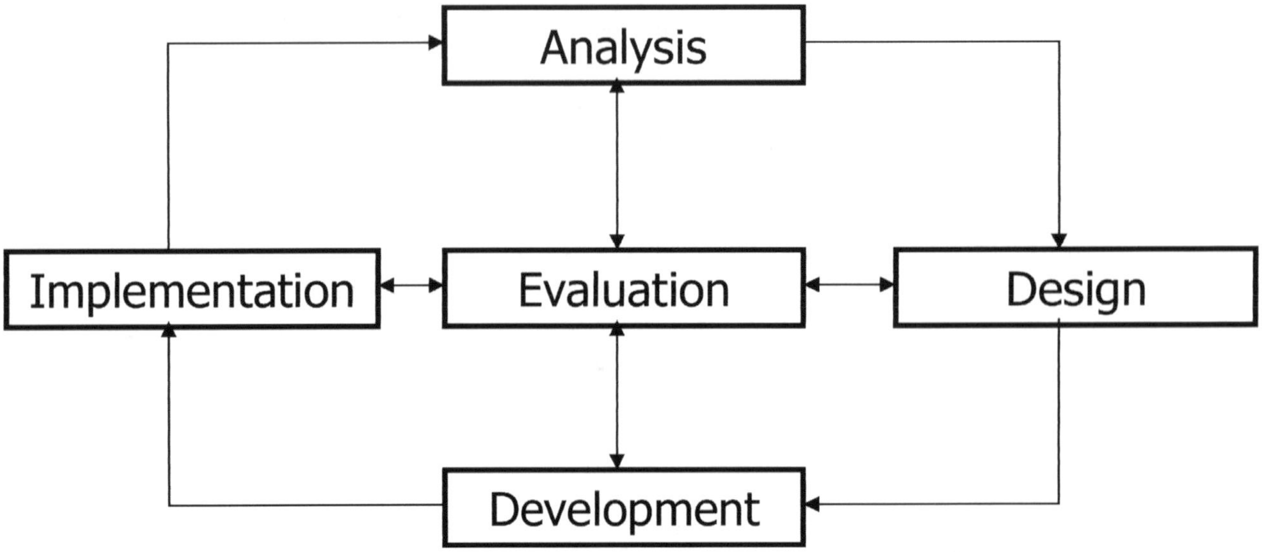

Figure 1. Flowchart for the ADDIE Instructional Systems Design Model

business goals, learner current capabilities, desired capabilities, and materials and resources. Part of this step also includes drafting of the overall program objectives, addressing all three areas – business, performance, and learning objectives.

An important aspect of the analysis step is to determine who should be involved at this stage. Typically these persons include a training specialist that will work with :

- Project sponsors (executives, senior management) – to discuss business goals
- Subject matter experts - to highlight existing resources and describe undocumented knowledge
- Software developer/programmer and Information Technology (IT) personnel – to address software programming and infrastructure
- Target audience representatives – to show current knowledge and skills and to serve as a good test of the materials later
- Training department leader – to discuss overall expectations and objectives and for day-to-day interaction regarding the training program

Primary steps within the analysis phase include:

- Assess the needs to determine overall goals and objectives (and instructional goals)
- Identify the target audience
- Determine the existing skills, knowledge, behavior, performance
- Determine the desired skills, knowledge, behavior, performance
- Identify the gap (performance gap)

- Discovery – general review of existing materials and resources
- Choose the instructional setting (overall)
- Initial technical infrastructure analysis and assessment
- Validate this phase
- Provide updated estimate of training costs

Some of the characteristics of the target audience may include: education and experience, background of the learners, experience in current or similar jobs, job performance requirements (compared to existing skills), cultural and language issues, perceptions, learning styles, the number and location of learners, motivation, mental or physical characteristics, and biases or special interests.

When determining the desired situation, the training specialist is effectively conducting an instructional analysis. Through this activity we identify the instructional goal and determine step-by-step what persons must accomplish to achieve the instructional goal. It is essentially a list of everything the learner must know at the conclusion of the course. The gap analysis is a critical step as this will be the focus of the instructional material. The discovery process can be used to identify existing materials and resources, collect information, analyze the information received, and confirm research to this point. A constructive discovery effort can quickly bring the training specialist "up to speed", places learning at the front end of the process, and reduces the time subject matter experts (SMEs) may need to invest. It can also save time and money backtracking. The initial technical infrastructure analysis should address the existing infrastructure and capabilities, the desired infrastructure and capabilities, and identify the gaps and needs. It is important that if the training program is likely to use e-learning components that the needed infrastructure is or will be made available.

Some of the typical questions to ask in the Analysis phase may include:

- What are the business needs driving the analysis?
- What are the goals and objectives for the course or program?
- How would you define success? (both for the learner and for the project)
- How will you measure success?
- Who is the intended audience?
- What do learners already know?
- What resources are available?

In the analysis phase of ADDIE, the learner analysis provides the starting point, business goals provide the final destination, and the instructional analysis provides step-by-step (high level) information to take the learners from beginning to end.

It is also a significant point to remember that in identifying instructional goals for the training program that they are achievable. It is easy to set lofty goals but much harder to achieve them. The project goals should be achievable and must also match the level of commitment from the sponsors (time, money, culture, etc.).

A key question to ask again near the end of the analysis phase is whether training will, in fact, meet the business need or if another solution is required.

DESIGN

The next phase in the ADDIE process is design. The training program and material design is based upon and utilizes the information gained and results of the analysis phase. The emphasis of the design phase is to design business, performance and learning objectives that form the basis for all of the instructional material. Instructional strategy is also a key component of the design phase. The primary steps in design are as follows:

- Design business, performance, and specific learning objectives (formal, specific ones). The objectives should be written using the "condition", "performance" and "criteria.
- Design instructional material using the objectives and other criteria or restrictions such as time limitations and available resources
- Sequence and structure learning objectives
- Plan/lay out instructional strategies
- Design performance test or assessment materials
- Determine entry behaviors and abilities
- Write the instructional design document
- Write the technical infrastructure plan document
- Specifically address security
- Validate

The business objective defines for the training project how the results of the instructional material will impact and benefit the company. Performance objectives then define the desired performance and behavior on the job as a result of changes from participation in the training. The learning objectives define what is taught through use of the instructional materials and can be measured or assessed at conclusion of the training.

When designing the objectives it is important to consider the condition, performance, and criteria. The condition addresses the following questions: "Under what circumstances will the behavior occur? Will they have access to any job aids or other materials? The performance aspect considers "What is the learner doing when successfully demonstrating achievement of the objective?" And finally, the criteria addresses things such as "What is the required quality, accuracy or speed of the performance? Objectives should be written to ensure they are specific, observable, measurable, achievable, relevant and track-able.

After the objectives are written, they should be grouped and sequenced in a manner to enhance learning. A competency map can be used as well to aid in this process. There are numerous options for sequencing the objectives, including: step by step, part to whole, whole to part, known to unknown, general to specific, concrete to abstract, simple to complex, facts to principles, principles to applications, most to least important, chronological order, orientate-enhance-synthesize, DEDICT (demonstrate, explain, demonstrate, instruct, critique, test) and IDEA (introduce, demonstrate, explain, apply) to name just some of the many possibilities.

The instructional strategy and tactics include both the structure of the training program and the methods and activities used. Basic decisions in terms of the course structure include whether it will be instructor-led training, self-study/self-paced or some combination of the two. The course structure also entails decisions regarding the use of traditional instructor-led classroom training, e-learning, and blended learning. The instructor-led classroom setting is a familiar, traditional learning environment. E-learning is all encompassing, in terms of computer, web, and technology assisted learning. For example, it may include Computer-Based Training (CBT), Web-Based Training (WBT), synchronous e-learning, web-based classrooms, self study guides and quick references, conference calls, video conferences, podcasting, and use of other new technologies such as wikis, blogs and more. Blended learning involves the use of both traditional training together with e-learning. It may also be a combination of both self-paced and synchronous learning. Many organizations are discovering that blended learning is often a highly effective solution to business training needs – for both technical and other types of training. The selection of media to be utilized as well as activities to teach and reinforce desired skills naturally follow selection of the instructional strategy - and is intertwined with use of the various possible e-learning tools. There are also a variety of tactics and methods that can be used to support instruction and learning. Methods and activities may include: group discussions, scenarios, mnemonics, modeling, drills, exercises, and applied practice to name just a few of the options.

A key step in the design phase of ADDIE is to design the performance tests and assessment materials. This includes assessment of both knowledge and application. A simple analogy is that a person, in order to secure a drivers license has to pass both a written and a driving test. An individual can know all of the rules of the road can still be a poor driver behind the wheel. On the other hand, a person can be skilled at the mechanics of driving, but if unaware of the rules of the road and driving laws, may also dangerous. To be a good driver you must have the knowledge, skill in application, as well as safe behaviors. So, in training it is important to assess knowledge, and application – and sometimes behaviors as well.

The design phase also includes determination of entry requirements (or entry behaviors) for each training course. This may include: experience, skills, knowledge, behavior, and performance that are required of learners to be able to begin the instruction. Acceptable equivalencies may also be written for the entry requirements. For example, a training course requirement may be that to be trained in PHA leadership, the learner must possess a four year college degree in a related field. An equivalency for this requirement may be ten years of hands-on operating experience.

A significant step in the design is to write the design document. It addresses the course structure and instructional strategies (e.g. methods, tactics). The design document should contain at least the following five elements: the overall learning approach, identify the structure and media to be used, cluster and sequence objectives, describe activities and exercises, and review the assessments. The written design document specifies the final training program. This does not have to be a thick report, but a concise document that can be used to build consensus and agreement before course development begins. It can also serve as a helpful quality check point.

For training courses involving e-learning or blended learning, it is similarly important to document the technical infrastructure plan and to address security. This concise document will serve the same function to infrastructure as the instructional design document does for the learning materials.

All of the work conducted in the design phase should be validated through formative evaluations, prior to proceeding to development. Rapid prototype development may be a useful tool in some circumstances, as outlined in the following section.

It is essential that throughout the design phase (and later the development phase) that the training developer/specialist thinks from a learner's perspective. Designing material while thinking from a subject matter expert's view, may result in an ineffective training program, skipping needed steps in the learning process and/or designing a course that is too complicated and frustrating for learners. The course designer/developer must view the training with "fresh eyes". This may be even more important in the case of working with an existing program to improve it or to make use of e-learning and blended learning.

In the design phase of ADDIE, we decide what we are teaching and also what we are not teaching. For example, leaders for PHA, LOPA (Layer of Protection Analysis) and SVA (Security and Vulnerability Analysis) studies often have to be familiar with the use of various software tools. One up front decision to be made in the design phase is whether the course will include learner training on the technical method and techniques only – or if it will also include training on the associated software program as well. These types of decisions should be carefully considered, as the most desirable and effective format for teaching the technical aspects of the technique may not be well suited for learning the software applications.

Course developers who are somewhat inclined to skip the instructional design phase often are the ones who could most benefit from its use. These individuals may not have the formal training or experience to skillfully make the necessary design decisions without systematically carrying out this phase. Naturally, the quality of the course, even with use of the ADDIE process, will still depend upon the skill of training designers.

Rapid Prototype Development – Optional

Some training companies or training specialists may suggest rapid development of a prototype for a more visual example of what is planned from the design phase. Originally, rapid prototype development arose as a result of significant timing constraints on the overall process. In an attempt to speed things along, rapid prototypes have been developed. Prototypes may be developed as soon as early in the ADDIE design phase, or in the initial development phase. Often rapid prototypes may be used in conjunction with e-learning and programming activities to get a "model" quickly to review. It can be helpful in a number of situations and may be advised by the training specialist too – when it can be constructively and effectively be used as part of the ADDIE process to aid in design and/or development. Prototypes can also be developed for traditional training materials as well, such as in the case of a prototype lesson plan or exercise. For classroom training and web-based classrooms for Virtual Learning Environments (VLEs) it could also include sample pages from the participant and instructor manuals. For e-learning such as CBT and WBT, the prototype could include use of pre-production storyboards to depict text, animation, images, layout and voiceover details. Blended learning prototypes can include a combination of the above materials, for instance story boards and sample pages from the manuals.

Straightforward courses may not use prototypes at all, or use only simple ones. However, with large and complex training projects involving e-learning and software development or programming, prototypes can sometimes become a helpful component. There are both potential advantages and disadvantages to rapid prototype development. The potential advantage is to develop a model early on that can be seen and used to quickly make changes – thereby saving time and money. The disadvantage is that by too quickly developing prototypes, you may bypass some of the desirable structure and systematic nature of the ADDIE process. In this way, the project could get caught up in endless iterations of prototypes without taking the time to properly identify objectives and design the training accordingly. In short, the project could "miss the mark" and end up delaying the training and increasing the overall cost. Rapid prototypes should, therefore, be utilized with due care.

DEVELOPMENT

Sometimes people want to skip right to the development phase (bypassing both analysis and design) of training. However, you do so at a big risk. It is difficult to develop effective training programs that consistently meet the learning, performance, and business objectives on an ongoing basis without information from the analysis and without the design document. It is analogous to moving across the country by simply one day getting into your car, driving to a location far away without a reason or goal; departing without preparing either the vehicle or yourself, leaving without knowing or checking out anything about the area (jobs, schools, amenities, communities), and traveling without any directions or maps. If you happen to get to your destination, and it all turns out well, it is by pure luck – and probably with many missteps along the way. It is possible for it all to go well? Yes, but the chances are low. A structured, systematic approach to training, starting with analysis and design prior to development is a superior approach – one that you can confidently use to develop a successful training program.

It is important to remember, however, that the ISD process is a tool to aid training programs and the training developers are not rigidly tied by it. It can be modified and streamlined if needed, especially for smaller and simpler courses. The primary steps in development are:

- Use results from the analysis and design steps to begin development of course materials
- Review existing materials
- Develop and integrate instructional strategies and choose the specific delivery system and methods
- Create a prototype
- Develop instructional materials
- Review and revise instructional materials
- Develop and/or program software tools
- Develop hardware and technical infrastructure, as needed
- Secure environment and access
- Review, revise and validate – prototypes, tabletop reviews, pilot training
- Final changes and quality reviews

Existing available resources are typically reviewed to keep from "reinventing the wheel" and to make the process as cost-effective as possible. However, a break from the past may be needed to create the most effective materials. While developing the materials and reviewing instructional strategies to be used, it is also valuable to consider different learning styles and alternate methods of teaching to reach all learners and keep the training interesting. Prototypes developed can come in a number of forms, such as lesson plans, presentation materials, participant manual samples, assessment materials, software tools, etc. This phase of ADDIE should result in the development of all instructional materials. This encompasses instructor materials (e.g. lesson plans, teaching aides), assessment tools, and participant materials. Lesson plans should be developed for use by trainers so that the course content can be consistently delivered, and meeting all objectives of the training. Lesson plans should include the learning objective, module content and timing, performance criteria and suggested instructional strategies (specifically including assessment).

Several techniques can be used to review, revise and validate the training materials developed. This can include reviews of prototypes, a tabletop review, and pilot training for example. In each of these it is important to view the course and materials from the learner's perspective. A review of prototypes can be used to get a visual model of the training deliverables. The tabletop review of all materials by SMEs, the training department leader, and training specialists together is a helpful step in the process as well. The materials should be reviewed for both completeness and accuracy.

The best test for newly developed or improved training materials is a pilot training session. A pilot training session is considered by some to be the final stage of development and by others as the initial stage of the implementation phase of ADDIE. In any case, it is a highly effective evaluation tool. The pilot (and/or beta test) should be run with a representative sample of learners. However, it is sometimes helpful to include individuals that may have recently taken somewhat similar training (or training from another source) but that still can provide a realistic learner's perspective. It may also be helpful to have an SME, training department leader, and/or any existing exceptional practitioners attend to identify helpful improvements in the training. It is important that all of these persons work together with the sole goal of improving the training without individual agendas, placing blame, or "ownership" issues. A pilot study can provide tremendous insight into the training that cannot be readily gained from the other review methods. Pilots can be used to ensure the material meets the project objectives and addresses the needs of the audience. In the pilot setting all aspects of the training can be tested. The course development team and trainer can check allotted times, content, learner engagement, clarity, difficulty, flow and balance, instructions, assessment tools, evaluation instruments, overall effectiveness, and whether the learners are able to achieve the stated course objectives. Based on the pilot study, any needed adjustments and improvements can be made to the training program.

Some individuals are tempted to skip the quality assurance steps and go right to implementation. However, it can be risky to launch an untested training program. They may do so because of a lack of knowledge (e.g. the ADDIE process), deadlines, budgets, overconfidence in the developer, or they simply don't value the QA process. However, some form of testing of the training program can be very helpful and a pilot program serves best.

In developing the training program materials it is essential that the program be designed and developed so that any qualified trainer can lead the training course. Otherwise it will be difficult to consistently deliver a quality training program that consistently meets expectations and the specified objectives.

IMPLEMENTATION

The final phase in the ADDIE model is implementation. Companies rely on training for improvements and reduced costs (improved profitability) in return for investments in the program. Training can be used to: ensure legal compliance and reduce liabilities; improve efficiency, productivity and reliability; train new people or train on new processes; and introduce the company and its safety culture, among other uses.

By following the ADDIE process you will have a high degree of confidence about the course when it is ready to launch. Great course material does not, however, guarantee a great launch or a great training program. Murphy's law still applies – though you should plan the course accordingly, to the extent possible. Think through and carefully consider reasonable backup plans as part of the course design and development. It is possible to write and launch a training course without the use of ADDIE or any ISD model, but it is done at much greater risk. The course may have the wrong focus (e.g. on symptoms rather than the problem) be unclear and confusing, with no or un-measurable objectives, lack critical content, not meet the business need, or simply not be very effective. Without a structured approach, you can only hope it goes well.

Primary steps in the implementation phase of ADDIE are:

- Use of a management plan – administration and logistics
- Train-the-trainer sessions if/as needed
- Assembly and use of a Training Management Plan (or Course Management Plan)
- Trainers review lesson plans and prepare
- Conduct the training and deliver the material
- Remediation for individual learners

- Application of the training by learners
- Adjustments and revisions

Often there are many details to be handled for the rollout of a significant training program. A management plan can be helpful to address the many details related to administration and logistics. For a classroom course, it may include course materials instructors, course schedule, classroom space and facilities, registration, travel, logistics, reporting and records. For e-learning courses it may also contain hosting, integration, access, learners' connections, learners' computers, security, available help, registration, logistics, report, and records, etc.

For some training programs it may be helpful or necessary to run train-the-trainer sessions prior to full rollout. In addition, a Course Management Plan (or Training Management Plan) should be assembled. It informs trainers of all factors related to a training program. It generally may include: the course description; target audience; directions to administer the course and score assessments; directions for guidance, assistance, and evaluation of learners; a list of all tasks to be instructed; a course map or sequence; program of instruction (how it is to be taught); a copy of all training material (training outlines, student guides, etc.); instructor and staff training requirements; and any other documents or information needed for instruction of the course.

Conduct of the training then involves the content, delivery, media, activities, assessments and everything needed to deliver the training. When conducting the course, it is important through assessments to identify learners with less than expected or desirable performance to receive additional training and remediation. This can be done to address application issues by training apprenticeships, including guidance and reviews. A key component in any successful training endeavor is for learners to apply their learning as soon as possible and as often as can be arranged following completion of the retraining. This is very important for retention, mastery and building of self confidence.

Implementation sounds like the easy part of the process, but it can be the hardest. Good trainers can make poorly developed programs work, and can make a good program great. Poor trainers can make neither work well. A well qualified trainer should be used to implement the training program. This may not be the best SME. Companies may move people that are excellent in one area to other areas where they may be unable to perform well. A good trainer should be versed in the art and science of teaching as well as in the subject matter. The trainer needs to utilize learning factors (knowledge, leadership, environment, involvement) together with learning theories to engage the whole brain (left brain - detail/technical, right brain – artistic, sensing). The trainer may also need to change leadership style during the course to best meet the learning styles and needs of the participants as they gain knowledge and abilities. The trainer also pays attention to the learning environment and changes that can deter from or enhance learning. Finally, the trainer needs to make use of a repertoire of involvement skills, including: flexibility, spontaneity, empathy, compassion, questioning, getting feedback, counsel, and positive reinforcement.

EVALUATION

The evaluation phase of ADDIE actually occurs throughout the entire ISD process – in analysis, design, development, and implementation as well. It is also performed after training program completion and can be applied as the learners return to their jobs. The purpose of evaluation is to gather and document learner performance in a course and, ideally, on the job. The overarching goal is to improve and make the training better. However, many companies do not invest in the evaluation phase or may put forth a very minor effort, though there are significant benefits to be gained. Without evaluation, training programs can rather quickly become outdated and obsolete. Sound evaluation does not require much in terms of expense, but does require some time. Companies that follow the ADDIE ISD model through the evaluation phase generally see much better results than those with haphazard or hit-and-miss approaches to training development and delivery.

Primary steps in evaluation are:

- Use of the results of the previous four steps to develop evaluation instruments that will be used to verify and validate that the training is meeting the need.
- Formative evaluations are conducted throughout the ADDIE process
- Summative evaluation is done at the "end"
- Evaluation results are fed back into the ISD process for improvement
- Success is evaluated against objectives (learning, performance, business)

There are four basic levels of evaluation often used – response, learning, behavior, and results. The first level is to get the target audience reaction to the instructional material. For formal training, this could be immediate or delayed training evaluations. The second level is a measure of what the target audience has learned from the instructional material. This can be in the form of written tests or other assessments to verify that the target audience has understood the instructional material and will be better able to apply it on the job. The third level is a measure of how the target audience performs back on the job. This can provide important feedback that the instructional material was effective as intended. The fourth level is a measure of the impact or results of the instructional material strategy on the business bottom line.

It can take a significant amount of time to measure changes in performance and behaviors and improvements in business goals. At times it may be difficult to develop instruments for all four levels, but it is important to consider how the training program is successful. The evaluations should be all encompassing – considering the course content, materials, course delivery, mechanisms, assessment, and evaluation mechanisms. Feedback and training program improvement are essential to an ongoing, relevant, quality training program meeting business needs.

TRADITIONAL INSTRUCTOR-LED, E-LEARNING AND BLENDED LEARNING

The structure or format of the training program should be chosen and designed to best provide and enhance the needed learning. The best structure can depend upon a number of factors to be considered.

Traditional Instructor-Led Training

Instructor-led training, as any structure or format, has both advantages and disadvantages. Generally, instructor-led training in a traditional "classroom" setting is useful for large scale training across a site or organization. It is personal, dynamic, and interactive. It facilitates the practice of skills, and learners can receive direct guidance from an expert trainer and/or available subject matter experts. In addition this format readily facilities direct observation by the instructor in practice exercises for the purpose of assessment as well. The instructor-led structure is helpful for training in highly complex and technical matters. The primary disadvantages to this format are the need to be at one location for each training session, incurring investments of time and, potentially, travel related expenses. In addition, the pace of learning is set for the entire class and the course must be scheduled when all participants can attend and remain focused on the learning objectives in the program.

E-Learning

"Electronic learning or E-learning (recently known as eLearning) is an all-encompassing term generally used to refer to computer-enhanced learning, although it is often extended to include the use of mobile technologies such as PDAs and MP3 players. It may include the use of web-based teaching materials and hypermedia in general, multimedia CD-ROMs or web sites, discussion boards, collaborative software, email, blogs, wikis, text chat, computer aided assessment, educational animation, simulations, games, learning management software, electronic voting systems and more, with possibly a combination of different methods being used." [3] It is a broader term than online/Web-Based Training (WBT) or Computer-Based Training (CBT) and includes the use of multimedia (text, audio, still images, animation, video, interactivity, etc.).

The advantages and disadvantages of e-learning vary with the specific format selected. Generally, e-learning can provide both interactivity and flexibility for the training delivery. Either self-paced or synchronous training may be used.

In self-paced training the learners proceed through the training individually, at their own rate of progress. In synchronous training, an instructor is actively involved in the training program, leading the group through a live presentation. In self-paced training the learning environment is "tutorial", using a web browser (or course reader) to use recorded media through CBT or WBT. The instructor is not actively involved in the training session, though a help desk may be provided. An advantage of self-paced training is that there is not a need for a central training location and the training can be delivered to the learners desktop at any time that is convenient.

In the case of synchronous learning, the pace of the course is set for the group in a virtual classroom. The course can be delivered via a live presentation using courseware or conferencing applications. An instructor can be utilized to lead the training sessions. There is no need for a central training location other than what may be necessary due to the existing infrastructure. Typically the courses are scheduled and announced in advance of the training and the training class size can vary significantly.

Some of the benefits of e-learning are the potential for greater productivity, timely delivery of training, flexible learning, interactivity and interactive help, and a lower overall cost per learner to name a few.

According to the American Society for Training and Development "Many organizations with high-performing learning functions are able to attribute clear cost savings, efficiency gains in the learning function, increased content reuse, and decreased costs of learning development and delivery to successful technology investments." [1]

Blended Learning

Blended learning may be a combination of traditional and e-learning. Blended learning can also be a combination of self-paced and synchronous learning. Many of the advantages of traditional instructor-led classroom training can be maintained while using a multimedia approach, as well as available new technologies and tools to enhance learning cost-effectively. Companies are now discovering that blended learning can gain the advantages associated with traditional, e-learning, synchronous and self-paced learning - resulting in an optimized training program and learning experience.

MEDIA AND NEW TECHNOLOGIES WITH APPLICATIONS IN TRAINING

There is certainly a wide array of media and tools that may be used in training programs through e-learning and blended learning formats. This includes traditional ones such as photos, models, videos, props, flip charts, videos and laptop based presentations for example. Multimedia training materials can generally consist of a combination of text, audio, still images, animation, video and interactivity. "Rich media" is often used for interactive multimedia. Advances in technology have also led to the use of Computer-Based Training (CBT), Web-Based Training (WBT), webcasts, podcasts, social media, and Web 2.0 applications. Today's communication and training needs to be relevant, interactive, portable, and immediate – to the extent possible. The current generation of web browsers have RSS (Really Simple Syndication) readers built in that can be used as a way to broadcast information. Just as in training the activities are varied to maintain interest and focus, the media utilized should be varied as well. Learner retention typically increases as the number of senses used in the training increase. Of the many new technologies and developments, Learning Management Systems, blogs, wikis, and podcasts (and potentially the second generation World Wide Web) appear to be best suited to aid in training programs.

Learning Management Systems

Learning Management Systems (LMSs) and Course Management Systems (CMSS) are recent developments. "LMS" is often used to refer to all of these types of systems. An LMS is a "high-level, strategic solution for planning delivering and managing all training events within an organization. It includes online, virtual classroom, and instructor-led courses." [3] They are a significant new tool for training, and as such, are worthy of further explanation here.

LMSs can manage administrative functions of training, and also the tools to deliver synchronous and asynchronous online training. Tools are provided for authoring content and virtual spaces for learner interaction (e.g. bulletin boards, chat). CMSs typically focus more on content, as compared to LMSs. These systems collectively give authors, instructional designers, and subject matter experts the means to create e-learning content more efficiently. The business problem solved is to create enough content just in time to meet training needs. An approach taken is to develop reusable content chunks and make them available throughout the organization. These systems help to eliminate duplicate efforts and facilitate rapid assembly of customizable content. All of this leads to lower training costs for the organization, while maintaining a quality program. Some of the features of an LMS or CMS may include (in addition to the usual course content): forums, content managing, quizzes with various types of questions, blogs, wikis, database activities, surveys, chat, glossaries, peer assessment, and multi-language support for the interface. LMSs and CMSs provide the overall learning environment for e-learning. Security features are also included, as well as interactive help. Some of the standard blocks within such systems are: activities, administration, calendar, course summary, courses, latest news, login, main menu, online users, people, recent activity, remote RSS feeds, search, section links and upcoming events. Static, interactive, and social course material can all be combined in this learning environment.

SUMMARY

There is a clear need and significant benefits that can be achieved through application of a structured, systematic approach to PSM, risk management, security and EHS training programs. The ADDIE Instructional Systems Design Model is a valuable tool that can be used to create and deliver training programs that can be confidently launched and achieve the desired business, performance, and learning objectives. It formally incorporates five steps in the overall training process, including Analysis, Design, Development, Implementation and Evaluation. The ADDIE model provides a process that can be followed and adjusted if needed to balance the cost, schedule and risks in training programs. Essentially, it can be adjusted to fit the desired point on a Time-Quality-Cost triangle to meet the specific business need, given all of the constraints.

The return on investment for use of the ADDIE process may be as high as a factor of ten for larger programs – and especially those with projects involving software development. Without use of an Instructional Systems Design (ISD) model such as ADDIE, if everything goes absolutely right, an adequate (but not effective) training program can result. However, in that same circumstance, if anything goes wrong the course will likely "miss the mark". Costs of redevelopment often significantly exceed costs associated with use of the ADDIE model.

The ADDIE ISD model can be successfully applied to new training needs or to significantly improve existing ones to produce superior content and ensure effective delivery. However, it is also important to have good trainers involved in the delivery phase.

Within the ADDIE process, traditional, e-learning and blended learning structures can be used to meet the business training needs. There are many new technologies that have been recently developed and others emerging that can enhance the learning experience in e-learning while increasing productivity. Blended learning has been shown to be possibly the best alternative, preserving many of the advantages of traditional instructor-led training with multimedia and promising new training tools and technologies.

Tools with promising applications in technical training include the use of new options in multimedia, social media, and Web 2.0. They can be used to meet training and other business needs in process safety, risk, security and EHS. Some of the specific new tools include wikis, blogs, Learning Management Systems (LMSs), and podcasts. The Learning Management Systems facilitate enterprise-wide, quality training programs that previously may have required significant investments and that can now be accomplished for a fraction of the cost. These tools and the host of available new technologies can be integrated into training programs through the ADDIE ISD process to take technical training to the next level.

REFERENCES

1. American Society for Training and Development, 2006. State of the Industry Report
2. International Society for Performance Improvement, www.ispi.org
3. www.wikipedia.org
4. Clark, Don. ISD Handbook.
5. Dick, Walter, Carey, Lou, and Carey, James O. Systematic Design of Instruction, 6th Edition. Allyn & Bacon, 2004.
6. American Society for Training and Development, www. astd.org

Dynamic Simulation / Dynamic Training
Overcoming The Graying Workforce Dilemma

Joseph L. Sims, Valero Energy Corporation, San Antonio, Texas; and
William F. Early, P.E., Early Consulting, L.C., Houston, Texas

Abstract

The aging workforce in the petrochemical industries today represents huge challenges for the future that require innovative responses. Considering that the 50+ age group is the fastest growing segment of the workforce, and recognizing that 11% fewer people were born between 1966 and 1985 than in the 20 years immediately following WWII, our industry faces many obstacles in maintaining a stable, well-trained workforce. When one recognizes that over 35% of workers in technical fields are 45 years and older, this problem becomes more alarming.

Valero Energy Corporation has committed to a corporate simulation project initially consisting of twenty-nine dynamic unit simulations. The program includes simulator / console training for Foxboro, Honeywell Experion / TDC 3000, and Yokogawa systems. As an extension of this program, Valero has embarked on a path intended to systematically address the problem of knowledge loss. In parallel to the simulator program, Valero is standardizing its operator, maintenance and craft training activities and taking additional steps to maintain subject matter expertise, capture institutional knowledge, and mentor the next generation. This paper will describe some of Valero's initiatives, and discuss the technical and commercial benefits obtained.

INTRODUCTION

The refining and petrochemical industry recognizes a growing problem with knowledge and experience loss through personnel retirement, as indicated in the abstract above. As knowledge and experience is set to leave the workplace, one of the greatest challenges is capturing that knowledge in a way that it can be transferred effectively to the next generation. As the workforce experience level diminishes, one of the greatest challenges, beyond the question of whether the next generations are sufficient in numbers to meet demands, is the question of whether the necessary knowledge base can be imparted effectively.

One response to these concerns is the development and implementation of dynamic simulators into the training environment. Though not a new phenomenon (note that surgeons practice on cadavers and/or models before removing your appendix), recent technological advances in dynamic simulation have allowed for a more effective and realistic training experience. Simulators have been used in diverse environments such as the military, medical industry, aviation, and of course the refining and petrochemical industries. The following paragraphs provide examples indicating how simulators can benefit different employee groups and departments within refineries.

Figure 1. Console User Manual

Operating personnel who work in a control room would benefit by being trained at a simulated control panel.... This type of training can be very effective in teaching employees correct procedures while allowing them to see the consequences of not following established operating procedures." (Dept. of Labor)

OPERATIONS BENEFITS

Console or board operators can benefit from dynamic simulations, as this is the group whose primary role most closely resembles the simulator environment. To enhance this training experience, the existing Distributed Control System (DCS) environment can be cloned.

Cloning enables the operator to recognize his / her environment, and demonstrate the ability to respond to a variety of failures which may be experienced during normal operations (though the failures are infrequent). The simulated failures typically include, but are not limited to, process and equipment malfunctions.

When these scenarios are captured in the simulation environment, it enables the instructor / subject matter expert (SME) to pass his/her knowledge on to the next generation within a controlled training environment. By constructing simulated experiences, operations personnel can learn from role-playing scenarios. In this environment, the stress factor, which is known to increase the likelihood of human error, can be experienced and potentially overcome (i.e., a potentially stressful scenario can be experienced, thus becoming a recognizable event which allows the console operator to use these experiences in real life).

When the simulator is coupled with an interactive training console, the instructor has the ability to monitor the trainee, and provide constant feedback throughout the training exercises. Most simulation software will also have the ability to score each movement made during the training exercise to ensure the trainee has responded to the failure correctly and within the time constraints set by the trainer (SME). By grading these response times, the system allows trainers to assess and certify trainees based upon training scenarios. Simulation scenarios can also be constructed to stretch the range of experiences normally encountered day-to-day in the real life environment.

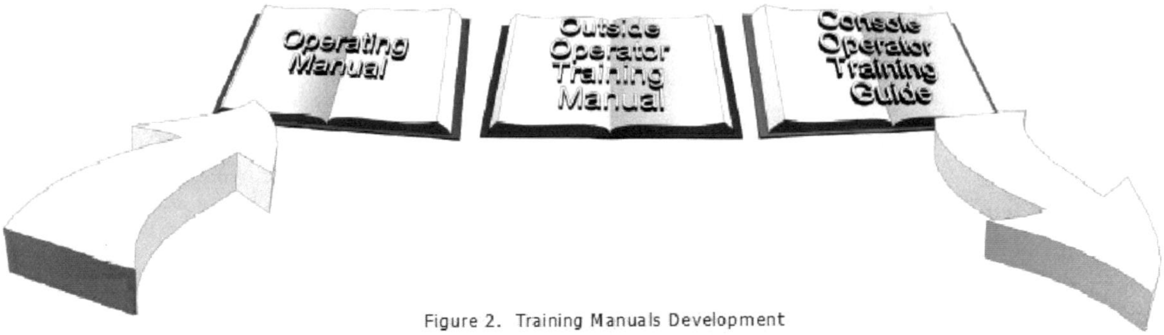

Figure 2. Training Manuals Development

PROCESS CONTROL BENEFITS

> ***"All plants with Advanced Process Controls (APC) should have an Operator Training Simulator to maintain operator skills."*** *(Hydrocarbon Processing, August 1999)*

Control Systems groups can benefit greatly from dynamic simulations. Valero uses the same hardware used in the process control system design to create a replicated environment. Control Systems will have the ability to make mock configuration changes and see the actual effects on the unit, before these changes are loaded on the live unit. There are several benefits to this approach, including loop tuning, identification of critical instrumentation, and setting alarm limits. This system will also be valuable when training new engineers who can interact with the simulator to learn what various control loops will do, how they interact with the process, and why certain loops must include redundancy for reliability or safety.

TECHNICAL SERVICES BENEFITS

Unit engineers also struggle with passing knowledge to the junior engineer and/or the next generation. Dynamic simulation software has the ability to display and print stream compositions, lab reports, cause-effect and unit modeling data. These attributes can effectively maintain and transfer the needed knowledge base.

Dynamic simulations also provide operations and engineering personnel a consistent tool for communicating difficulties in process units. Operations, safety, and reliability issues can be reviewed, with potential solutions being tested off-line before risking capital.

TECHNICAL TRAINING BENEFITS

Technical trainers struggle with ensuring operating procedures are accurate and certified annually as required by the Process Safety Management (PSM) regulation. This includes content accuracy and the ability to document steps for all operating phases within their facilities. By having dynamic simulators in place, Operations and Training can prove what is written and documented within the standard operating procedures (SOP's) are valid. This platform will also allow technical trainers and operations personnel to simulate management of change (MOC) modifications to procedures, and validate the expected effects of the proposed change before commissioning the MOC.

From 1976-1996, the largest 100 accidents in the hydrocarbon-chemical processing industry cost $7.52 Billion in losses; Operator error accounted for 21% of these events at an average of $75 Million per loss." *(J&H Marsh & McLennan, Loss Control, 1997)*

Figure 3. Trainer / Instructor Manual

Although the stress experienced by console operators is reduced with this learning environment, it can still exist in view of their comfort level with operating the process unit.

Some hands-on field experience with each unit modification will still be necessary to bridge the real or perceived knowledge gap.

INDUSTRIAL SAFETY BENEFITS

Another interesting benefit that dynamic simulators have to offer is a more structured look into root cause analysis (RCA). With a dynamic simulator, the RCA investigation team will be able to replay the events of any incident related to process plant operations in order to aid the investigation team. Simulating an incident to study what went wrong on the process unit, gives the investigation team an opportunity to test several possibilities and develop correct countermeasures. The actual scenario then becomes a future training scenario so the knowledge is captured to prevent reoccurrence. Scenarios identified in a process hazard analysis (PHA) similarly could be tested to determine the probability of an event being experienced, or the severity that might be anticipated.

In reactive mode, this will allow the investigation team to recapture the actual unit conditions prior to the incident and monitor the process through the incident. This in turn will provide a more thorough look into the events that occurred from the console operator perspective. Even in the most severe accidents, simulator modeling might enable incident re-enactment.

PLANNING AND ECONOMICS BENEFITS

Many of today's modeling tools do not provide an actual gauge of how the process unit will react when a new feedstock is introduced. With Valero's dynamic simulator process report tool (Material Balance) section, engineers will be able to model the actual process variable fluctuations during feedstock changes/additions. This will allow planning and economics (P&E) groups the ability to maximize throughput and unit conversions rather than making a best estimate guess.

With the use of dynamic simulations in the P&E planning phase, refiners can better anticipate and understand how their units will react to different weight percentages of sulfur, percentages of metals, and or other foreign materials in feedstocks (whether benign or malicious).

Simulators will allow for maximization of unit conversions/throughputs (hence margins). Through better tuning of the process unit, and possibly avoidance of unit upsets runaways, or catastrophic failures can potentially be prevented.

PROCESS SAFETY BENEFITS

The training benefit from the use of dynamic simulators is obvious in terms of meeting the requirements of the PSM standard. The benefits include addressing the following specific requirements:

(g) Training.

(1) Initial training. (i)Each employee presently involved in operating a process, and each employee before being involved in operating a newly assigned process, shall be trained in an overview of the process and in the operating procedures as specified in paragraph (f) of this section. The training shall include emphasis on the specific safety and health hazards, emergency operations including shutdown, and safe work practices applicable to the employee's job tasks.

(2) Refresher training. Refresher training shall be provided at least every three years, and more often if necessary, to each employee involved in operating a process to assure that the employee understands and adheres to the current operating procedures of the process. The employer, in consultation with the employees involved in operating the process, shall determine the appropriate frequency of refresher training.

(3) Training documentation. The employer shall ascertain that each employee involved in operating a process has received and understood the training required by this paragraph. The employer shall prepare a record which contains the identity of the employee, the date of training, and the means used to verify that the employee understood the training.

(i) Pre-startup safety review.

(iv) Training of each employee involved in operating a process has been completed.

(o) Compliance Audits.

(1) Employers shall certify that they have evaluated compliance with the provisions of this section at least every three years to verify that the procedures and practices developed under the standard are adequate and are being followed.

Operators trained using a simulator have the equivalent of 6 years of "On-the-Job experience".
(Power Engineering, May 1992)

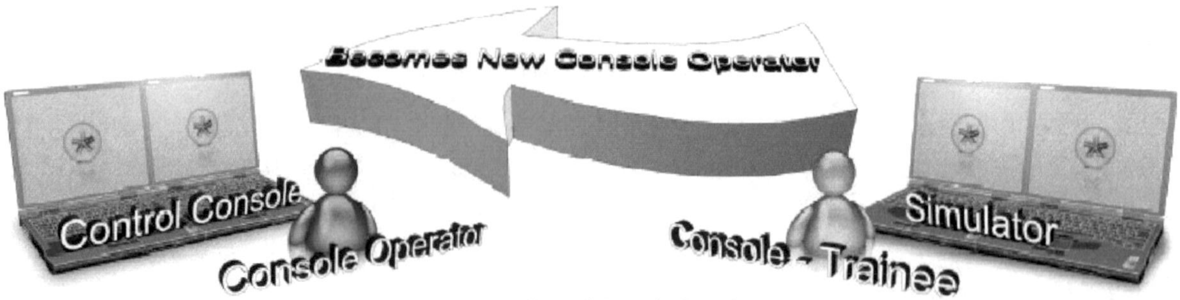

Figure 4. Trained Console Operator

CAUTIONS

Both simulations and direct hands on experience share some limitations. First, it makes little sense to immerse someone in a simulation or a real experience without some basic receptors to appreciate the information being conveyed in the simulation scenario training. Second, it is true in all learning environments that people need to be motivated to learn. However, motivation can be enhanced by making the simulation highly realistic. This is one reason that the Valero simulator workstations will appear very much like the actual control room user interfaces.

SUSTAINABILITY

The sustainability of the capital investment (in the Simulators) is critical. Changes to process chemistry and the technology will have to be maintained within this environment in order to assure personnel are using evergreen information for training. When the integration of changes is incorporated into the sites' MOC program, the accuracy of the process safety information being used for training is assured.

Process control engineers, technical service engineers, operations personnel and site leadership are always trying to improve unit performance and reliability. This can present significant changes in the dynamic model, resulting in challenges for maintaining the model. Train-the-trainer courses will assure that program administrators understand the complexity of the dynamic simulator model and how the model is to be maintained.

Beyond the question of model sustainability, simulated DCS screens maintained off-site represents an independent layer of protection against a major loss of the DCS system. When one recognizes that information exchange with the basic process control system (BPCS) does not have to be one-way, improved overall BPCS sustainability is achieved.

CONCLUSIONS

Refinery leaders, operations, engineering and human resources personnel are responsible for identifying and capturing a company's vital knowledge and imparting that knowledge to the next generation of employees. Attending leadership training classes, reading applicable books, even talking with or interviewing experienced leaders cannot substitute for first hand experience. Dynamic Simulators provide a vehicle to capture knowledge and transfer that knowledge to the next generation of refining employees.

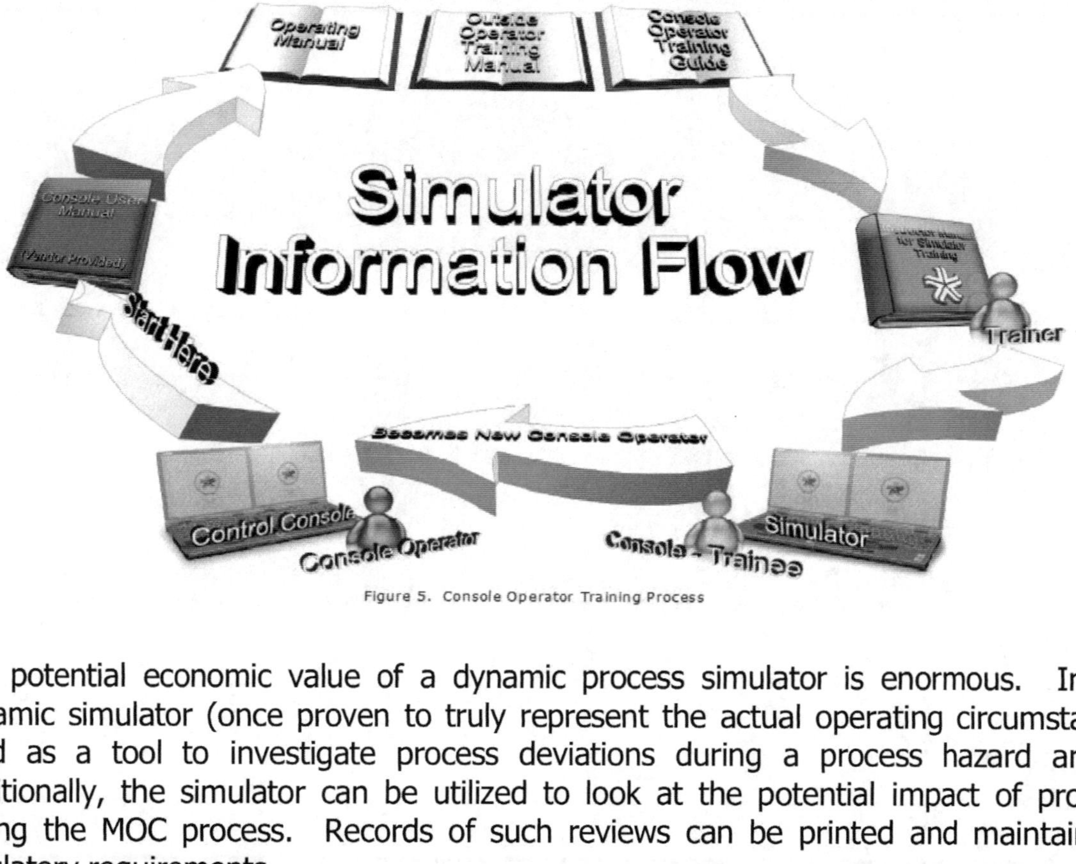

Figure 5. Console Operator Training Process

The potential economic value of a dynamic process simulator is enormous. Interestingly, a dynamic simulator (once proven to truly represent the actual operating circumstances) can be used as a tool to investigate process deviations during a process hazard analysis (PHA). Additionally, the simulator can be utilized to look at the potential impact of process changes during the MOC process. Records of such reviews can be printed and maintained to satisfy regulatory requirements.

Dynamic simulators create a stable knowledge transfer environment for all affected departments. They allow the review of changes being made to a process (or anticipated to be made) using a team approach without affecting the real time process operations. Thus, dynamic simulators are an effective tool in overcoming the graying workforce dilemma.

Acknowledgements: *The co-authors would like to thank Miya Maysent – Valero Corporate Director of Organizational Development, Ken Powers – Valero Manager Corporate Technical Training, Jennifer Kester – Manager Training and Development, Darril Wilburn Organizational Development Specialist, Pepper Wilson – Organizational Development Specialist and Grace Mortell – Valero Corporate Training & Development Specialist for their support of the dynamic simulator program and this paper. Thanks are also offered to Dr. Vish Mantri, P.E., President of Alpha Sim Technology, Inc., for his contributions to an effective dynamic simulator program.*

THE CASE FOR COMPUTER BASED TRAINING: MAKING IPS TRAINING LESS PAINFUL

Kevin L. Klein, Solutia, Inc., Springfield, Massachusetts

ABSTRACT

Training is an important part of the ANSI/ISA 84.00.01-2004 requirements to ensure the people designing, maintaining, and operating Safety Instrumented Systems are competent and the standard is being consistently applied throughout the Company. As a user of the standard, Solutia has implemented training using several different training methods including a set of Computer Based Training (CBT) presentations to provide initial introductory training and refresher training. This paper discusses the objectives of the Computer Based Training, the CBT development process, and the effectiveness of these techniques.

INTRODUCTION

How many companies have the time and resources to train the whole company on a new initiative using instructor lead training on a budget of $20,000? Solutia did it with only one subject matter expert, a consultant, a trainer, and Computer Based Training (CBT). Do you want to know how? Just read on. In this paper, I discuss the issues with rolling out a technical program using CBT with reduced resources while still maintaining effective training.

INSTRUMENTED PROTECTIVE SYSTEM (IPS) TRAINING PROGRAM

When Solutia decided to migrate to the new edition of ANSI/ISA 84.00.01 issued in 2004 Functional Safety: Safety Instrumented Systems for the Process Industry Sector, we knew we had a significant task ahead of us. Training was a major concern since we have 11 major Chemical plants to be trained in four different countries. The first task was to determine who needed to be trained, how they would be trained, and what they needed to be trained on. We identified five main groups needing training: IPS Experts, Process Engineers, Maintenance personnel, Operations personnel, and Management. We then identified the training requirements for each group. See Chart 1 for more detail.

Chart 1 – IPS Roles and Training

	Knowledge Required	Job requirements	Training Required
IPS Experts	Detailed Understanding	• Detailed understanding of ANSI/ISA 84.00.01-2004 • Detailed understanding of Corporate requirements • Training on IL verification calculations.	• Instructor Lead expert training with exam. • Hands on training with another expert • Training on Site specific Requirements
Process Engineers	Understanding of basics	• Detailed understanding of Process Hazards • Basic understanding of ANSI/ISA 84.00.01-2004	• Overview CBT • Instructor Lead Process Engineer training • Training on Site specific Requirements
Maintenance	Overview of program with understanding of role-specific requirements	• Detailed understanding of Testing and inspection requirements • Basic understanding of IPS design • Basic understanding of Bypassing requirements	• Overview CBT • Maintenance CBT • Training on Maintenance procedures • Training on Site specific Requirements
Operators	Overview of program with understanding of role-specific requirements	• Detailed understanding of IPS operation requirements • Basic understanding of Bypassing requirements • Detailed understanding on operation of IPS	• Overview CBT • Operations CBT • Training on operating procedures • Training on Site specific Requirements
Management	Overview of program with understanding of role-specific requirements	• Basic understanding of IPS program • Basic understanding of Bypassing requirements	• Overview CBT • Management CBT • Training on Site specific Requirements

The next task was to determine how to train the groups. In school, many text books follow the spiral technique for training in which a concept is taught several times at different points in the text. This is done to help the student better retain the subject matter and to give the student time to absorb the concepts. Due to the complex and foreign nature of the subject, we decided to cover the subject three times with increasing detail at each stage to help ensure retention of the concepts.

An overview CBT was developed to introduce the student to the subject. The CBT was designed for all personnel and provides basic information on the program. This training covers the:

- Reasons for the training,
- Overall Protective System Philosophy change,
- Basics of Instrumented Protective System (IPS) Lifecycle.

The overview CBT introduces the concepts of:

- Protective and Non-Protective Systems,
- Instrumented Protective Functions (IPF),
- Instrumented Protective Systems (IPS),
- Integrity Level (IL),
- Layers of Protection Analysis (LOPA),
- And Protective Requirements Specification (PRS).

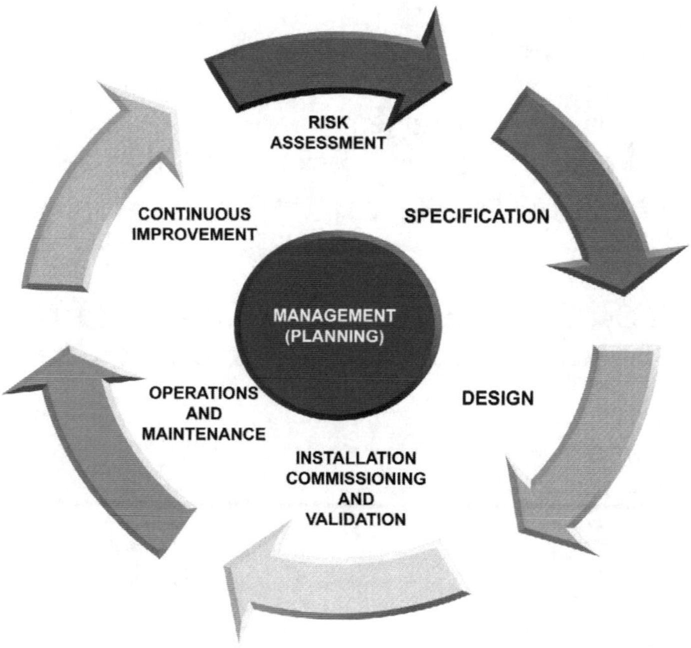

Figure 1 – The IPS Lifecycle

We included graphics to illustrate the slide text to help the students learn concepts and retain information. The coverage of each concept is introductory in nature. The final slide informs the student of additional role-specific training to learn more.

Role-specific training focuses more on the specific job responsibilities for the role and reinforces the concepts presented in the Overview CBT. Role-specific training covers the work processes and expectations for three disciplines: Management, Maintenance and Operations. Since the IPS Experts and Process Engineers are provided in-depth classroom training, a CBT was not developed for these students. The Management CBT would be the next CBT to be developed since many of the concepts and information being presented would also be presented from the role specific viewpoint in the Operations and Maintenance CBT.

The focus of the Management CBT is on the responsibilities of IPS Management in supporting the lifecycle and approving IPS bypassing. With this CBT we were able to build on the information presented in the overview. The following concepts were critical for Management to know:

- Verification of Lifecycle steps
- Performance Tracking
- Failure Detection and Response
- Fault Tolerant Bypassing
- Procedural Bypassing
- Non-Fault Tolerant Bypassing
- Maintenance and Operations roles,
- Management of Change Requirements, and
- Assessment of existing systems designed to the previous Recognized and Generally Accepted Good Engineering Practice.

This is the longest of the CBT courses at 25 minutes for the presentation.

The Maintenance and Operations CBT would be developed together. The Maintenance CBT focuses on the Maintenance Activities the maintenance personnel need to perform including:

- Reporting Errors and Failures
- Bypassing
- Use of approved Equipment
- Inspection and testing
- Management of Change

The Operations CBT focuses on the activities the operator needs to know to maintain the reliability of the Instrumented Protective Systems including:

- Reporting procedure errors, trips and component failures
- Response to trips
- Bypassing

- Detecting and responding to failures
- Management of Change

To build on the concepts introduced in the Overview, this training includes a more in-depth explanation of protective systems. Because maintenance activities should follow the IPS Lifecycle, we introduced a Maintenance Workflow to help drive home the parts of the IPS Lifecycle important for Maintenance Personnel.

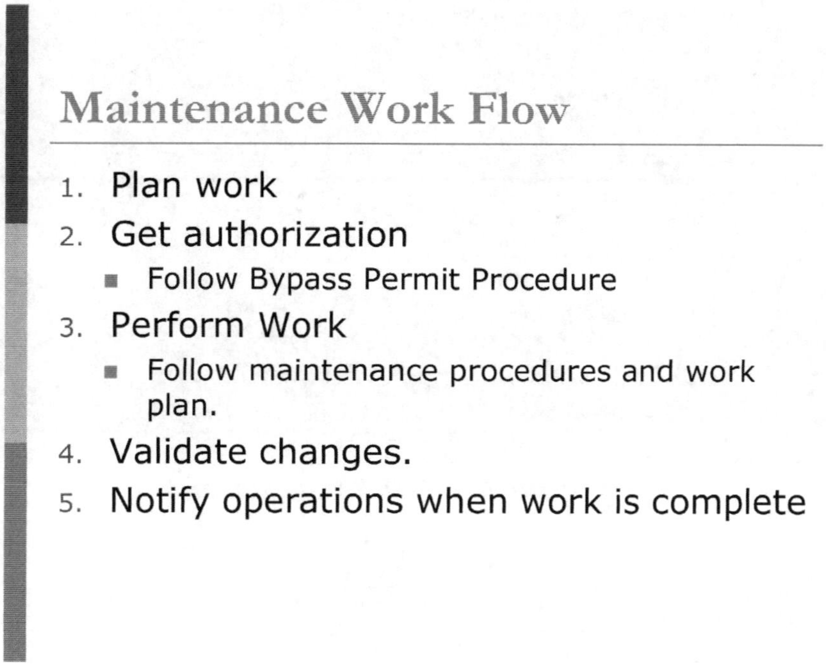

Figure 2 – Maintenance Work Flow

Finally, site specific classroom training is conducted through site based trainers. Additional practical and hands-on training is provided once the CBT and classroom training is completed.

WHAT IS COMPUTER BASED TRAINING?

You know those Power Point slides you generate for all those excellent presentations and training classes? Well, CBT is simply a way to record your verbal presentation of those slides once and enable other people to view the training at their convenience. It may sound easy, but there is a lot of preparation and forethought which goes into the presentation. Once the CBT course material is developed and a test is added, the course is assigned to an individual's training plan. When the student is ready to take the training, they log on to the CBT from their computer. From the main menu, the student chooses the course they want to take. See Figure 3.

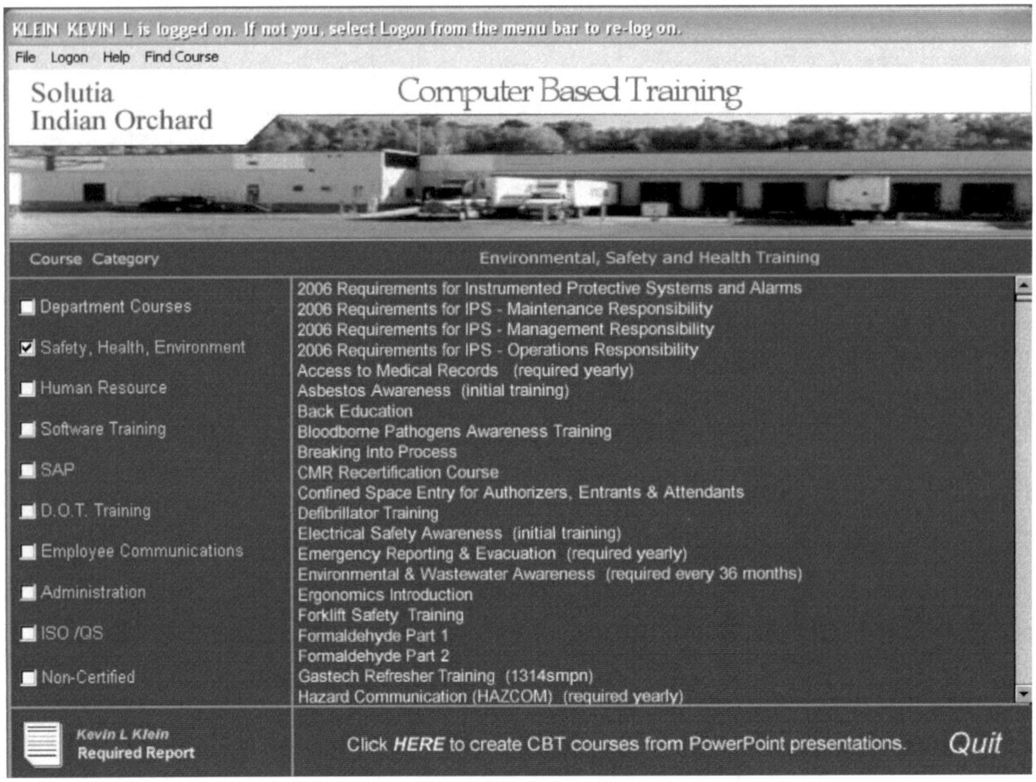

Figure 3 – CBT Main Menu

Figure 4 – CBT Slide

The student then chooses to take the course for reference or for credit. The difference is when the course is taken for credit the student will get a test at the end to verify they learned the main points of the training. In Reference mode, the student has the ability to freely navigate to specific pages, whereas in Credit mode the training is taken in a linear path and each lesson page must be completed before advancing.

The student then watches and listens to the CBT course materials. At the end of each slide the student chooses to either advance to the next slide or listen to the current slide again. If the student chooses to quit the program the program before completion, it will set a bookmark indicating how much of the CBT was completed. The next time the student runs the CBT, the course will start from where the student left off.

At the end of the presentation the student takes a test consisting of multiple choice and true/false questions. When the student answers the question, the CBT program gives feedback on whether the student got the question correct and provides additional review if the question is answered incorrectly, as shown in Figure 5.

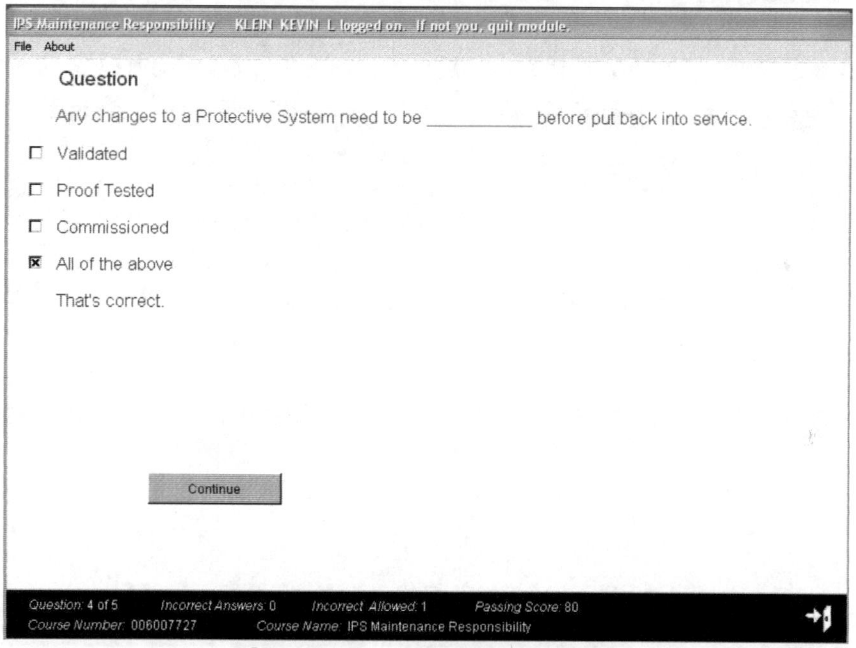

Figure 5 – Test Question

Finally the program grades the test and determines whether the student passes the course and gets credit for the course or not. The test scores and completion is recorded by the program and sent to the HR system that records and tracks training for each individual.

Before the CBT program exits, a slide is displayed with the name of the creator of the CBT and contact information so the student can contact the creator to ask questions that may arise while taking the CBT.

WHY CBT VS. TRADITIONAL CLASSROOM?

With this size and number of Solutia's Manufacturing facilities, rolling out training on a complex subject to the majority of the employees would be a daunting task involving many trainers and a scheduling nightmare. Many of the employees being trained are working shifts, so training must be scheduled for off hours as well. This is where CBT works well. It can deliver a consistent message to many employees and is available at their convenience. Other styles of training considered were classroom and video conference training. With the limited resources available for this effort, these options were not practical for Solutia Inc. An added benefit to the CBT delivery method is the ability for the student to take the CBT again at their own pace and at any time, without the need of an instructor.

With classroom training, you have the advantage of seeing the students and are able to answer questions and clarify the information in the presentation as needed during the presentation. The material can also be tailored to the specific audience. You can provide demonstrations as well. A disadvantage of classroom training is that the pace of the delivery of the material is planned to meet the needs of the whole class. Whereas the CBT training can be played at the students pace and each slide audio can be replayed, allowing the student greater opportunity to gain clarification of the information presented.

See Chart 2 for a summary comparison of CBT and classroom training's advantages and disadvantages.

Chart 2 – Training Method Comparison

	Advantages	Disadvantages
CBT Training	• Consistency • Available 24 hours a day • Automatically records completion of class • No Instructor needed to deliver the course • Cost-effective delivery • Self-paced • Tireless • Immediate feedback to Student	• Time to develop and record the narrative • Student cannot get questions answered immediately • Initial cost • Can not be updated and changed quickly • Requires computer skills to access
Classroom Training	• Immediate feedback for student and instructor on understanding • Can mix up slide presentations with demonstrations. • Very flexible and can be modified for the specific audience • Student can get questions answered during the training.	• Instructor needs to be available for training • Manual record keeping of class attendance and completion of class. • Less consistent message due to variability and flexibility.

As with the classroom instructor lead training, the CBT training can involve multiple senses and this can lead to greater retention of the new knowledge. On average, when people receive training, people remember:

10% of what they read,
20% of what they hear,
30% of what they see, and
50% of what they hear and see.

Based on the number of sites and people requiring training, CBT was chosen for Overview and Role-specific training. Solutia has used CBT successfully for many EHS training areas, such as HAZCOM, Lockout/Tagout, Emergency Reporting and Response, Electrical Safety Awareness, etc. Some training is required to be viewed yearly, some on a three year cycle, and others on a five year cycle.

Acknowledging the complexity of the subject and the time available to work on the CBT slides and narration, we decided to hire a IPS expert consultant to help define the content and scripting for the CBT.

DEVELOPING THE CBT

When we decided to develop the CBT, we developed an outline for the three disciplines containing all the information we felt each audience would need. From the outline we estimated the number of slides needed to cover the subject. Based on two minutes per slide on average and close to 30 slides, the training class for each discipline could have lasted an hour or more. Research studies have suggested that the average adult can "listen with understanding" for approximately ninety minutes, and "listen with retention" for approximately twenty minutes. So we had a problem. How do we cut the size of the training module and continue to present all the information? When comparing the content of the three training outlines, we found concepts that were repeated in each presentation. The decision was made to take the common information and make an Overview CBT for all students, which would be followed by role-specific CBT targeted at discipline needs and responsibilities. This allowed us to present the concepts from two different perspectives and also keep the presentation length to 20 - 30 minutes. The CBT narration must be carefully developed to ensure consistency between courses and clarity of the message.

Unlike a presentation you give to an audience, you will not get any feedback from your audience when you give the presentation, so you need to anticipate the needs of the audience when you develop the slides. It helps to give the slide presentation at least three to four times to a live audience. This will allow you to obtain advance feedback from your target audience. A word of warning to those who like those animated clip art. They are cute and attract our attention for sure, but generally that is not what the presentation is about, so my advice is to leave them out and just use the still clip art.

Next, you need to script the presentation being careful to be consistent with the concepts and making sure the terminology has been adequately defined. Remember the audience cannot ask any clarifying questions when CBT format is used. The length of the narrative for any given slide needs to limited to 60 - 120 seconds depending on the amount of action on the slide. If the narrative is too long, the audience will get bored and start to wander from the subject you are presenting or start working on something else. Also remember that each sentence your write you will need to read without a breath. Depending on how good your lungs are, you may need to split up some of the sentences to allow the sentence to be read all the way through without breathing.

Once you have your PowerPoint slides and narrative, you are ready to start recording. This is the hard part. First, the person narrating needs to read the script smoothly and have a pleasant tone. We can all read, but the narrator must also maintain voice volume level, narrative pace and even plan when to take breaths. If you want a good final product, work with your company's training group. They probably have done this before and know how to read the narrative without losing the audience. You could have the best presentation in the world, but as with a live presentation most of what the audience will remember is how you presented the information. So it is critical to get the right voice. Think about those television commercials or radio announcers you have heard. If the voice is annoying, you change the channel and miss the message. With safety presentations, you do not want them to change the channel on you.

Plan to read through the part several times before starting to record and then you need to do several takes. Don't try to read the whole narrative for the slides all the way through. We have all seen the blooper shows where the professional actors and actresses mess up their lines and have to re-do a part of the show. The nice thing about recording is you can delete the sentences you mess up and re-record them.

Next the words and slides come together to tell the story about your new program and you're done. Right? Not so fast. To verify the student watched the presentation and learned the important concepts and major points, you need to develop 5 to 10 multiple choice questions. As with any test, you should define the pass/fail criteria. The CBT records the test score and successful completion of the training into the student's employee training records.

THAT'S A TAKE

As you can see, there is a lot of up front work to putting together a CBT. But when compared to conducting on site training once every two to three years, the time is well spent.

REFERENCES

1. Bixler, Brett and Bergman, Terri, "Selecting and Implementing Computer-Based Training", National Alliance of Business, pp. 4

2. Summers, Angela (2007), "Guidelines for Safe and Reliable Instrumented Protective Systems", CCPS/ AIChE

3. ANSI/ISA, ANSI/ISA84.00.01-2004 Part 1– Functional Safety: Safety Instrumented System for the Process Industry Sector, The Instrumentation, Systems, and Automation Society, Research Park, NC, 2004

Conduct of Operations - the Glue Between Training and Operational Performance

Donald K. Lorenzo, P.E.
ABS Consulting
Knoxville, Tennessee
dlorenzo@absconsulting.com

Lee N. Vanden Heuvel
ABS Consulting
Knoxville, Tennessee
lnv@absconsulting.com

ABSTRACT

A consistently high level of human performance is a critical aspect of any process safety program. As the complexity of operational activities increases, there must be a commensurate increase in the formality of operations to ensure safe, reliable, and consistent performance of critical tasks.

Workers are initially and periodically trained in the organization's policies, procedures, and practices. Conduct of operations addresses the day-to-day implementation of that training so that operational and management tasks will be executed in a deliberate and structured manner. Workers at every level are expected to perform their duties with alertness, due thought, full knowledge, sound judgment, and a proper sense of pride and accountability.

The Center for Chemical Process Safety created ***Risk Based Process Safety*** *as the framework for the next generation of process safety management. Conduct of operations is a new element in that framework. This paper describes how it will serve as the "glue" that connects excellence in training to excellence in operational performance from a safety, environmental, and economic perspective.*

1. INTRODUCTION

Read any incident investigation report, and almost certainly, one of the recommendations will be to train or retrain workers in some operations, maintenance, or administrative procedure. Yet in many cases the workers are already well-trained in the subject procedure and have met all company requirements, including demonstrations that the training was understood. How, then, is it possible that well-trained, qualified workers fail to execute the procedures as required?

Some will claim that human error is the cause, and indeed, we all make mistakes. But in high-reliability organizations, isolated human errors should not result in serious incidents. Excellence in training does not automatically result in excellence in execution, and the "glue" that is often missing is a management system governing the conduct of operations.

A serious incident spurred one company to implement a management system for the conduct of operations that produced world-class quality and productivity results while compiling an outstanding safety and environmental record. In the context of that incident, this paper describes some of the key elements that should be incorporated in a management system for formalizing operations to produce excellent results.

2. THE INCIDENT

The company manufactures a variety of plastic resins that are stored in large atmospheric tanks. The tanks are heated to keep the resins fluid so they can be pumped into trucks and railcars for shipment. The vent on each tank is also heated to prevent the condensation and solidification of vapors that could plug the vent. Once each 12-hour shift, operators check the temperature of the condensate from each steam tracing loop to ensure the systems are working properly.

A scheduled shutdown was approaching, so some additional tasks were required in preparation for the outage. An operator was taking the temperature readings as part of his routine rounds, and he noticed that the temperatures were unusually low. Unfortunately, before he could troubleshoot the situation, he left the plant due to a personal emergency. He informed the relieving operator of the low readings, but nothing was done to correct the situation. (We believe another operator had incorrectly isolated the heat tracing in anticipation of the shutdown, working ahead to get that task out of the way.)

For the next 22 shifts (the next 11 days), only two other valid temperature readings were taken. Unfortunately, the apprentice operator who took those readings did not communicate them to her supervisor, and he did not review her log sheets, so no action was taken. On every other shift, the readings were skipped (no values were entered on the log sheet) or were "pencil whipped" with typical values. (The false entries were all the same value and were all entered by the same operator.) Meanwhile, resin vapors were freezing in the tank vents, which restricted their size.

The day of the outage arrived, and the unit was shut down. The procedure specified that the storage tank inlet valve should be closed and that the rundown line should be blown clear with compressed air. However, any material blown back to the unit would have to be disposed as scrap. Operators saw no reason to scrap perfectly good product, so their practice was to close the unit outlet valve and blow the product forward to the storage tank. During past shutdowns, with fully open tank vents, no apparent harm had occurred. However, with the tank vent restricted, the surge of 80-psig air split a side seam of the tank. The tidal wave of hot product surged over the dike walls and flooded the employee parking lot, where the resin froze and encased the parked cars.

3. A NEW APPROACH

The traditional response to such an incident is to dismiss or punish those deemed derelict in their duties and to retrain those who remain. However, previous application of this approach had failed to have a lasting effect. Retraining and fear of discipline had not resulted in the operational excellence the company sought.

While investigating the practices of high-reliability organizations, the company uncovered a concept called "conduct of operations," "formality of operations," or "operational discipline" by various sources. Conduct of operations (*operations*) is the execution of operational and management tasks in a deliberate and structured manner, and it is closely tied to an organization's culture. Conduct of operations institutionalizes the pursuit of excellence in the performance of every task and minimizes variations in performance. Workers at every level are expected to perform their duties with alertness, due thought, full knowledge, sound judgment, and a proper sense of pride and accountability (Refs. 1 and 2).

Conduct of operations applies to all work activities, not just those of the operations department. Thus, all workers, employees, and contractors are included. Each work group defines the framework of controls necessary to ensure that tasks for which it is responsible are performed reliably. The manager of each work group is responsible for the conduct of operations in his group, but overall responsibility rests with the facility manager. The human resources group is often involved with the process because it includes fitness-for-duty, progressive discipline, salary, bonus, and retention decisions.

The output of this activity is a policy describing the organization's overall expectations for worker conduct and specific procedures and goals for implementing those policies. A defined framework of controls should be established that implements a defense-in-depth strategy to ensure that process operations remain within safe operating limits. Workers must then be trained in the policies and procedures so that they understand the goals and expectations, the lines of authority, and their personal accountability. They must apply good reasoning and judgment (founded upon a sound process safety culture) in all situations, but particularly when action is required in situations not specifically addressed by policy or procedure.

4. DEVELOPING THE *OPERATIONS* SYSTEM

The root causes of the incident highlighted several specific areas that needed improvement: communications, the use of qualified workers, the verification of task performance, and following written procedures.

One of the first failures in this chain of events was a breakdown in communication between work groups. Someone took it upon himself to isolate the steam tracing without communicating it to the operators. This type of failure frequently occurs during nonroutine activities such as startups, shutdowns, construction, and partial plant outages. The new *operations* system required written communication between work groups to minimize the potential for future misunderstandings. The new system encompassed a wide range of documents, including purchase orders, batch sheets, work orders, and safe work permits. The facility also (1) adopted standard forms to help ensure that each work group received all the information it needs and (2) assigned specific contact points to help ensure that documents are properly routed. To handle abnormal situations, designated individuals were appointed to handle emergency communications and to notify appropriate external work groups, authorities, or stakeholders.

The next failure was a breakdown in communications between workers. The worker who had to leave the plant did not effectively communicate the plant status to the relieving operator. Everyone depends on communication to exchange information, and reliable communication is essential to reliable operation. The information must be transmitted and received accurately, completely, and in a timely manner, and it must be understood. For radio communications, the company adopted a new protocol that required that (1) the sender identify the recipient and himself, (2) the sender transmit the message clearly and concisely, (3) the recipient repeat the message back to the sender, (4) the sender confirm or retransmit the message until it is repeated back correctly, (5) the recipient report completion of the task or any difficulties encountered, and (6) the sender acknowledge the report. The use of public address or paging systems for critical communications was forbidden unless an independent means was available to confirm that the message was received and understood.

The new *operations* system also included a structured method for management to communicate guidance and short-term information to the operations staff. For example, operators may need to know the schedule of special activities during a plant shutdown, or they need to be aware of recent changes to policies, procedures, equipment status, production runs, and so forth. The system for issuing timely orders to operators (1) defined a standard format for orders, (2) identified who has authority to issue such orders, (3) distinguished between daily and ongoing (standing) orders, (4) described how orders should be posted and acknowledged, and (5) described how orders should be canceled or superseded. The system also specified that urgent matters be orally communicated to workers in a pre-job or pre-shift briefing. For less urgent matters, a prioritized list of documents was provided for workers to review during their next shift, so everyone would get the same message, even if they were away from the plant for several days or weeks.

There was also a breakdown in communications between shifts. Before the incident, there was no formal shift turnover, and written logs were completed at the discretion of the operator. After the incident, workers were required to maintain a written record of operations in a structured logbook to ensure the oncoming workers understood the status of all equipment and work activities when they took control. Logs were established for each key operating position, and written guidance or a standard form for recording information was developed. The logs were required to be recorded in a timely manner, and the entries had to be legible. At shift change, the outgoing shift was required to brief the incoming shift, in addition to noting its final entry and transferring the written logs to them. Supervisors periodically reviewed the logs for accuracy and completeness to emphasize their importance – both for intershift communication and for building a useful database of performance information.

A fundamental breakdown that contributed to the incident was the failure to monitor equipment status. Despite the communication failures, there were 22 more opportunities to detect and correct the situation. However, the operators failed to maintain a current knowledge of the status of their equipment and support systems. The new *operations* system placed great emphasis on operators both monitoring the information available in the control room and physically inspecting their equipment on regular tours or rounds. Local readings were used to confirm the accuracy of remote instrument readings and reveal incipient problems that require attention, such as abnormal configurations, odd noises, or unusual smells. The field log sheets were modified to clearly indicate the acceptable range for each parameter so that abnormalities could be spotted immediately.

The apprentice operator's failure to recognize the implications of the low condensate temperatures suggested that workers were being assigned to tasks for which they were not fully qualified, and their work was not being closely reviewed and supervised. The new *operations* system established a more formal protocol for on-the-job training. In particular, requirements for supervision of trainees were established along with clear criteria for advancing trainees through the program.

Supervisors also failed to monitor the work of qualified workers. While a qualified employee should require significantly less supervision, regular checks on the quality and completeness of their work are essential. The fact that log sheets could be left blank or "pencil whipped" with impunity suggested that supervisors had effectively abandoned their responsibilities. The new *operations* system included frequent, unannounced reviews on all shifts to help ensure that operations were conducted in accordance with best practices. The system primarily relied on peer inspections, but supervisors and upper-level managers were expected to be visibly involved on a routine basis. Supervisors were required to ensure that rounds were conducted as scheduled and that logs and other paperwork were properly completed. Persons conducting routine safety and housekeeping audits were also expected to check on ongoing maintenance work activities to ensure that (1) the proper permits were in place, (2) the permits specified appropriate conditions, and (3) workers were following procedures and conforming to permit requirements.

The coup de grace for the storage tank resulted from a failure to follow written procedures. The tank should have never been subjected to plant air pressure, and with a restricted vent, its rupture was inevitable. A cornerstone of the new *operations* system was a "condition of employment" requirement to follow the written procedures. However, management emphasized that strict adherence was not blind adherence – the ideal was "thinking" compliance, where the worker was expected to execute the procedure as written unless doing so would create an adverse condition. In that case, the worker should stop and get permission to deviate from the standard procedure and initiate the process to change or update the procedure, as appropriate. When checklists were provided with a procedure, the worker was expected to have the checklist at the actual work location and check off tasks as they were completed. When independent verifications are specified, they must be performed – no matter how experienced or reliable the worker.

Of course, the complete *operations* system was not derived from the findings of just one incident. Other key topics included:

- Assigning adequate resources, so workers were not overburdened with work, nor fatigued by excessive overtime
- Adhering to safe operating limits and limiting conditions for operation, regardless of the economic incentives to "ride it out" and avoid a process shutdown
- Controlling access to operator work areas to minimize distractions
- Formalizing equipment access protocols, so there was a clear "owner" responsible for maintaining the equipment in a safe state at all times
- Maintaining good housekeeping to instill a sense of ownership and pride in the workforce, as well as to reveal incipient problems

- Maintaining clear labeling and good lighting to minimize human errors when selecting equipment or taking readings
- Maintaining equipment and tools in good working order so they provide reliable information to the workers
- Emphasizing observation and attention to detail so warning signs of imminent failures were noted during rounds
- Promoting a questioning/learning attitude so workers' troubleshooting skills were always improving
- Training workers to recognize hazards and emphasizing lessons learned from incidents throughout the plant, the company, and industry
- Training workers to self-check and peer-check so human errors could be identified and corrected before any harm resulted
- Establishing standards of conduct that promoted constructive cooperation with zero tolerance for disruptive behaviors
- Maintaining workers' fitness for duty so impaired individuals would not be assigned to safety-critical roles
- Correcting deviations immediately, so workers were aware of management's ongoing interest and attention to unsafe actions and conditions

The result was an ambitious program that set out to convert the trained workforce into an effective workforce for whom excellence was the only acceptable performance standard.

5. IMPLEMENTING THE *OPERATIONS* SYSTEM

Like all major initiatives, implementation of the *operations* system had its share of opponents and resistors. Previous initiatives had come and gone with various managers, so the workforce was generally jaded about its prospects for success. However, the incident and its aftermath had prompted a corporate review of the plant's economics, and there was a real possibility that the plant would be downsized, sold, or closed. The situation created a climate for change and improvement, but management leadership was required.

Management signaled its intention to break with past practices by not firing those directly responsible for the incident. Management then engaged the workers to define measurable performance goals for all key aspects of the *operations* program. These metrics were based on a wide array of information, including:

- Number of incidents with operations issues as root causes
- Number of shortcuts identified by near misses and incidents
- Number of incidents during which safe operating limits were exceeded
- Number of human-machine deficiency and near-miss reports
- Number of nuisance and always-on alarms
- Number of nonroutine and emergency maintenance work orders

- Number of incomplete shift logs or reports
- Number of missed surveillance rounds
- Number of work orders attributed to equipment abuse
- Average time to complete repairs on the human-machine interface
- Number of housekeeping audits and their scores
- Average time to resolve off-normal findings
- Number of inspection deficiencies related to labeling
- Number of inspection deficiencies related to lighting
- Number of incidents caused by a lack of self-checking or peer-checking
- Percentage of overtime hours
- Absenteeism
- Average time required to complete required reading
- Number of manager inspections of work locations
- Number of unplanned shutdowns
- Number of unplanned safety system activations
- Number of disciplinary actions
- Percentage of workers failing random substance abuse tests

Management also scheduled regular reviews of the *operations* program itself, in addition to the metrics, to identify and correct problem areas before they impeded progress.

A few areas showed immediate success, and management seized upon them to celebrate success and build "can do" momentum. After a year, the plant's economic performance had improved to the point that talk of sale or closure faded. This significantly improved morale, and success began to feed on itself. However, management continued to invest energy in the initiative, rather than abandoning it as a completed project. A financial incentive program was implemented that incorporated many of the *operations* metrics, so workers had a tangible stake in continuous improvement. After 5 years, the plant's economic, safety, and environmental performance was among the best in the company, and it was being selected as the site for several new product lines. The plant's program became a model for other sites to emulate when the company implemented *operations* programs globally.

6. CONCLUSION

Workers must be initially and periodically trained in an organization's policies, procedures, and practices. However, having well-trained workers does not alone ensure excellent operating results. Conduct of operations addresses the day-to-day implementation of that training so that operational and management tasks will be executed in a deliberate and structured manner. It is the "glue" that binds good training to a consistently high level of human performance. Any company can adopt the principles of an *operations* program, but it will only be successful if the company's culture embraces and reinforces the pursuit of operational excellence. The Center for Chemical Process Safety included it in the framework for risk-based process safety because of its proven value as a management system in high-reliability organizations.

7. REFERENCES

1. *Conduct of Operations at Nuclear Power Plants, DS347*, International Atomic Energy Agency, Vienna, 2005.
2. *Conduct of Operations Requirements for DOE Facilities*, DOE Order 5480.19, U.S. Department of Energy, Washington, 2001.

Guidelines for Identifying, Evaluating, and Selecting Safety Instrumented Functions

Russell A. Ogle and Andrew R. Carpenter
Exponent
185 Hansen Court, Suite 100
Wood Dale, IL 60191
rogle@exponent.com

ABSTRACT

For any given process hazard, the engineer has at his disposable three main categories of safeguards: operator actions, safety instrumented functions or SIFs (e.g., trips and interlocks), and physical safeguards. Careful consideration must be given to the interactions between these three components of safety. In an optimal situation, these three elements work together in concert to achieve the goal of controlling the hazard. However, our experience in chemical process accident investigations has shown that sometimes both engineering and operating personnel have a poor understanding of the system of safeguards dedicated to protecting a particular process unit. This situation can arise because the safeguards were implemented on a piecemeal basis instead of being the result of a systematic hazard analysis. Engineering and operating personnel need to understand the interaction between operator intervention, SIFs, and the physical safeguards.

INTRODUCTION

A considerable number of guidance documents on evaluating the effectiveness and reliability of safety instrumented functions (SIFs) are available to the process engineer [ISA 84, 2004; CCPS, 1993; IEC 61508, 1998]. There is very little guidance, however, on identifying, evaluating, and selecting SIFs for controlling a particular process hazard. This paper attempts to fill this gap by presenting a methodology based in part on chemical process safety guidance documents [e.g., CCPS 1998 and MacDonald 2004]. We examine common hazards found in three types of process systems:

- Process vessels
- Fluid transfer equipment
- Heat transfer equipment

A methodology is presented for each type of process equipment that would identify, evaluate, and select SIFs. The identification of hazards is based in part on the concept that an accident is an uncontrolled release of energy or hazardous chemical(s). The goal of an SIF is to detect and prevent or control the release of energy or hazardous material. The identification of potentially useful SIFs follows naturally from this concept. The evaluation and selection of SIFs explicitly considers the potential benefit of operator intervention and physical safeguards. We illustrate the use of this methodology with both hypothetical examples and case studies of accidents that could have been prevented with a single SIF.

A brief note on the terminology used in this paper:

- **Interlock**: An interlock is a device or system that, in response to a predetermined condition, initiates or prevents a predefined action (CCPS, 1993, p. xx). "Interlock" is considered to be an older, less precise term for an SIF [CCPS, 2001, p. 3].
- **Safety Instrumented Function (SIF)**: The SIF is a combination of sensors, logic solver, and final control elements with a specified safety integrity level that detects an out-of-limit (abnormal) condition and brings the process or equipment to a functionally safe state without operator intervention, or by initiating a trained operator response to an alarm [CCPS, 2001, p. 262].
- **Safety Instrumented System (SIS)**: The SIS is a combination of sensors, logic solver, and final control elements with a specified safety integrity level that performs one or more safety instrumented functions [CCPS, 2001, p. 263]. The SIS is an assembly of SIFs. It is frequently referred to as the emergency shutdown system or the safety shutdown system [Gruhn and Cheddie, 1998, pp. xi-xii].

BACKGROUND

This paper attempts to answer the question, "Where on process equipment should one place safety instrumented functions?" This deceptively simple question has no easy answer. There is a wealth of reference material available to the process engineer that addresses subjects like hazard analysis and risk assessment for chemical process systems in a broad and far-ranging manner. And there are specific references that present methods for designing and analyzing the reliability of safety instrumented systems. For example, the Center for Chemical Process Safety (CCPS), a technical committee of the American Institute of Chemical Engineers (AIChE), has developed a risk analysis process called Layer of Protection Analysis (LOPA) [CCPS, 2001]. For a given accident scenario, LOPA explicitly considers the degree of protection offered by operator intervention, SISs, and physical safeguards.

But the topic of how to select locations for SIFs is rarely discussed. It is often implied that the selection of SIFs is the outcome of a formal process hazard analysis conducted near the end of the design effort (e.g., after the piping and instrumentation diagrams have been completed). This implies that a key layer of protection is being built into the process at a late stage of the design effort when it is more difficult to accommodate the inevitable changes that occur during the design process. A better practice is to select the layers of protection during the earliest phases of the design process (conceptual design stage) [Marszal 2006]. The approach presented here is appropriate for conceptual design and is based on a preliminary hazard analysis technique.

SELECTING SAFETY INSTRUMENTED SYSTEMS BASED ON PROCESS HAZARD ANALYSIS

The fundamental goal of chemical process safety is to ensure that the energy and hazardous chemicals contained inside the equipment stay inside the equipment. A chemical process accident can be defined as an uncontrolled release of energy or hazardous chemical(s) from the process equipment. The types of energy most frequently encountered in a chemical process system are: mechanical, thermal, and chemical. Common categories of hazardous chemicals encountered in industrial processes include reactive (flammable, explosive, thermally unstable, etc.), toxic, and corrosive. The severity of the consequences due to the hazardous release is an important factor in determining the nature and characteristics of safeguards used.

The process engineer uses hazard analysis techniques to identify accident scenarios, to evaluate accident consequences, and to identify safeguards or layers of protection to mitigate these consequences. Three of these layers of protection are critical alarms and operator intervention, SIFs, and physical safeguards. In the selection of the SIFs, it is important to consider the potential benefits or limitations imposed by operator intervention and physical safeguards. The conceptual procedure described below can be applied to any type of process equipment. It is based on the most common hazards encountered in a process operation. However, no attempt has been made to comprehensively consider all hazards that one could possibly encounter (e.g., ionizing radiation hazards and infectious agents are not discussed). Furthermore, mechanical and electrical hazards specific to machinery are not considered.

Conceptual Procedure for Selecting Process Equipment SIFs

The conceptual procedure for selecting process equipment SIFs is outlined in Table 1. The procedure can be easily documented in a tabular or worksheet format. It can be implemented by a single individual or in a group setting.

The first step in the procedure is to define the process system of interest. By analogy with thermodynamics, a control volume boundary is drawn around the equipment item to facilitate the identification of mass and energy flows that can enter, exit, or accumulate within the item. Normal operation of the item must be described because it is used later as a reference point.

The second step is to identify the key process variables. From the perspective of process control, the typical process variables are flow rate, level, pressure, temperature, and composition. Key process variables are those variables that can change in the course of normal or abnormal operation. Next, the controlled, measured, and manipulated variables are listed. Finally, the design value and both lower and upper operating limits are specified.

The third step is to specify the operational deviations that could result in an exceedance of the lower or upper operating limit. Operational deviations are then categorized into one of two categories: safety related and non-safety related. Safety related operational deviations are those that could cause a failure of containment resulting in an uncontrolled release of energy or hazardous chemical(s). These safety-related operational deviations are candidates for SIFs.

The fourth step is to identify an SIF that could act to stop an operational deviation. To be effective, the SIF must be capable of three functions:

- The SIF must be capable of detecting the operational deviation, i.e., the process measurements must be capable of detecting the exceedance of the operating limit.
- The SIF must be capable of stopping the operational deviation, i.e., the final control element must be capable of stopping or controlling the exceedance of the operating limit.
- The SIF, upon detection of the operational deviation, must deliver the appropriate command to stop or control it by actuating the final control element.

If the current design of the equipment item lacks either the process measurement or the final control element needed to control the operational deviation, one should consider adding this to the design requirements.

The final step is risk assessment: evaluate the proposed SIFs and select those that offer the desired level of risk reduction. The consequences and likelihood of a hazardous release must be determined. Then the potential risk reduction of operator intervention and physical safeguards should be considered. If operator intervention and physical safeguards cannot achieve the desired level of risk reduction, add the SIF with an appropriate safety integrity level (SIL).

Table 1. Procedure for Selecting Process Equipment Safety Instrumented Functions

1. **Describe the process system of interest.**
    a. Select the process equipment item to be considered.
    b. Draw a control volume boundary around the item.
    c. List the inputs and outputs of mass or energy that cross the control volume boundary.
    d. Describe normal operation.
    e. List the types of energy or hazardous chemical(s) entering, leaving, or accumulating within the item.
2. **Identify the key process variables and the equipment operating limits.**
    a. List the controlled, measured, and manipulated process variables.
    b. Specify the design value, and the lower and upper operating limits for the equipment in terms of the process variables.
3. **Determine the consequences of operational deviations.**
    a. What operational deviations could cause an exceedance of the lower or upper operating limit?
    b. Determine if an exceedance of the lower or upper operating limit could result in a failure of containment that would, in turn, result in a release of energy or hazardous chemical(s).
4. **Identify SIFs to control the operational deviation.**
    a. For each exceedance of an operating limit that could result in a release, consider a SIF.
    b. Can the measured variables detect the operational deviation? If not, consider adding a process instrument.
    c. Can an existing final control element prevent the operational deviation? If not, consider adding a final control element.
    d. What is the control logic for the SIF?
5. **Evaluate the proposed SIFs and select those offering the desired level of risk reduction.**
    a. For the specified operational deviation, determine the consequence and likelihood of a hazardous release.
    b. Consider the risk reduction offered by operator intervention and physical safeguards.
    c. If the desired level of risk reduction is not achieved, add the SIF.

The use of this procedure is illustrated with a simple example that uses three types of process equipment:

- Process vessels

- Fluid transfer equipment
- Heat transfer equipment.

A pump is used to transfer liquid through a heat exchanger and then to a distillation column. Figure 1 below is a block flow diagram of the system.

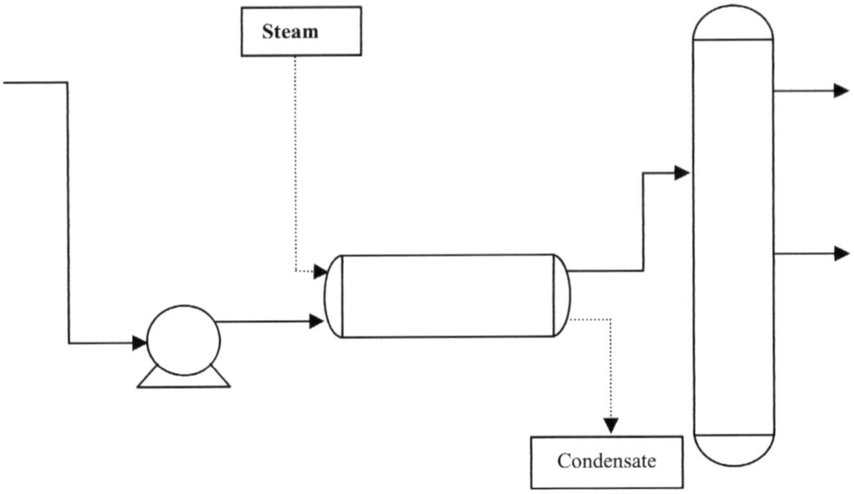

Figure 1. Block Flow Diagram

It is assumed that the liquid is a flammable liquid. Thus, the process engineer must manage the fire and explosion hazards posed by the liquid.

SAFETY INSTRUMENTED FUNCTIONS FOR PROCESS VESSELS

Process vessels include atmospheric storage tanks, pressure vessels, and vacuum vessels. The characteristics that typically define their function are the materials contained in the vessel, the operating pressure and temperature of the contents, the flows into and out of the vessel, and specific construction details (e.g., materials of construction). The key process variables are level/inventory, flow in/out, pressure, and temperature. Normal operation can be characterized between two extremes: a steady inflow and outflow through the vessel with an essentially constant inventory of material, or periodic filling and discharging of the vessel. A process vessel is an energy storage device. The forms of energy within the vessel include mechanical (pressure), thermal (temperature) and chemical (composition) energy.

The safety objective is to keep the energy and hazardous chemical(s) inside the process vessel. Operational deviations that threaten the safety objective include overflow, excessive pressure or vacuum, and high or low temperature. Depending on the magnitude of the deviation, each of these could result in a failure of containment. Hence, each of these deviations is a candidate for an SIF.

Consider the distillation column in the example. An overflow of a flammable liquid can have catastrophic consequences if the overflow is not easily diverted into a safe containment. One way to prevent an overflow hazard is to install a SIF to close the vessel inlet to prevent overfilling. The process engineer should consider the consequences of the SIF action on both upstream and downstream equipment. For example, if a level SIF is used to protect the distillation column in this example, then one must consider what happens to the pump and heat exchanger. In this case, it is recommended that in the event of a high level trip on the distillation column, the pump would be tripped off-line and the steam supply would be closed. This is illustrated in Figure 2.

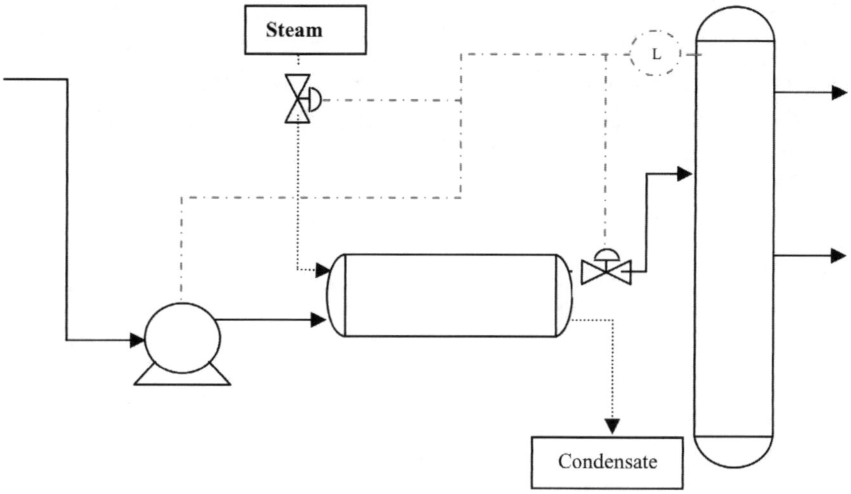

Figure 2. Level Control SIF

Two recent accidents involved the overflow of a flammable liquid from a process vessel. The BP Texas City accident involved an overflow of flammable liquid from a distillation column that resulted in explosion and fire [Baker, 2007]. The Buncefield storage facility accident was caused by an overflow of flammable liquid from an atmospheric storage tank [Powell, 2006]. Both of these accidents could have been prevented with functional high-level SIFs.

SAFETY INSTRUMENTED FUNCTIONS FOR FLUID TRANSFER EQUIPMENT

Fluid transfer equipment includes pumps, blowers, and compressors. The characteristics that typically define their function are the materials transferred; the operating flow rate, pressure drop and temperature; and the operating limits of the source and receiving vessels. The key process variables are flow rate and pressure.

Normal operation is a steady transfer of fluid from a source vessel to a receiving vessel. Fluid transfer equipment store little energy; they are work producing devices. The work produced by fluid transfer equipment is transformed primarily into mechanical energy (pressure, flow). Compressors, however, can transform a significant portion of their work into thermal energy (temperature).

The safety objective is to keep the energy and hazardous chemical(s) within the process equipment. The effect of operational deviations on fluid transfer equipment requires consideration not only of the device itself but also of the potential impact on the source and receiving vessels.

Consider again the distillation column example. In the event of a loss of flow into the pump, two potential consequences are excessive pressure in the heat exchanger due to the loss of cooling provided by the product stream through the heat exchanger, and reverse flow from the distillation column due to its hydrostatic head. The potential consequence of reverse flow is overfilling the source vessel. Employing an SIF to close an isolation valve on the steam line and to close an isolation valve on the product line inlet to the pump can mitigate these hazards. This SIF is illustrated in Figure 3.

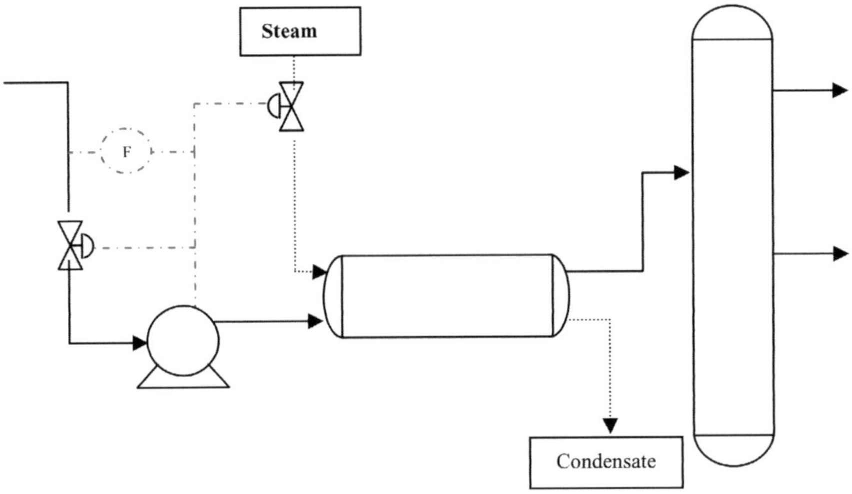

Figure 3. Flow Control SIF

Another hazard associated with fluid transfer equipment is unintended forward flow while the device is idle. A source vessel filled with concentrated hydrochloric acid drained and spilled its entire contents when the acid dissolved the pump. The pipe between the source vessel and the pump did not contain an isolation valve. If an isolation valve had been installed with an SIF to close the valve when the pump was not running, the accident would have been prevented [Ogle, 2001].

SAFETY INSTRUMENTED FUNCTIONS FOR HEAT TRANSFER EQUIPMENT

Although there is a wide diversity of design concepts for heat exchangers, they are in essence very simple devices. Consideration is restricted to devices that passively exchange energy from one fluid stream to another; combustion systems such as furnaces and boilers are excluded. The characteristics that typically define the function of heat exchangers are the heating or cooling duty, operating temperatures and pressures of the process streams, and specific construction details. The key process variables are flow rate, temperature, and pressure. Normal operation is a steady transfer of heat from a hot stream to a cold stream. The forms of energy within the exchanger are mechanical (pressure, flow rate) and thermal (temperature).

The safety objective is to keep the energy and hazardous chemical(s) within the process equipment. Operational deviations that threaten the safety objective include excessive heat transfer rates, unexpected phase changes, excessive temperatures (i.e., too hot or too cold), and overpressure.

In the distillation column example, excessive steam flow could result in a high-temperature hazard for the product stream entering the distillation column. A high-temperature SIF is recommended to close an isolation valve on the steam. This is illustrated in Figure 4.

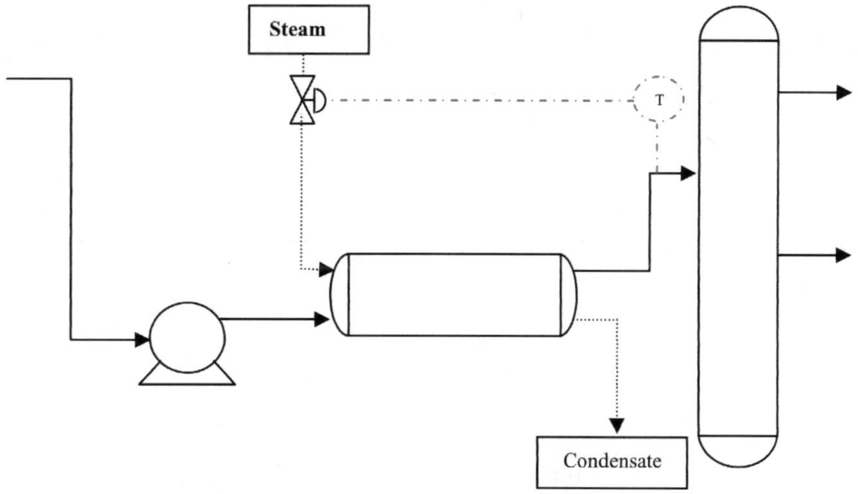

Figure 4. Temperature Control SIF

CONCLUSIONS

SIFs are an important safeguard for process equipment. A methodology has been presented in this paper for identifying opportunities to incorporate SIFs. The methodology was demonstrated using examples for process vessels, fluid transfer equipment, and heat exchange equipment.

REFERENCES

1. ANSI/ISA-84.00.01-2004 Functional Safety: Safety Instrumented Systems for the Process Industry Sector (Parts 1, 2 and 3), ISA-The Instrumentation, Systems and Automation Society, Research Triangle Park, NC (2004).

2. Baker, et. al., The Report of the BP U.S. Refineries Independent Safety Review Panel, (2007).

3. CCPS, Guidelines for Design Solutions for Process Equipment Failures, American Institute of Chemical Engineers, Center for Chemical Process Safety, New York (1998).

4. CCPS, Guidelines for Safe Automation of Chemical Processes, American Institute of Chemical Engineers, Center for Chemical Process Safety, New York (1993).

5. CCPS, Layer of Protection Analysis: Simplified Process Risk Assessment, American Institute of Chemical Engineers, Center for Chemical Process Safety, New York (2001).

6. Gruhn, P. and H.L. Cheddie, Safety Shutdown Systems: Design, Analysis and Justification. ISA-The Instrumentation, Systems and Automation Society, Research Triangle Park, NC (1998).

7. IEC 61508-1 Functional Safety of Electrical/Electronic/Programmable Electronic Safety-Related Systems, International Electrotechnical Commission, Geneva, Switzerland (1998).

8. Macdonald, D., Practical Hazops, Trips and Alarms. Elsevier, Amsterdam, (2004).

9. Marszal, E., "Dispelling Myths about SIL Selection," INTECH, p. 66 (November 2006).

10. Ogle R., Morrison D., "Investigation of an Acid Spill Caused by the Failure of an Air-Operated Diaphragm Pump," Process Safety Progress, Vol. 20, pp. 42-49 (2001).

11. Powell, Taf, The Buncefield Investigation – Third Progress Report. Buncefield Major Incident Investigation Board, London (2006).

Equipment Safety Manuals - A User's Expectations

Harold W Thomas
Air Products & Chemicals, Inc.
Allentown, Pa

David Deibert
Air Products & Chemicals, Inc.
Allentown, Pa

David C. Arner
Air Products & Chemicals, Inc.
Allentown, Pa

David Weir
Air Products & Chemicals, Inc.
Hersham, United Kingdom

ABSTRACT

With the recently approved IEC 61511 and ANSI/ISA-84.00.01 standards, process industry companies are being challenged to determine the level of performance required from their safety instrumented functions (SIF) and to verify that the design, installation, operation and maintenance employed is sufficient to ensure that the required performance is sustained until the SIF is decommissioned. As a practical matter, user companies must self-certify or user approve their SIF equipment, since they are responsible for the correct operation of the SIF in the operating environment, not just individual components as manufactured. To accomplish this, user companies need lifecycle assistance from equipment manufacturers, irrespective of hardware or software certification status related to IEC 61508.

One of the best means to communicate the necessary information is via a safety manual. IEC 61508 requires that equipment manufacturers provide a safety manual with any product claimed to comply with IEC 61508. Although required for product certification, it is the collective authors' current experience that quality and consistency in safety manuals is lacking. Part of this is due to IEC 61508 not providing a clear concise template with minimum requirements. The safety manual requirements are scattered throughout the rather complex standard.

This paper seeks to provide a user's perspective of what essential information is needed from equipment manufacturers to not only comply with IEC 61508, but also to provide enough information to the User of the product to allow the essential balance of reliability and safety in both SIF and instrumented protective function applications.

INTRODUCTION

A Safety Manual is a compilation of installation, configuration, and mechanical integrity requirements for equipment approved for safety instrumented systems (SIS) applications. The equipment manufacturer should provide a safety manual for any product claimed to comply with IEC 61508, Functional Safety of Electrical/Electronic/Programmable Electronic Safety-Related Systems. The safety manual specifies the user's responsibilities for equipment installation and operation required to maintain a particular safety integrity level (SIL) as defined by IEC 61508.

The information contained in the Safety Manual is essential for users to design and maintain the SIF that achieves a specified SIL. When performing SIL verification work, the average probability of failure on demand (PFDAVG) will change as a function of equipment failure rates, subsystem architecture, how the equipment is configured to respond to various diagnostics, the equipment proof test interval, proof test effectiveness, and how long it takes to respond and repair identified SIF equipment faults. No manufacturer can provide a PFDAVG for the final design, installation, operation, and maintenance that exists in the field. This is the user's responsibility. The manufacturer is responsible to provide the fundamental technical information for that which they have control, enabling users to efficiently and consistently do their job.

This paper addresses information that should be provided by manufacturers to support the user. The data expectations have been sub-categorized as basic, restrictions of use, and mechanical integrity guidance / information. The paper ends with an overview of a work process our company is implementing to facilitate obtaining this essential information.

BASIC INFORMATION

Boundary Diagram of Covered Equipment: Each safety manual is expected to provide a clear definition of what is included and what is not included within the equipment boundary in the manufacturer's analysis. The user's analysis must cover the devices, components, and interfaces needed for the equipment to operate according to the specification. A pictogram is a straightforward and understandable means of providing this information.

As an example, assume a manufacturer is providing a smart positioner capable of performing partial stroke testing. To the user, the equipment is simply one part of the final element. The high diagnostic coverage claimed for the actuation subsystem does not necessarily translate into high diagnostic coverage for the final element. The authors' experience is that the IEC 61508 boundary often does not include the coupling mechanism between the actuator and the position or the solenoid operated valve required for valve actuation. Figure 1 shows the typical boundary diagram of a final element from a user's perspective.

Another example of the equipment boundary limitation is exclusion of the process connection in most manufacturer analysis, e.g., the interface of transmitters with the process. In some processes, impulse lines can plug thus increasing the dangerous failure rate. It is important to know whether the probability of plugging is included in the manufacturer information or not. Suppose a transmitter can be purchased with a capillary and remote diaphragm seal. The choice of the process connection can certainly impact the performance of the sensor, but may not be included in the information provided by the manufacturer. Figure 2 shows three different equipment boundary diagrams. Boundary 1 is what the user must address and in most cases, is not being addressed by the manufacturers. Too often, the IEC 61508 analysis boundary is the only meaningful information being provided by manufacturers. This boundary is indicated by boundary 3.

A well designed safety manual should clearly denote the analysis boundary. The boundary interfaces, required support systems, and external diagnostic requirements should be described. The restrictions regarding the support systems or communications should be clearly listed. These assumptions are essential to properly understand the "handshake" between the components on opposite sides of the interface. The proper specification of the communication interface is essential, because it ensures that the "handshake" conveys important diagnostic data, without which, diagnostic coverage is impossible. In the case of programmable electronic systems (PES), the supplier may provide either one or multiple boundary diagrams (corresponding to the various subassemblies and components on which they depend i.e. racks, terminations, power supplies, etc.)

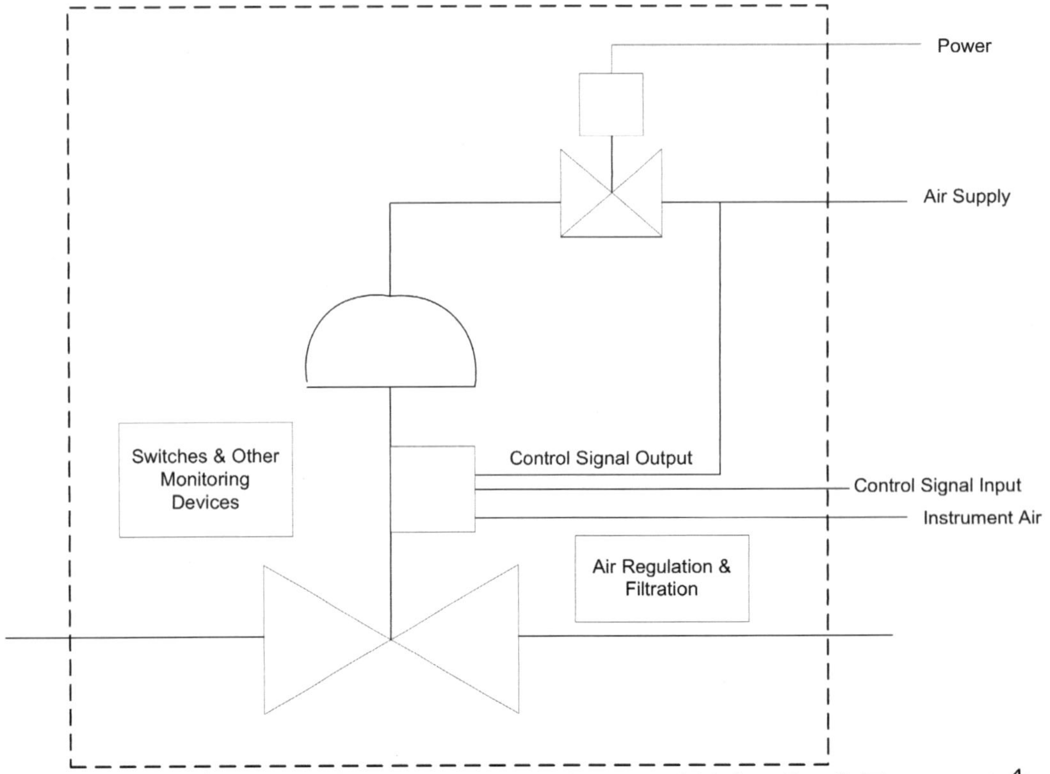

(Excerpted from CCPS PERD Remote Actuated Valve Draft Taxonomy[4])

Figure 1. Example Boundary Diagram – Remote Actuated Valve

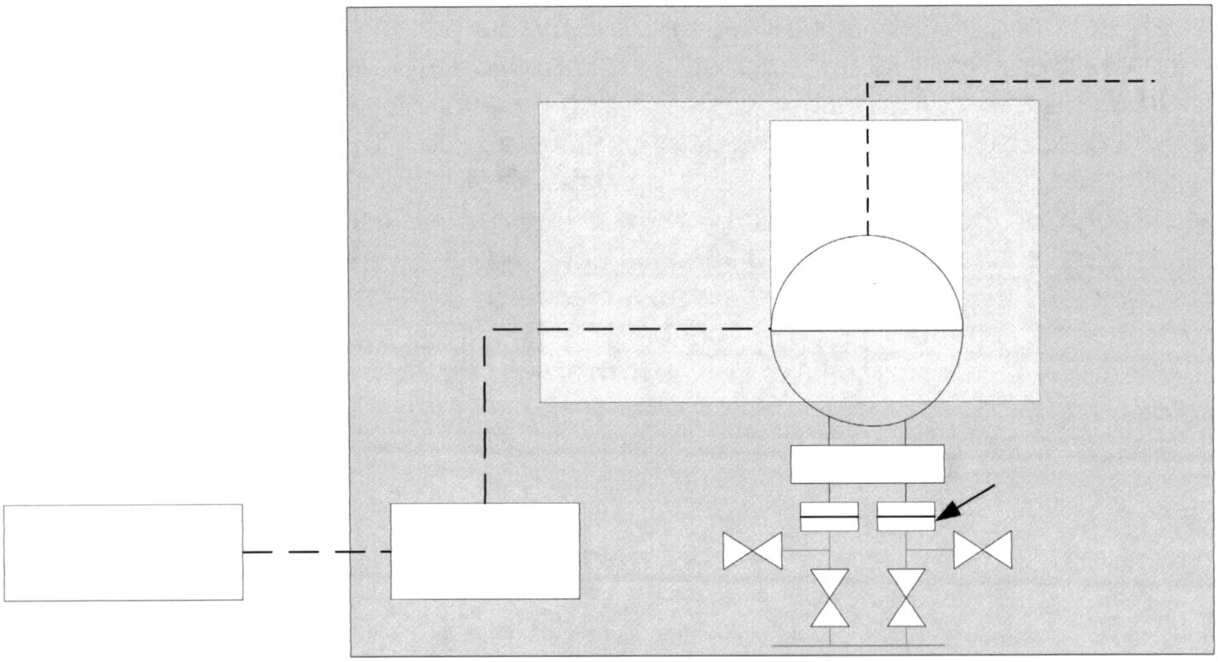

Figure 2. Transmitter: Possible Boundary Diagrams

Failure rates and failure modes: The Safety Manual should contain the estimated failure rates for the equipment failure modes that adversely impact the operation of the SIF equipment. The analysis should be based on a Failure Mode and Effects Analysis (FMEA) using appropriate failure rates. For equipment included as part of a fault tolerant subsystem, an estimate of the beta factor is used to account for common mode failure.

The user needs adequate information to design a SIF so that its PFD_{AVG} is less than or equal to the target value established by the hazard and risk analysis. The PFD_{AVG} is a function of the equipment proof test interval and the required time for on-line repair. In addition, to predict plant reliability, users also need data to determine the rate of spurious equipment operation. This requires data for failure modes related to degraded failures as well as complete loss of function type failures.

Table 1 documents a complete list of potential failure modes for a remote actuated valve as determined by a functional analysis as part of the Center for Chemical Process Safety (CCPS) Process Equipment Reliability Database (PERD) initiative. Depending on the process application, some of these failure modes are dangerous, causing a failure to operate on demand. Other failure modes are safe causing spurious equipment operation. The failure modes listed are applicable to the boundary diagram of a remote actuated valve as depicted previously in Figure 1.

In general, the supplier needs to provide the information on "how" it fails, e.g., loss of function, spurious operation, and degraded operation. Along with the failure frequency, the user can determine whether the equipment is acceptable for the application. It is the user's responsibility to ensure that the manufacturer's assumptions are appropriate for the specific application, especially assumptions regarding whether a particular failure mode is dangerous or not.

As an example, consider a SIF initiated on high pressure. If a transmitter output were to drift high (undetected failure) or go to its saturated high state (detectable failure), the transmitter failure would be safe rather than dangerous. Conversely, if the drift or saturation were to the low state, the transmitter failure would be dangerous. For a low pressure SIF, the failure mode would have the opposite effect. As shown in this example, the effect of the failure mode is application dependent. If we just used dangerous and safe (as many manuals report today), we will either overstate safety while understating reliability, or vice versa.

The user should include the operating environment impact in the evaluation of the failure modes, as well as the configuration for detected failures when determining the failure rates. Manufacturer's currently do not supply process application data. It is an unfortunate outcome of the IEC 61508 certification process that most manufacturers' stopped providing observed failure rates. Most manufacturers should provide significantly better documentation of the fundamental requirements for the safe and reliable operation of equipment. In this way the user may better apply equipment in safety applications.

In the case of PES, the author's experience is that some failure modes and associated frequencies are more complex and indeterminate than claimed by the manufacturer. It is possible for some failures to have one or multiple effects. If there are multiple effects of a single fault (such as one fault rendering an entire 8 analog input card failed), the failure rate should consider the cumulative effect. The user can negate dangerous effects through diagnostics and fault tolerant architectures. Knowledge of the equipment failure modes and effects is critical to designing a safe and reliable plant.

Fault Tolerance: The user may implement fault tolerance because either it is required to achieve a certain level of performance or because the user requires higher reliability. Any equipment fault tolerance requirements should be specified in the equipment safety manual.

Some equipment is provided with diagnostics that depend (via supplier firmware) on fault tolerance to achieve reliable performance. For example, the diagnostics detecting a failed point on a module may be used to automatically switch the equipment to a backup module. The frequency of individual point failures may be small, but the effect of common mode issues on the module failure rate may be high. The impact is that all inputs/outputs on the module are affected more frequently, hurting overall reliability. Certain settings may appear to affect only reliability; however, the diagnostics afforded by the redundancy is often critical to maintaining the needed PFDAVG.

For the case where the user is required to implement fault tolerance by using redundant equipment, the safety manual should include adequate information for the user to be able to perform the calculations properly.

Table 1. Remote Actuated Valve Failure Mode Table

Failure Modes	Failure Classification
Complete Failures	
Spuriously fail to closed position	Potentially dangerous as a hazard demand
Spuriously fail to open position	Potentially dangerous as a hazard demand, not as a standby protective SIF
Fail to close on demand	Dangerous (SIF related when SIF closes on demand)
Fail to open on demand	Dangerous (SIF related When SIF opens on demand)
Frozen Position (Modulating Service)	Potentially dangerous as a hazard demand, not as a standby protective SIF
Valve Rupture	Potentially dangerous as a hazard demand, not as a standby protective SIF
Seal/Packing Blowout	Potentially dangerous as a hazard demand, not as a standby protective SIF
Degraded Failures	
Reduced Capacity	SIF related when open on demand and protection requires a defined capacity
Seat leakage	May be SIF related. Depends on user application
External Leak	Potentially dangerous as a hazard demand
External Leak - Body/Bonnet	Potentially dangerous as a hazard demand
External Leak - Packing/Seal	Potentially dangerous as a hazard demand
Fugitive Emission	Potential environmental / health issue
Controlled variable high	Potentially dangerous as a hazard demand
Controlled variable low	Potentially dangerous as a hazard demand
Fail to hold position	Potentially dangerous as a hazard demand
Unstable control (hunting)	Potentially dangerous as a hazard demand
Responds too Quickly	SIF related dependent upon process application safety time
Responds too Slowly	SIF related dependent upon process application safety time
Excessive Noise	Potential health issue

(Excerpted from CCPS PERD Remote Actuated Valve Draft Taxonomy[4])

Diagnostic coverage percentage for each failure mode: When reporting diagnostic coverage, the manufacturer should state whether the analysis assumes that the equipment automatically takes the equipment to the safe state or initiates an alarm output contact. For fault tolerant architectures, it is necessary to know when only a channel is taken to the safe state versus the entire subsystem. If the diagnostics require the installation of additional equipment outside the manufacturer's boundary diagram, it should be clearly explained in the safety manual. For instance, a transmitter may have the ability to detect certain faults and drive its output either to full scale or to zero, but the vote to initiate the SIF must be incorporated in the logic provided by the user that is outside of the manufacturer's transmitter boundary. Configuration requirements should be clearly defined in the safety manual and should be understood by those implementing the equipment. Deviation from the configuration requirements should be justified.

Operating mode that data applies to: Failure rate data for a single piece of equipment will be different depending upon its operating mode. The four fundamental operating modes in the process industry include:

- Continuous Operation
- Demand Mode Operation
- Cyclical Operation
- Batch Operation

Failure rate data obtained from cyclic testing should not be used for equipment that is continuously energized and may not cycle for months or years, i.e. the typical SIF that is in demand mode service.

Special performance capabilities: It was shown earlier that equipment has several failure modes and the application often dictates when a particular failure mode is safe, dangerous, or immaterial. For instance, using the failure modes for valves (Table 1), we can contrast the failure modes, 'Fail to Open' and 'Seat Leakage.' In some cases, seat leakage is not a significant concern, while in others it is a major concern. The majority of safety manuals we have seen to date have not generally made this distinction. Only recently have we seen some safety manuals start to address seat leakage as a failure mode, but the treatment is still inadequate.

One user company reported that every time they tested particular fuel gas isolation valves that came with a specific BMS service approval listing, the valve leaked. The company reported removing them from service and replacing them with ball valves that have consistently met the leakage requirements.

Special requirements such as level of seat tightness, time to close, time to open, etc. are very important to process industry applications. In addition, these special requirements must be capable of performing to this level of rigor over the entire interval between proof tests.

RESTRICTIONS OF USE

Environmental Limits: The equipment failure rate is impacted by the range of ambient conditions (i.e. indoors, outdoors, temperature, humidity, Alaska, Sahara Desert, Gulf Coast, etc.) to which it is exposed. The failure rate provided by the manufacturer should be valid for the range of environmental conditions claimed for the equipment, otherwise, the safety manual needs to make any limitations clear. Some flow meters may be affected by piping stress or vibration. If so, limits must be established as well.

Utility Limits: Similar to environmental limits, equipment is impacted by the quality of the utilities provided to the equipment by the user. Utility requirements, such as electrical power and/or input signal quality, should be clearly specified with permitted operating limits, such as maximum and minimum voltage, maximum ripple levels, surge suppression requirements, etc. For mechanical devices, such as control valves, instrument air quality specification should address the maximum and minimum operating pressures, filtration, moisture content, etc.

Process Limits: Just as with ambient environmental limits, the actual process severity (e.g., the process operating temperatures, pressures, corrosiveness, fluid phases, metallurgical / process compatibility, etc.) can impact the equipment failure rate. For instance, some electronic transmitters will not last very long in high pressure hydrogen service unless the sensor diaphragm is gold plated. In some instances, the manufacturer may have excellent information regarding potential process impact. In other cases, the manufacturer may simply not know. It is important that the equipment be proven to work in the operating environment prior to use as SIF equipment. Also, it is important that the manufacturer not overstate equipment capability and acknowledge the process conditions for which the failure rate data is relevant.

Configuration Requirements to Achieve Performance Results: When equipment has physical settings or configurable parameters that can be changed, the safety manual should specify the impact of the configuration on the integrity and reliability for the various options available. For instance, a configuration setting that would change the failure direction of fault detection would have a significant impact on appropriate failure rates to use in an analysis.

One of the strongest attributes of programmable equipment today is flexibility. From a manufacturer's perspective, the same equipment can be used to achieve a variety of different desired outcomes. Although this flexibility appears to be a benefit, it comes with a double edge sword regarding SIS. Increased opportunities to commit configuration errors can cause the SIF to fail to operate and therefore to fail to achieve the required PFD_{AVG} for a SIF.

It is incumbent upon the manufacturer to explicitly state what diagnostic settings are required to achieve specific SIL and spurious trip rate. It may be possible to optimize safety and reliability by judicious selection of the configuration settings for different applications, but only if the fundamental impact of the configuration settings are clearly communicated from the manufacturer to the user via the Safety Manual.

As an example, the safety manual may dictate that a certain diagnostic be turned on, however, that diagnostic is programmed in firmware by the supplier to generate both an alarm and an automatic response that shuts down the process. In certain applications, users may want the flexibility to keep the diagnostic, and generate the alarm which reveals the fault, but allow time for the operating / maintenance people to repair before causing a shutdown. The available options and impacts of specific choices must be clearly communicated.

MECHANICAL INTEGRITY GUIDANCE / INFORMATION

Useful Life of Equipment: No equipment lasts forever, although the failure rates provided by manufacturers make it seem like they will last for generations to come. That's because the IEC standards make the assumption that the failure distribution is exponential, meaning the equipment is not in its early life or wear out period (equipment is not used beyond its useful life where the hardware failures are random). In reality, equipment does wear out and age-related mechanisms result in an ever increasing failure rate. To meet the useful life assumption, the mechanical integrity program must address wear out so that the frequency of inspections, tests and preventive maintenance is sufficient to maintain the equipment in the "as good as new" condition and to identify when it is time to replace the equipment.

Not having or considering this information can lead to misuse or abuse of the information provided. One FMEA report that was obtained for a combustible gas detector stated a useful life of 3 years. This information is significant when establishing the mechanical integrity program. In today's world, a number of companies are trying to extend proof test intervals. Without intelligent consideration of useful life, there is a danger of serious mistakes when formulating mechanical integrity programs.

There is a product stewardship expectation on the manufacturers to not present information in a self serving manner that can be confusing. In the example above, the report was intended to aid the users by calculating and reporting PFD_{AVG} for various proof test intervals. Unfortunately, they did not fully think through what they were doing and **reported PFD_{AVG} values for up to 5 years, 2 YEARS BEYOND THE EQUIPMENT'S USEFUL LIFE**. In this case, it was an honest mistake. When the faux pas was brought to their attention, they immediately took steps to institute improved quality assurance measures in their technical writing work process.

In complex systems, a relatively short useful life may be due to specific components. As a practical matter, users need major pieces of SIS equipment to last considerably longer than 10 years. As such, when a relatively short useful life is determined (one major control system manufacturer listed its useful life as 10 years), it is necessary to identify the weak link(s) and address them with preventive maintenance. Figure 3 represents a form that the manufacturer can fill out, listing the component useful lives in increasing order until an acceptable life can be obtained by addressing the weaker components with an appropriate maintenance program.

Figure 3. Useful Life Limitation Information

Notes:

1. Table only needs to be completed for components/parts with a useful life less than 30 years.
2. Components/parts are to be listed in order of increasing useful lives
3. Column three is intended to document what degraded of loss of function occurs if the component fails, taking NO credit for diagnostics.
4. Column four is to document one of the following four classifications of failure:

   Dangerous Undetected, DU

   Dangerous Detected, DD (documents whether the equipment has diagnostics that detect the failure)

   Safe Undetected, SU

   Safe Detected, SD (documents whether the equipment has diagnostics that detect the failure)

Component	Useful Life (years)	Effect on equipment function (note 3)	Failure Classification (note 4)
•			
•			
•			
•			
•			
•			
•			
•			
•			
•			
•			

***Documented Proof test procedures*:** The Safety Manual should advise all procedures required for a test to detect known potentially "dangerous failures" as identified by the product FMEA. In addition, the procedures should document pass/fail criteria. Any tools required to perform the proof testing should be documented as well as the expected skill level of those doing the work. As some tests are more comprehensive than others, the effectiveness of each test should be documented, i.e. what percentage of the failure rate can be detected if the test is performed correctly. See Table 1 for example failure modes that are potentially dangerous for remote actuated valves.

When automated diagnostics are employed to improve the equipment's SIL, it should be recognized that the diagnostics themselves may fail. As such, the Safety Manual should provide appropriate guidance for the mechanical integrity program. Examples of appropriate guidance are:

- Document the diagnostic probability of failure at the end of some reasonable lifetime such as 50 years,
- Document the diagnostic probability of failure as a function of time,
- Provide practical procedures to test the diagnostics and a basis for proof test interval, or
- Instruct the user to return the device to the manufacturer for factory "test" to certify the "as good as new" condition.

Depending upon where in the safety life cycle the equipment is, the appropriate answers and guidance for the above may be different. For instance, during factory acceptance testing, it is possible to more heavily weigh actual testing of the diagnostics, whereas, once the equipment is installed and operational, this may be impractical. Knowledge of probability of diagnostic failure as some function of time is essential.

If the equipment has been designed with on-line testing capability, proof testing is still required as it is generally impossible to completely test the entire SIF on-line. As part of any test or inspection procedure, whether it is on-line or off-line, there should be pass/fail criteria and guidance on how to repair or replace the product, components or parts, as applicable. Once again, any tools and special skills required need to be specified.

For systems with multiple components it may be possible to test each component, however, without definition of acceptable criteria for testing the interfaces (overlap) between these components, the test may not constitute a full proof test, thus lowering the proof test effectiveness to an unacceptable level.

Finally for any equipment where firmware upgrades are possible in the product, procedures need to be provided with any needed tools identified and the expected skill level of those doing the upgrade should also be specified.

USER WORK PROCESS

This paper has documented a number of information items the user expects from manufacturers in their safety manuals. Currently, it is not likely that these expectations will be fully met. To assist moving closer to the ideal, the following work process illustrated in figure 3 and audit checklist in table 2 has been formulated and is being implemented to facilitate communication with our suppliers and to help develop long term relationships where both parties (user and supplier) are committed to continuous improvement.

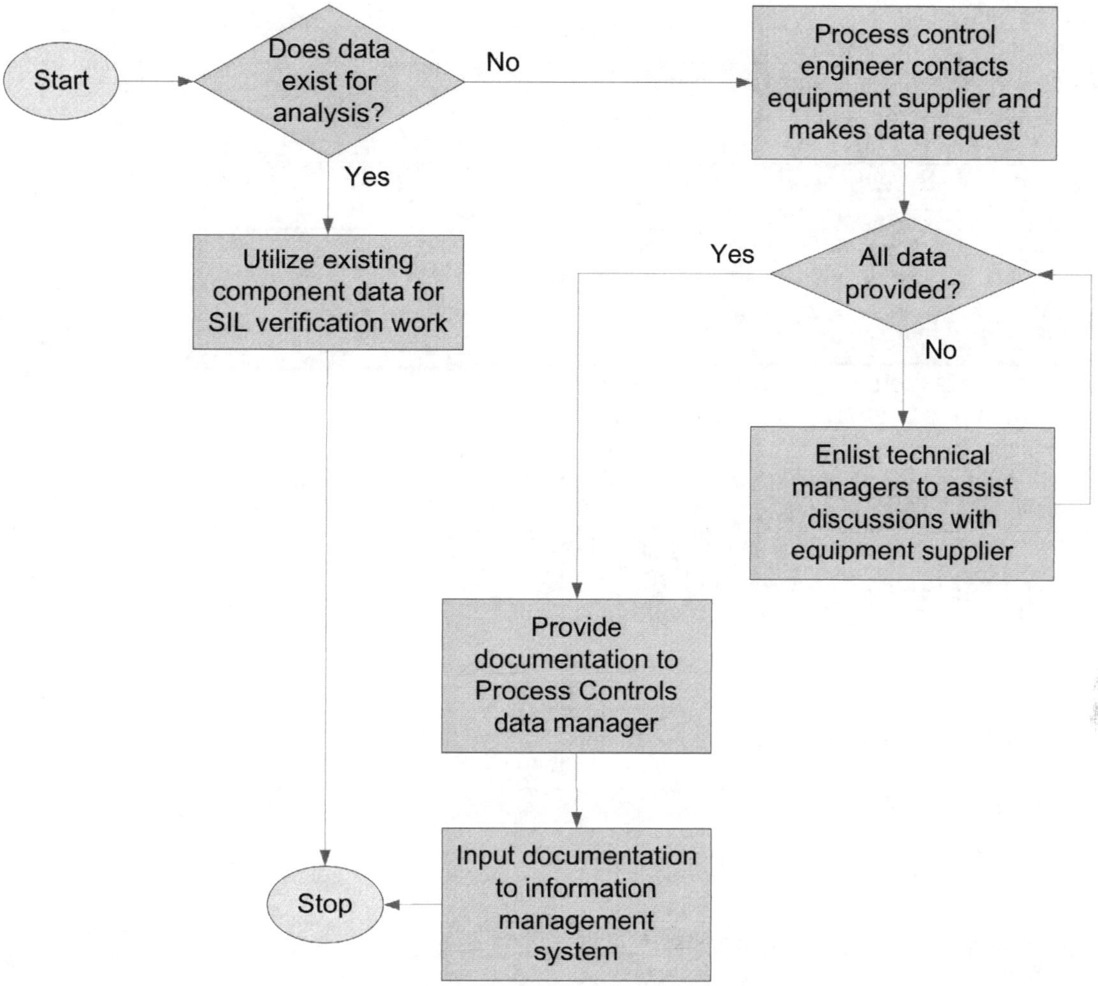

Figure 3. Equipment Reliability Data Procurement, Storage, and Maintenance

The check list in Table 2 is intended **to be filled out by the user** as part of an audit of information available for equipment intended for use in SIS service. It is recommended that it be shared with the manufacturer to help them understand and provide the necessary information in the most expeditious and cost effective manner possible.

Table 2. Safety Manual Audit Checklist

General	Yes/No	Auditor Notes
Have failure rate numbers been documented for applicable failure modes?		
Has a clear boundary been established		
Architecture (Hardware) Requirements		
Have the architecture requirements (i.e. 1oo1, 1oo2, 2oo3, etc.) for the intended SIL been clearly defined?		
Have environmental limits been clearly stated (or a reference pointing to another document)?		
Have process application limits been clearly stated (or a reference pointing to another document)?		
Hardware/design constraints adequately defined?		
Configuration Requirements		
Have required parameter settings to achieve a specific SIL claim limit or PFDavg been clearly documented?		
Are performance degradation consequences clearly defined for each parameter not properly configured?		
Software Considerations		
Is type of software (Fixed, Limited Variability, Full Variability) employed in the system documented? If Yes, document type of programming language on the line below; ________________________ NOTE: Full Variability Programming Language is not permitted in SIS service.		
Diagnostic Coverage		
Are all diagnostics receiving credit in the determination of the defined SIL claim limit clearly documented?		
Mechanical Integrity Requirements		
Have all proof test procedures required to detect known potentially "dangerous failures" as identified by the FMEA of the product been documented?		
Are procedures to repair or replace the product documented?		
Have the tools to accomplish applicable inspection/testing/repair been documented?		
Have the expected skills to accomplish inspection/testing/repair been documented?		
Has the proof test interval(s) used to estimate the documented SIL been clearly stated?		
Has the fundamental information necessary for the user to calculate alternate PFDavg based upon different proof test intervals been provided?		
For fault tolerant systems operating in a degraded state, has the allowable repair time been documented as measured from when a diagnostic is first alarmed until the system is returned to 100% health?		

Has each failure of the diagnostics employed been classified as detected and/or undetected?		
If failure of the specific diagnostic is detected, how is it detected?		
For those failures of specific diagnostics that are undetected, have appropriate proof test procedures been documented?		
For those failures of specific diagnostics that are detected, have allowable repair times been documented?		
Management of Change Requirements		
Does manufacturer have QA program in place capable of supporting replacement in kind and potential upgrades to their equipment necessary to support MOC?		
Have the procedures to accomplish firmware upgrades been provided?		
Have the tools to accomplish firmware upgrades been provided?		
Have the expected skills to accomplish firmware upgrades been documented?		
Quantified Performance		
Has the detailed FMEA been made available for review?		
Does the Safety Manual document the estimated failure rates (or reference other documents containing this information).		
Have BETA factor(s) used in estimation of SIL been documented?		
Have the basis of equipment specific beta factors been documented?		
Have product useful lifetime limits been documented?		
Does the failure rate data allow calculation of spurious trip rate when all safety manual requirements are implemented?		
Does the failure rate data allow calculation of PFDAVG and spurious trip rates for different proof test intervals/repair times/different diagnostic settings that satisfy safety manual requirements?		

CONCLUSION

In order to meet the intent of the current standards and have designs that meet the required process application performance criteria for our plants, it is necessary to obtain safety related information from our suppliers. This paper has looked at what one user expects from manufacturers in their efforts to comply with these industry standards.

The importance of the information that should be contained in a safety manual was shown, supported by examples where relevant.

Looking into the future, there will be an increasing need for users and manufacturers to work together to do a better job of supplementing the predictive data being provided via validation of actual performance in a way that addresses all causes of failure; hardware, software, random, systemic, human error, etc. This information needs to be fed back into the on going safety lifecycle (i.e. defensible quality data to support SIL verifications, improved designs, optimization of mechanical integrity program, etc.) so that continuous improvement can be made.

REFERENCES

1. IEC 61508, *Functional safety of electrical/electronic/programmable electronic safety-related systems*, 1998.

2. ANSI/ISA 84.00.01-2004 (IEC 61511), *Functional Safety: Safety Instrumented Systems for the Process Industry Sector.*

3. CCPS Guidelines for Safe Automation of Chemical Processes, 1993.

4. CCPS PERD Remote Actuated Valve Taxonomy (December 2004).

5. CCPS PERD Transmitter Functional Analysis (November 2001).

6. Thomas, H. W; *Meeting the SIS Standard's Challenge – Users and Manufacturers Working Together*, Presented at the Valve Manufacturer's Association Technical Seminar, Houston, TX, March 2005.

7. ISA-TR84.00.04, Guideline *on the Implementation of ANSI/ISA 84.00.01-2004 (IEC 61511).*

8. Thomas, H. W, Arner, D.C., Deibert, D.A., and Roposh, C.: *Approved Equipment for SIS Service; Past, Present and Future*, Presented at ISA Expo, Houston, TX, October 2006.

9. Goble, William; EXIDA Report, Safety Manual Requirements, No.: R02-08/01 R210 Version V2, Revision R1.1, October 15, 2003

Sustainable SIS/LOPA Technology in Dow

Greg Schultz

The Dow Chemical Company
Plaquemine, Louisiana
gkschultz@dow.com

ABSTRACT

The Dow Chemical Company has written safety standards for critical instruments, which have been in effect for many years. Like many in the chemical industry, Dow's implementation of these standards has evolved into a Safety Instrumented Systems (SIS) approach. In 2000, Dow introduced Layers of Protection Analysis (LOPA) as the primary means of determining the need for and the integrity level of SIS. At that same time, Dow began to adopt globally the principles and methodology of industry standard IEC-61511 (later ISA S84-2004), even before it was finalized.

Dow first applied the industry standard to new projects. In 2002, as part of its overall process safety management strategy, Dow chose to use LOPA in its existing plants and to apply SIS protection layers as needed to manage risk. Applying new standards in existing plants and incorporating the "technology" of SIS for future projects in a way that is sustainable presents many challenges.

This paper describes how Dow has utilized Six Sigma methodology to implement SIS technology in new projects and existing plants. Specific areas of discussion will include tools, work processes, benchmarking, and organization.

1. Introduction

In 2006, Dow publicly announced 2015 Sustainability Goals reflecting the principles of Responsible Care®. The commitment to local communities includes operating excellence as measured by several key EH&S indicators. Process safety incidents, for example, will be reduced by 75% and the severity rate by 95% based upon a 2005 baseline. Overall, these goals are focused on strengthening community relationships, improving product stewardship and innovation, and reducing Dow's global footprint.

Mechanical integrity (including safety instrumented systems (SIS)) is a key target area for ensuring the Sustainability Goals are met. By the end of 2006, Dow had made measurable progress toward the implementation of the industry standard for safety instrumented systems in existing plants. Globally, 20,000 SIS protection layers have been identified using Layers of Protection Analysis (LOPA). The assessment of the existing instrumented protective systems in Dow's 600 plants is underway, with overall progress measured at 50%. attention has turned from viewing the implementation of SIS and LOPA as a company "initiative" to one of sustainable technology, by "institutionalizing" it as the way work is normally done.

2. Background

Looking back as far as 1982, Dow has had a Loss Prevention Principle (LPP) for critical instruments, covering the basic requirements of identification, testing, and documentation. In 1996, the concept of safety integrity level was introduced as the first step toward quantifying risk reduction. In 2000, even before the international industry standard IEC-61511 was finalized, Dow incorporated the objectives, requirements, and principles and renamed the LPP "Safety Instrumented Systems".

To support the revised LPP, tools were developed based on the concepts of independence, reliability, integrity, and probability of failure for SIS. LOPA was introduced as the primary means of analyzing process hazards and determining the need for and integrity level of SIS. A SIS Work Process, modeled after the life cycle approach of IEC-61511, was created to define roles and responsibilities surrounding SIS and LOPA. Small groups of coaches were trained to facilitate LOPA studies, calculate probability of failure on demand, develop software code, and determine proof test intervals. Then, starting in 2001, all new capital projects began using the new methodologies and tools. This initiative represented a sizeable effort for Dow.

Even while the first capital project teams were starting to implement the new SIS/LOPA technology for new construction, a decision was made to adopt LOPA as the preferred hazard analysis methodology for process hazard analysis (PHA) in all existing plants in Dow. After all, the communities surrounding the plants expect outstanding process safety performance from all assets, regardless of age. With more than 600 plants globally, the size and complexity of this effort increased by an order of magnitude. The plants represent different process technologies, construction eras, and operational philosophies. Feedback from internal stakeholders during 2003 identified significant issues and concerns. Clearly, the approach for implementing SISs as part of new capital projects would require significant modification if it were to be adopted for use in existing plants.

3. Value Proposition

There are four questions always clamoring for attention with most safety-related projects--what will it achieve?, how much will it cost?, are the people resources available?, and what is the schedule? In short, the value of the chosen approach must be clearly articulated. The mechanical integrity benefits of implementing SIS in Dow are in the areas of compliance, consistency, testing intervals, and incident reduction. Many countries require chemical production plants to have mechanical integrity programs in place which include instrumented safety systems. Dow's past critical instrument programs were subject to interpretation, thereby resulting in variable implementation. The new standards offer data-driven techniques for extending test intervals. Most important is the safety benefit of reducing process safety incidents.

By implementing SIS/LOPA in all plants, Dow will ensure that risk is managed consistently and in conformance with corporate criteria. Sustainable process safety performance improvement will be achieved by eliminating potential incidents caused by faulty or non-existent instrumented protective systems. Past diverse practices will be harmonized and aligned with best industry practices.

4. Using Six Sigma to Implement SIS/LOPA

Six Sigma is a disciplined, data-driven approach and methodology for eliminating defects in any process[1]. The fundamental objective of the Six Sigma methodology is to improve processes and reduce variation. There are two basic approaches—MAIC (Measure, Analyze, Improve, Control) for improving existing processes and Design for Six Sigma (DFSS) for innovation and new processes. Both approaches require measurement and analysis of data and determination of causal effects, in order to make measurable improvements. Dow has utilized Six Sigma for many years to make all areas of the company better.

With the decision to implement SIS and LOPA in all Dow plants, the required changes to processes and practices were not merely incremental ones. There was no existing process to implement, nor one even similar to modify. Therefore, Design for Six Sigma (DFSS) was chosen as a means of developing an innovative approach to implementing SIS and LOPA in Dow. Data were gathered internally using interviews and Voice of Customer sessions to define the critical customer requirements and to guide the development of the new process. External benchmarking was valuable in understanding Dow's approach relative to other premier chemical companies.

An "umbrella" approach was chosen as the best way to implement the DFSS methodology for the identified critical customer requirements of reduced incidents, reduced costs, and reduced cycle time. The DFSS project oversaw and provided guidance to twenty-five MAIC project teams charged with improving various aspects of SIS and LOPA. For example, one project resulted in the development of an implementation guide or "cookbook" containing examples of best practices. Another project provided guidance and role clarity for the proper documentation of LOPA scenarios. A probability of failure on demand (PFD) calculation tool was developed, as were guidance for using certified instruments and partial stroke testing devices. The projects covered everything from training to assessing to bypassing—even resourcing. Six Sigma enabled the resolution of many SIS/LOPA issues by providing a structured methodology which delivered sustainable results.

5. Benchmarking

One of the most important of the MAIC projects chartered under the DFSS "umbrella" effort was an external benchmarking study on SISs implementation. As questions from individual plants and the Technology Centers which support those plants began to grow in number and complexity, those in Dow charged with implementing

SIS/LOPA looked internally and externally for answers. In 2003, Dow benchmarked with eight other global companies to compare hazard analysis approaches and results. Each company's approach was different, but the effort involved and, for the most part, the results obtained were quite similar. Simultaneously, a re-validation of the bases for LOPA was conducted internally. Two fundamental changes resulted from this Six Sigma project.

First, the LOPA methodology, being "an order of magnitude" tool, was found to be slightly conservative when used to analyze toxic hazard scenarios. The methodology was modified and calibrated to eliminate this defect. Tools were updated, guidance developed, and training conducted.

Secondly, the need for a systematic approach to evaluate and implement SIS requirements in existing facilities was recognized. ISA 84.01-2004 requires an assessment of existing hardware and software for SIS according to the following statement: "For existing SIS designed and constructed in accordance with codes, standards, or practices prior to the issuance of this standard, the owner/operator shall determine that the equipment is designed, maintained, inspected, tested, and operating in a safe manner"[2].

Although Dow had made the strategic decision prior to 2003 to apply the LOPA risk assessment process to existing facilities, a systematic approach to assessing and qualifying the acceptability of existing hardware and software had not yet been implemented. The Six Sigma project led to the classification of those logic solvers, software revisions, and instrumentation, which were viewed to have an excellent safety record and meet the essential elements of the ISA 84.01-2004 requirements. These identified technologies were compiled into a list available to instrument and process automation designers, called the "Dow Good Engineering Practices" (DowGEP). Assessment tools were created to document the evaluation of each SIS. A work process was developed to guide a team of process, safety, control, and instrumentation experts through an assessment of existing instrumented protective systems to determine conformance with safety requirements for SIS.

6. Implementing the Requirements into Existing Facilities

Another Six Sigma project was launched to determine the most effective and efficient approach to conducting the assessment work on an existing plant. Since best practices had not yet been developed, the first assessments were time-consuming and labor-intensive. This Six Sigma project developed a database of plants, schedules, and resources. A "dashboard" was developed to display and track company progress and estimate resource requirements. A supply/demand forecast was developed to address the defect of forecasting difficulty and to ensure timely completion of the LOPA studies and SIS assessments. It became apparent as a result of this project, that the work involved with these assessments would challenge the existing planning process.

While a plant-by-plant approach to implementing the "existing facility" clause for SIS was effective, it was not very efficient, nor was it likely to be consistent. Rather than train a different assessment team for every plant, two dedicated implementation teams were formed to supplement the existing effort. The data analysis suggested that one team of process control hardware and software and instrumentation experts could support the North American plants, and a second team the European plants. Other global areas could be covered by these two teams or other teams formed as needed. The teams are comprised of Dow employees and some Dow retirees. Soon contractors may be added to the teams to match work load demand. The feedback from the plants who have used these teams has been very positive. The teams are successfully leveraging the work processes and tools, assisting the plants in identifying and closing gaps found during the assessments, and are helping to assimilate SIS/LOPA technology in Dow plants. By assessing the hardware, software, instrumentation, and operating discipline of each safety instrumented system, the plant is assured of operating protection layers consistent with Dow's minimum requirements and in conformance with the "existing facility" clause of ISA 84.01-2004. Additional benefits include reliability improvements and better testing intervals.

7. SIS/LOPA Issue Resolution

For the past six years, SIS and LOPA have been in rapid development inside and outside of Dow. As with many new technologies, there are issues to overcome, details to define, and questions to answer. A SIS/LOPA issue resolution process (see Figure 1) was developed in 2004 to facilitate getting questions answered. Coaches were called upon to help those in the plants or on the project teams with the common implementation concerns. A technical resolution board (TRB), comprised of subject matter experts was active in developing solutions to the uncommon problems. A management-level steering team was charged with providing oversight. The persons responsible for implementing the resolution process (within the project teams and support groups) are credited with developing SIS/LOPA as a technology within Dow.

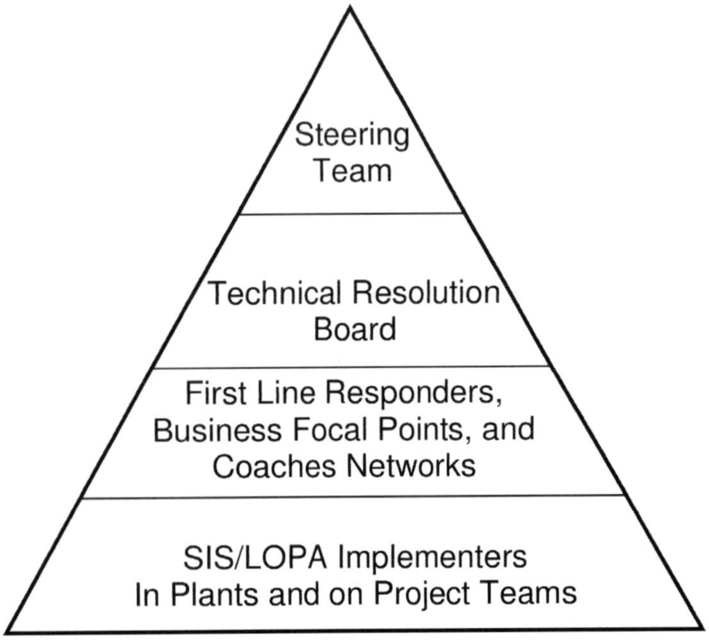

Figure 1: Dow's SIS/LOPA Issue Resolution Process

8. SIS/LOPA as Sustainable Technology

A Six Sigma project was chartered in 2005 to move SIS/LOPA from an "initiative" to a "sustainable" technology. Role clarity, assigned ownership, knowledgeable resources, available information, and training were defined as attributes of sustainability. Three defects were identified—work practices not finalized, support structure not in place, and feedback mechanisms ineffective. The pre-project defect rate was determined to be 50%.

The Six Sigma team used tools from the MAIC and DFSS processes (such as the Voice of Customer tool) to solicit input from internal stakeholders relative to the defects. These data were analyzed to determine the root causes for the high defect rate. A multi-generation plan was developed to organize solutions in a time-based approach. Thirteen changes, implemented over a two-year time frame, is expected to eliminate 80% of the defects resulting in SIS/LOPA being a sustainable technology in Dow.

The cornerstone of this project is that SIS and LOPA technology must become fully integrated as the way work is done in three existing discipline areas—Process Safety, Process Control, and Instrumentation—in order to be sustainable. These three technology areas must own, nurture, and manage change to SIS/LOPA until and after it is mature in Dow. These three areas will develop their own subject matter experts and implementation coaches. In concert with each other, they will solicit and solve stakeholder problems and develop or purchase new tools. A Technical Steering

Committee will manage the "white spaces" between the three discipline areas. (see Figure 2).

Additionally, SIS/LOPA training materials are being re-formatted based upon learning objectives. A documentation library including guidelines of SIS best practices are under development. Commercial tools and software are being evaluated for long-term use. A process for addressing frequently asked questions and for managing internal communications is being developed, as well.

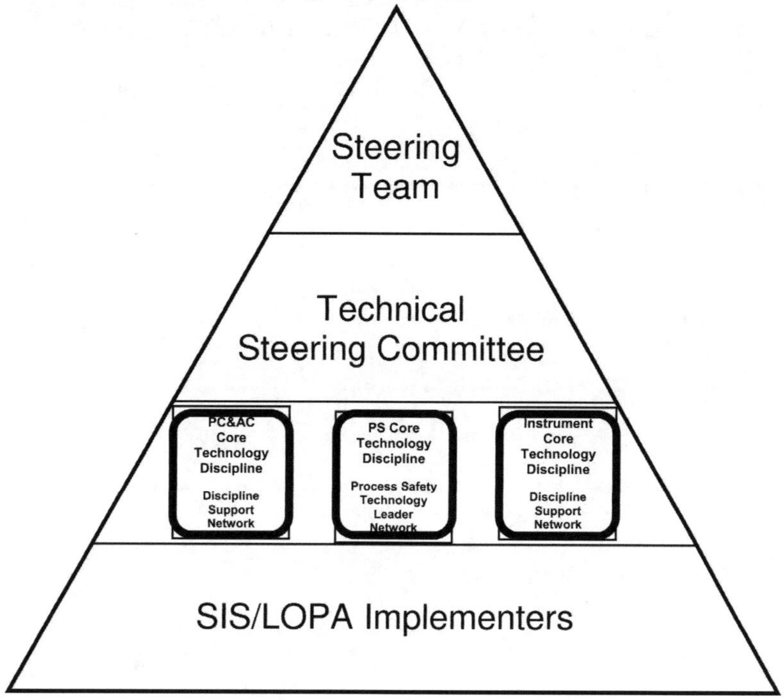

Figure 2. Sustainable SIS/LOPA Technology in Dow

"PC&AC" is Process Control and Advanced Control
"PS" is Process Safety
"Discipline Support Network" includes coaches and project leaders

9. Summary

Implementing SIS and LOPA technology in Dow has been underway for more than six years. Much of that time was spent defining the process which is now being implemented. Substantial completion of the assessment of existing plants and implementation of needed improvements will take several more years, dependent upon risk tolerance and turnaround schedules. The technology of SIS and LOPA will be owned and managed by the existing disciplines of Process Safety, Process Control, and Instrumentation. These disciplines will incorporate the tools and training from the DFSS project and the MAIC projects to eliminate the defined defects. In the next few years as the assessments and gap closure within exiting plants is completed, SIS/LOPA will cease to be looked upon as an initiative, and will be incorporated as a sustainable technology into the way work is done in Dow.

10. References

1. iSixSigma.com

2. ANSI/ISA-84.00.01-2004 Part 1 "Functional Safety: Safety Instrumented Systems for the Process Industry Sector"

CONTINUOUS IMPROVEMENT IN SAFETY INSTRUMENTED SYSTEMS

Angela E. Summers, Ph.D., P.E.
SIS-TECH Solutions, LP
Houston, TX 77034 USA
Phone: 281-922-8324

ABSTRACT

The historian John Lewis Gaddis defines strategy (Alden 1996) as "the process by which ends are related to means, intentions to capabilities, and objectives to resources." The goal of the process industry is no incidents, no harm to people, and no damage to the environment. This goal can only be reached by going beyond compliance and establishing safety as a strategic business value. Developing a safety culture requires that values and policies are converted to practices and behaviors. The policy of "Safe operation is our mission" must be backed with the authority, resources, and tools necessary to achieve the policy. Everyone from the front-line operator to the board of directors must understand and believe that safe operation is a core operating principle.

Many factors affect performance expectations and the decisions made to attain them, such as the economy, market trends, and technology, along with legal and political issues. A strong safety culture expects ownership and accountability for safe and reliable performance of the process equipment over its life. Management should support continuous improvement to ensure existing equipment is designed, maintained, inspected, tested, and operating in a safe manner.

GOOD ENGINEERING PRACTICE

IPS design is a team effort requiring knowledge and skills from a variety of disciplines. Some of these skills may be rarely used, but when required, high levels of expertise must be available. Massive restructuring and mergers have become commonplace within the process industry. Corporate technical departments that once served as clearinghouses for new technologies and new ideas no longer exist. Responsibility has shifted in many companies from specialized corporate personnel to site personnel, who often find it difficult to maintain the required expertise. In some facilities, there is an increasing reliance on manufacturers and other third party contractors to provide the missing expertise. At the plant site, numerous departments and many different individuals may be responsible for various, yet limited, aspects of safe operation.

Without strong oversight, it is easy for simple slips of reason, errors, loss of knowledge, loss of skill, or bad behavior to become the predominant cause of process safety incidents. A safety and reliability driven culture can only exist when the owner/operator documents the information, practices and procedures necessary to achieve the attributes important to safe and reliable operation. The goal is to maintain the IPS equipment in an "as good as new" condition throughout the process equipment life. To accomplish this, random failures and systematic errors that cause IPS equipment to fail must be identified and corrected, as early as possible.

The protective management system recommended in the CCPS/AIChE *IPS Book* encourages the early identification and correction of failures and errors. It provides a framework for the implementation of quality control processes for the development, implementation, and management of a risk reduction strategy that reduces the risk below the owner/operator risk criteria. The protective management system is intended to ensure:

- Protective functions and required risk reduction are identified for each mode of process operation,
- IPSs are designed, installed, commissioned, and validated to meet the design basis,
- IPSs are inspected, tested, maintained, and operated to ensure the equipment operates in an "as good as new" condition,
- The process is adequately protected during periods where the process is being operated with a known IPS failure,
- Changes to the IPSs are evaluated through a management of change process,
- Access to the IPS is controlled administratively and physically, and
- IPS equipment failures and the occurrence of process demands are tracked and periodically assessed to ensure prompt response and resolution of any identified inadequacy.

The continuous improvement phase of the protective management system, as shown in Figure 1, ensures that everything ties back together and continues the lifecycle. Performance is monitored and gaps between expected and actual performance are closed in a timely manner. Investments are made to reduce risk further when practical. Continuous improvement evolves the design and management practices to ensure relevance to current needs, consistency with good engineering practices, and the installed equipment is designed, maintained, inspected, tested, and operating in a safe manner.

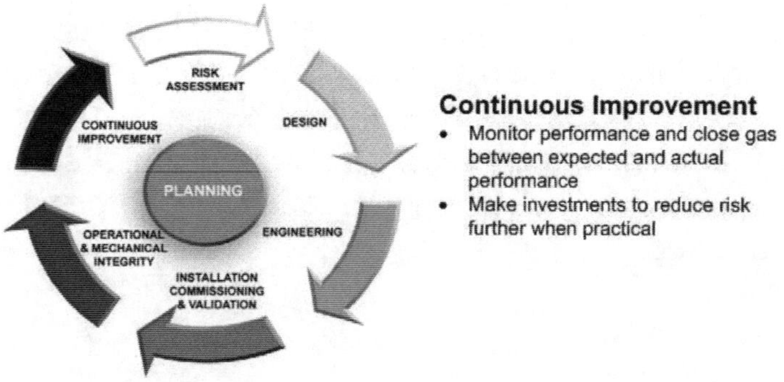

Figure 1. Continuous Improvement in the Lifecycle

FINDING THE BALANCE

Balancing safety and production goals can be a tenuous, delicate and complex act. It is undeniable that safety and production are compatible. It is indisputable that investments in safety yield long-term benefits. However, these benefits are not as obvious nor do they produce the rapid results associated with production investments, which generally have a high certainty of providing a measurable, positive effect within a short time frame.

For protection and safety, many of the benefits are less tangible. When successful, the IPS is blamed for a process outage; when it fails, it is blamed for the incident. The hazard and risk analysis describes the hazardous event prevented by the operation of each instrumented protective function (IPF). When an IPF operates as required, the IPF should be given credit for the event avoided by its successful operation, including potential fatalities, injuries, environmental releases, equipment damage, and financial losses. Also, the IPF should be credited when its design prevents a safe IPF equipment failure from taking spurious action on the process.

Figure 2 provides a summary of the decision making process, illustrating how available resources must be allocated across safety and production goals. Decision makers often have defensive filters which affect the receipt and interpretation of information (Reason 1990). Today's business climate puts pressure on personnel in a variety of forms, such as production forecasts, budget cuts, resource reductions, or colleague retirement.

In the absence of a strong safety culture, production and budget pressure can result in a culture of denial where the decision maker's defensive filter refuses to acknowledge any evidence that does not support production or budget plans. Risk assessment can become skewed, with credible safety recommendations and concerns being dismissed without appropriate consideration. Erroneous assumptions concerning equipment and procedure robustness lead to complacency and an acceptance of increased risk. Often, this is done in the absence of dependable documentation, information, and data, or a rigorous mechanical integrity program.

Good engineering practices should be applied in preventing process safety incidents. Internal practices should be benchmarked against those of market sector peers or other process industry companies. Periodic gap analysis should be conducted to determine if existing equipment is designed, maintained, inspected, tested, and operated in a safe manner. Based on observed performance and benchmarking information, action plans for improvement should be developed and implemented.

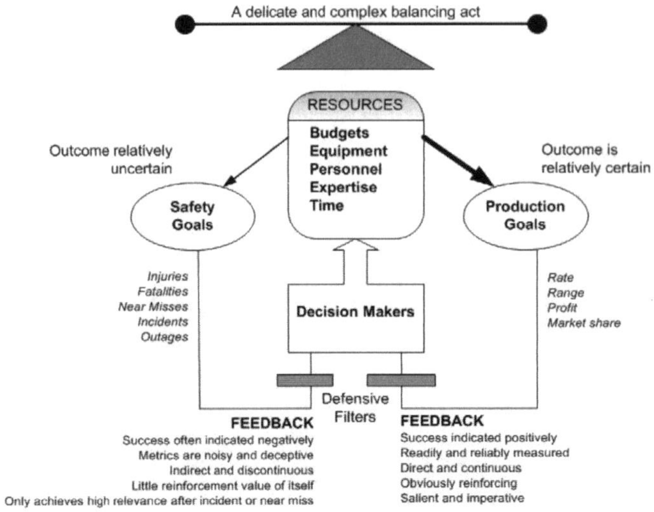

Figure 2. Overview Illustrating the Complexity of the Decision Making Process.
(adapted from Reason 1990)

UNDERSTANDING HISTORY

A series of catastrophic chemical incidents occurred during the 1970s and 1980s. These incidents are so legendary that they are often referenced by city only: Flixborough (1974), Seveso (1976), Mexico City (1984), and Bhopal (1984). They were catastrophic incidents that awakened the world to chemical industry risk. These incidents are summarized in *Lee's Loss Prevention in the Process Industry*.

The first process safety regulations were issued in Europe in the 1970s in direct response to the impact on the communities of Flixborough and Seveso. Nearly a decade later, the tragedy caused by the Mexico City explosion and the Bhopal chemical release resulted in process safety regulations being issued in the United States and many other countries. Industrial societies responded worldwide by publishing numerous codes, standards, and practices on a variety of process safety topics.

Unless you understand history you are doomed to repeat it. Despite these incidents occurring more than three decades ago, the errors and root causes for these incidents still exist today. Trevor Kletz discussed this problem in his book, *Lessons from Disaster: How organizations have no memory and accidents recur* (Kletz 1993). He presents numerous cases where an incident occurs and is repeated just a few years later. Kletz finds that organizations have poor memory due to many factors, such as insufficient failure investigation, inadequate communication and distribution of investigation findings, lack of information retention and little training concerning previous events.

A safety culture does not rely on balance sheet improvements to justify IPSs. It understands that the potential for incidents is an inherent part of the process design and that, without focused effort, incidents invariably occur.

BENCHMARKING CURRENT STATUS

The owner/operator should understand how internal practices compare with recognized and generally accepted good engineering practices. This is often referred to as benchmarking and establishes the owner/operator's position with regard to industrial and market peers. This is often the most painful part of the continuous improvement process, as it tends to shed light on the shortfalls and inadequacies of the protective management system as a whole.

Some owner/operators will likely find that their design and management philosophy is out of alignment with various aspects of good engineering practices. When unacceptable risk is identified, an action plan should be established with short-term and long-term measures sufficient to reduce the risk below the owner/operator risk criteria. It is well proven that investments to improve safety and reliability of the process operation yield long-term economic returns.

DEFINING GAPS

An increasing number of owner/operators are finding themselves operating within a regulatory framework which often does not provide prescriptive requirements. Instead, the requirements are a moving target based on the somewhat fuzzy concept of "good engineering practice." Keeping up and complying with the basic requirements concerning process safety and IPSs can seem taxing enough. How does an owner/operator move forward with continuous improvement when it seems that even the immediate goals are a moving target? To some, it may seem challenging enough just to maintain the status quo; let alone to embrace more changes.

Continuous improvement operates on the principle that finding failures and errors is the beginning of a learning process. Minimizing their reoccurrence requires an understanding of how these failures or errors developed. Continuous improvement, as well as the application of good engineering practice, should be viewed as an on-going process rather than an endpoint. Work processes, information, resources, and skills should be analyzed periodically to identify weaknesses that limit performance and to recommend improvements, as necessary.

Information systems, whether computerized or manual, should provide personnel with up-to-date information in a format that is easy to understand. Information should be revision controlled, yet accessible.

Many factors affect IPF performance expectations and the designs used to attain them, such as the economy, market trends, and technology, along with legal and political issues. A strong safety culture expects ownership and accountability for safe and reliable performance of the process equipment over its life. Management should support periodic evaluation of the existing equipment to determine that it is designed, maintained, inspected, tested, and operating in a safe manner.

Changes in operability, functionality, reliability, or maintainability expectations may require the implementation of different design/or management practices. Proof test, failure investigation, alarm, trip, and audit reports provide valuable insight into personnel and management system performance. Operating excellence requires that the root causes of unacceptable process reliability and equipment performance be identified and resolved. Improving equipment mechanical integrity requires a culture that values maintenance.

A need for improvement may be identified through various management system activities. Continuous improvement processes should include periodic examination of overall available information to identify, trend, and correct systematic problems. A gap analysis (stage 5 functional assessment) is performed to compare the observed IPS performance to the expected performance. The gap analysis should determine that:

- Equipment is operating according to design intent,
- Safety, operating, maintenance, and emergency procedures are appropriate for competency and risk reduction expectations,
- Hazard and risk analysis or management of change recommendations are addressed in a timely manner, and
- Training of relevant personnel is adequate for current expectations.

Significant issues may be identified during the analysis. Management system failures are often reflected in multiple performance metrics. Systematic problems may be identified, such as poor adherence to policies, procedures, and practices or insufficient inspection and preventive maintenance. If IPS equipment is not maintained, it is likely that other equipment is suffering from the same inadequate maintenance. The cumulative maintenance deviations, whether intentional or unintentional, may cause a breakdown of multiple protection layers.

Team effort is often needed to evaluate IPS requirements and performance. Some organizations establish a formal structure where identified personnel participate as site representatives on a core team. The core team evaluates changes in the good engineering practices and makes recommendations for modifying internal practices.

Whenever work processes are modified, a shift in emphasis often leads to changes in the way team members perceive the process, its associated risks, various protection layers and IPSs.

This shift may result in recommendations for additional risk reduction or IPS. These recommendations and other continuous improvement efforts complete the lifecycle moving the process toward safer and more reliable operation.

DETERMINING PATH FORWARD

The key aspect of continuous improvement is charting the course to achieve it. Over time, various options will be presented to upgrade hardware, software, or human interface systems. Proposed changes should be reviewed using a management of change process to identify how the change affects other functions or systems. Areas for improvement should be addressed with an action plan, which typically prioritizes recommendations based on consequence severity and risk gap.

Action plans should define objectives, milestones, and timelines. Action plans should be periodically assessed to determine whether there is a need to accelerate the schedule or broaden its objectives. For example, a planned IPS upgrade may be accelerated when the manufacturer withdraws support for critical equipment. To be successful, action plans should be communicated to affected personnel so they understand and commit to them.

Implementing upgrades aimed at improving long-term operational effectiveness takes time to complete, depending on the complexity and degree of change involved. As the IPS is changed, operating plans and targets should consider any additional risk borne by the process during the transition. Once the design basis changes are underway, the operating and mechanical integrity basis should be reviewed and needed revisions implemented.

There are many barriers to improvement, including:

- Poor data integrity and quality,
- Poor information availability and consistency,
- Lack of broad understanding of facts and procedures,
- Poor or missing internal practices and procedures,
- Poorly understood compliance expectations,
- Inadequate revision control or notification of changes, and
- Lack of comprehensive training on data, information, procedures, and practices.

To overcome these barriers, personnel must be provided with more than just another initiative or mandate for change. Continuous improvement must be part of an organization's culture, beginning at the highest management level and continuing to the front-line operator. A continuous improvement culture requires that all personnel understand the importance of following approved practices and procedures. Personnel should feel that safe and reliable operation is an institutional value and that they won't lose their jobs or be held back for speaking out. Front-line personnel must believe that continuous improvement is supported by all levels of management. Everyone should understand that employment is conditional on safe work performance.

A continuous improvement culture requires that all personnel understand the importance of following approved practices and procedures. To succeed, personnel must be aware of the potential risk and be committed to do what is necessary to maintain and continuously improve operational and mechanical integrity. The path forward encompasses many detailed tasks, but generally includes the following:

- Assign responsibility and hold personnel accountable,
- Audit to ensure practices and procedures are followed,
- Question norms and reduce risk further when practical,
- Integrate business and process safety goals,
- Track performance, address bad actors, and celebrate success, and
- Learn and remember.

A CORPORATE RESPONSIBILITY

An organization's culture is ultimately driven by what management indicates is important; what is measured; and what is rewarded. Safe operation can only be sustained where it is recognized that the direct costs of an incident represent the tip of an iceberg. Hidden from view are the indirect costs and long-term business damages resulting from unsafe operations. When the true cost of an incident is understood, it becomes very clear that being cost effective is much more than simply today's budget.

Success requires that personnel believe that investment in reducing risk further is encouraged and rewarded. Market leaders recognize that this investment provides benefits that far outweigh its costs. Operating excellence seeks to prevent incidents, because it is good for business and it is the right thing to do.

REFERENCES

1. Alden, Edward, "Comment And Analysis: Five Fraught And Futile Years - Why America must align aims and reality," www.ft.com, Financial Times (2006).

2. Excerpted from CCPS/AICHE, *Guidelines for Safe and Reliable Instrumented Protective Systems*, New York (2007).

3. Kletz, Trevor, *Lessons from Disaster: How Organizations Have No Memory and Accidents Recur*, Institution of Chemical Engineers, United Kingdom (1993).

4. Mannan, Sam, *Lee's Loss Prevention in the Process Industries,* Volumes 1-3, Elsevier Butterworth-Heinemann, United Kingdom (2005).

5. Reason, James, *Human Error*, Cambridge University Press, Cambridge United Kingdom (1990).

Continuous Improvement of SIS - Panel

Angela Summers
SIS-TECH Solutions
Houston, TX 77034
asummers@sis-tech.com

Kevin L. Klein
Solutia
Springfield, MA 01151
klklei@solutia.com

D. C. Arner
Air Products and Chemicals, Inc.
Allentown, PA 18106
ARNERDC@airproducts.com

Greg Schultz
The Dow Chemical Company
Plaquemine, LA N/A
gschultz@dow.com

and

Dallas Green
Rohm and Haas
Croydon, PA 19021
dallas_l_Green@rohmhaas.com

Abstract:

This panel session will begin with a brief presentation by each panelist on their company's activities related to ISA 84.01-2005/IEC 61511 for new and existing SIS. These presentations will be followed by a question and answer session to allow audience members to gain better insight into the challenges of moving forward and continuously improving SIS performance.

"Safety and Regulatory Compliance of Reconditioned Equipment"

Robert D. Baker
Retired; Safety Consultant to Emerson Process Management
1521 Southport St.
The Villages, FL 32162

Responding to challenges of seemingly unending reductions in capital and maintenance budgets, the process industry has increasingly turned toward the purchase of lower cost, recycled equipment including salvaged control valves and instrumentation. Although it may be acceptable from a functional perspective, depending on equipment age, repair history, application severity and other factors, this "reconditioned" equipment may be out of compliance with safety standards, or with manufacturer's specifications as originally designed to applicable industry codes, for safe use in hazardous locations.

If the use of these devices does not meet an original equipment manufacturer's technical specifications for the process application, there are possible implications with respect to the Process Safety Information and Mechanical Integrity elements of OSHA's Process Safety Management (PSM) Standard (29 CFR 1910.119) [1] and EPA's Risk Management Program Rule (RMP) (40 CFR 68) [2]. Another possible implication exists with respect to OSHA's Hazardous (Classified) Locations Standard (29 CFR 1910.307) [3].

This manuscript describes the issues and offers recommendations buyers can use to increase their confidence that salvaged, remanufactured, refurbished or new-surplus equipment meets plant design safety standards and regulatory requirements.

Investigations into process plant explosions and fires have resulted in regulatory and investigative agencies issuing reports, fines and recommendations. Plant managers are responsible for assessing whether their site programs are adequately addressing safety requirements, yet the potential issues associated with reconditioned and new surplus equipment may not be known and/or included in such assessments. Two issues are:

- Mechanical integrity of piping system components such as control valves
- Maintaining hazardous (classified) location approvals of electrical/electronic equipment

Potential equipment or compliance problems may arise when purchases are made from third party salvagers who recondition and resell used control valves and/or used and new surplus instrumentation. These third party sellers often do not have access to the original manufacturer's specifications and standards used during new product design of process instrumentation and control valves. Yet the following common marketing claims may confuse industry:

- "meet or exceed factory specifications"
- "meet and even exceed OEM testing standards"
- "remanufactured to like new"
- "fully reconditioned to OEM specifications"
- "remanufactured to original manufacturers' specifications and tolerances"

Despite these claims, salvagers often do not have the information needed to guarantee the restoration of equipment to full compliance with an original manufacturer's design attributes.

The question that begs answering is: "when standards and approvals are required to better ensure the safety of new equipment when introduced into the marketplace, why would not the same requirements be required whenever such products are salvaged, refurbished, reconditioned, remanufactured or repaired over their life cycle?"

Most original equipment manufacturers (OEM) of process instrumentation can offer solutions to assist plants in identifying and abating potential safety-related risks and regulatory non-compliance. In addition, implementing recommended solutions may potentially lower rates for property, liability and business-interruption insurance while simultaneously increasing reliability and process uptime.

This manuscript's information on used and new surplus control valves and instrumentation should:

- Assist plant management and technicians in identifying their chemical processing units' potential regulatory non-compliance issues

- Help maintenance, operations and engineering recognize a potential need for more frequent implementation of management of change (MOC) evaluation processes for reconditioned equipment that may visually look like replacement-in-kind, but may not be technically equivalent to be considered as an acceptable replacement-in-kind.

DEFINITIONS

To avoid confusion, the following terminology will be used in the context of this discussion.

Reconditioned refers to any form of *salvaged, refurbished or remanufactured* equipment. Unlike an end user repairing its own control valves or instruments, transfer of equipment ownership has typically taken place during a reconditioning process with limited, if any, traceability associated with the equipment's prior application, environmental conditions, handling, maintenance history, use of OEM parts during repairs, etc.

New-Surplus refers to *unused* current or obsolete equipment, which:

- May or may not still be in the original packaging
- May have aged significantly since being purchased and inventoried by an end user, or by a third party salvager / distributor not authorized by the OEM
- May have been previously installed, calibrated and subsequently removed and repackaged without a plant or process unit being started up
- May have had multiple ownership transfers without having actually been used

More traceability typically exists for new-surplus equipment than for reconditioned equipment, but it is still usually limited. More specifically, for new-surplus electrical or electronic instruments subjected to the scenarios outlined in the bullets above, the opportunity exists for unknown changes or hidden damage to occur during the typically lengthy time and extensive handling between original manufacture and subsequent resale.

MECHANICAL INTEGRITY (MI)

Mechanical integrity refers to the ability of equipment to maintain its original design integrity over its entire life cycle (i.e, to resist the loss of containment of chemical processes throughout operational and design maximum process pressures and temperatures). Loss of containment not only presents potential onsite safety issues, but also potential offsite health/safety and environmental issues, especially if a released hazardous process is carried beyond the confines of the plant site.

The mechanical integrity (MI) element of the PSM Standard[1] and RMP Rule[2] may apply in several ways to the use of reconditioned control valves:

- Paragraphs 1910.119(j)(1)(ii) and 68.73(a)(2) include valves when applying mechanical integrity to process equipment: "Piping systems (including piping components such as valves)"

- Paragraphs 1910.119(j)(4)(ii) and 68.73(d)(2) cover inspection and testing: "Inspection and testing procedures must follow recognized and generally accepted good engineering practices"

- Paragraphs 1910.119(j)(6)(i) and 68.73(f)(1) include quality assurance: "In the construction of new plants and equipment, the employer shall assure that equipment as it is fabricated is suitable for the process application for which they will be used"

With respect to the "information pertaining to the equipment in the process", these standards apply in the following ways:

- Paragraphs 1910.119(d)(3)(i)(F) and 68.65(d)(1)(vi) include: "Design codes and standards employed"

- Paragraphs 1910.119(d)(3)(ii) and 68.65(d)(2) state that: "The employer shall document that equipment complies with recognized and generally accepted good engineering practices."

Since the PSM Standard and RMP Rule require that equipment covered by the regulation be suitable for the existing process application, and that the equipment design, operation, and maintenance must conform to recognized and generally accepted good engineering practices, a detailed assessment of reconditioned piping system components, such as control valves, would be beneficial any time the valves are applied in PSM and RMP-covered processes. These assessments could include:

- Verification that control valves, when purchased after reconditioning, continue to meet all OEM design specifications as designed in accordance with the appropriate ASME pressure class standard.
- Similarly, verification that existing control valves, when repaired, continue to meet all design specifications in accordance with the appropriate ASME pressure class standard.

Development and execution of assessments better prevents potential loss of equipment integrity and loss of containment (LOC) of toxic and flammable materials.

Market Background

Historically (early 1990's), the market for salvaged/reconditioned control valves originated in shallow water offshore and onshore oil and gas facilities. However, over the last few years the installation of this reconditioned process equipment has significantly expanded into onshore chemical process and refining facilities due to declining maintenance budgets and financial pressure on small, locally engineered capital projects.

The closing of many chemical, paper, and other industrial processing facilities, built during the hey-day construction period of the 1970's through the 1990's feeds this phenomenon with recycled equipment of varying ages and quality. So along with the typical worn-out valves regularly recycled from plant boneyards, suppliers and/or service providers now often have an ample supply of salvaged control valves to recondition and resell.

Subsequent to reconditioning, these control valve assemblies are often repainted to look like new, and often *still bear the original nameplate applied when manufactured new, or contain a salvager's re-applied nameplate that is marked or stamped with valve ratings that infer continued compliance with the valve's originally designed pressure class.* But does a potentially aged, reconditioned valve still comply with, and has it been recertified to, all of the OEM's specifications as originally designed in accordance with the ASME standard?

Standards Background

As with any piping component, a control valve is a pressure-retaining device. Control valves are designed by the original manufacturer in accordance with the ASME B16.34 standard, to ensure integrity for the appropriately designed pressure class and compliance with the designed piping system pressure class when installed. The ASME B16.34 standard is the recognized and generally accepted good engineering practice for new control valves.

An important, but often overlooked, element of valve design, as referenced in ASME B16.34 paragraph 6.1.7 is "Additional Metal Thickness."[4] Unlike cylindrical shapes such as piping, additional wall thickness is designed into valve bodies and bonnets in order to handle the additional stresses occurring from:

- Assembly loads
- Actuating (closing and opening) loads
- Shapes other than circular
- Stress concentrations

Process application and age have a major impact on a control valve's life-cycle and its integrity. Erosive and/or corrosive applications have a greater impact on body wall thickness than the gradual, time-based effects of surface oxidation (steel castings).

Thus, sustaining the design parameters referenced by ASMBE B16.34 paragraph 6.1.7 is a critical element in maintaining a control valve's pressure integrity. If a designed, minimum wall thickness is required for new control valves off the assembly line, why wouldn't they continue to be required for a reconditioned control valve?

Many individuals incorrectly perceive hydrotesting as the sole indication of control valve integrity. Upon reviewing ASME B16.34, one will find that a shell test (hydrotest) is required, but is an additional element in meeting design specifications in accordance with the B16.34 standard, specifically paragraph 6.1.7.

Gary Icenogle, retired Engineering Department Manager of a major global manufacturer of process equipment including valves, was consulted about the standard's wall thickness and hydrotest (shell test) requirements. Icenogle gained an in-depth understanding of B16.34 during his 28 year membership on ASME B16.34 Subcommittee N, beginning in 1971 when appointed to American National Standards Committee B16 / Subcommittee 15 (subsequently renamed Subcommittee N). So he was extensively involved in the Subcommittee's ongoing development and revisions of the B16.34 standard as we know it today.

Icenogle:

"When ascertaining the pressure containment integrity of a valve, the appropriate data and application of the data is critical. With the exception of repair entities associated with the valve OEM, the typical source of wall thickness data and shell test requirements would be the ASME B16.34 standard, which gives valve design requirements to meet a specific pressure-temperature rating. It addresses minimum wall requirements as well as pressure testing requirements per pressure Class.

A sophisticated salvager, remanufacturer, refurbisher or non-OEM repair shop may measure a salvaged valve's wall thickness and compare it to the required minimum wall thickness of ASME B16.34 Table 3 or the thickness calculated per the equations of Appendix VI. These entities' assumption is that if their measured wall thickness is greater than the table or calculated value, then the valve is capable of the pressure ratings for that valve size and pressure Class.

However, this is an incomplete application of the B16.34 standard that produces a potentially significant safety risk, since the minimum wall thickness described in paragraph 6.1.1 "*Wall Thickness*" and listed in Table 3 and Appendix VI are ONLY for cylindrical shapes. Other than bar stock bodies, most control valves are by no means cylindrical throughout, other than at their inlets and outlets. By using only Table 3 and/or the equations, a salvager will have ignored the exception paragraphs identified in paragraph 6.1.1 and detailed in paragraph 6.1.7 "*Additional Metal Thickness.*"

The purpose of the shell test, typically referred to within the industry as a "hydrostatic test" or "hydrotest", is to ensure that the pressure boundary is capable of containing pressure in a static condition (as the term "hydrostatic test" implies). In other words, the shell test is performed to identify whether the pressure boundary has any material defects that could allow the fluid contained within the valve to leak to atmosphere. *However, the shell test is not intended to be the sole requirement that defines the pressure containing capability of a valve.*

The requirements within ASME B16.34 for a valve's pressure rating are to have adequate wall thickness per paragraph 6.1 *and* to pass the appropriate shell test of 1.5 times the rated maximum cold working pressure per paragraph 7.1. *Neither of these, if one is done without the other, will ensure that the pressure boundary is adequate to maintain a valve's original design pressure rating.* Both are required, along with meeting the rest of the standard."

Equivalency Claims

The equivalency claims often made by third party reconditioners have incorrectly advanced industry perceptions that their reconditioned control valves are always equivalent to new control valve specifications. But there is evidence indicating otherwise.

A major OEM maintains a measurements database of used valve body wall thickness (valves obtained from the plant demolition market and/or end user bone-piles), documenting the number of such valves required to be scrapped or requiring body wall restoration. Without access to manufacturer specifications, third parties may not identify such deficiencies and some reconditioned valves, which would potentially require de-rating of their pressure/temperature capability, may unknowingly be installed in piping systems where the valve's pressure class rating is no longer compatible.

Again, logic would indicate that an old used control valve body would need similar wall thickness as a new valve. And with the significant retirement and demolition of plant assets, a salvaged, reconditioned valve may already be in excess of 20 years old. Thus, *an end user should consider including the certification of control valve body wall thickness as an integral part of its mechanical integrity program,* especially when implementation could increase plant safety and demonstrate ongoing plant compliance to regulatory standards, but could also simultaneously increase operational reliability.

The obvious solution for a mechanical integrity program is to have reconditioned or repaired control valves individually certified that they continue to meet the original manufacturer's specifications as designed to ASME B16.34. This can be accomplished by specifically requiring the measurement, verification and certification of wall thickness and performing the appropriate hydrotest.

These additional requirements could easily be incorporated into a plant's maintenance and turnaround specifications for repaired or reconditioned control valves. An OEM or OEM-authorized facility can measure wall thicknesses with available non-destructive testing technology, and then compare the results to the latest revision of the OEM's casting drawings.

The ability to meet such a specification already exists today with most OEMs, with minor additional time required for measurement, testing and issuing a certificate of conformance (COC). This is similar to the Fitness For Service evaluation process that is used to restore and re-certify a pressure vessel before application in a service for which the design basis of the vessel cannot be confirmed.

Importance

Loss of equipment integrity presents potential safety, property and environmental issues. The results of PSM audits by OSHA have consistently demonstrated that MI is a PSM element receiving a large number of citations at most facilities, and in some cases has been the last PSM element to be fully addressed. And, often the Mechanical Integrity element of PSM has been a difficult element for many facilities to implement.[5]

In a recent case, the U.S. Chemical Safety Board (CSB) raised issues about a site's mechanical integrity programs for equipment, where one of its incidents involved a valve. Review of other OSHA citations reveals cases addressing mechanical integrity of other types of equipment. Some of these citations were classified as "egregious willful violations" or as "willful violations" with each violation accompanied by a significant fine.

Assessment of fines should be considered, but the potential financial impact from civil actions arising from a serious incident resulting in personal injury or death could potentially far exceed the investment required to meet regulatory standards..

HAZARDOUS (CLASSIFIED) LOCATIONS

Similar to the situation with reconditioned control valves, over the last several years there has been an increasing number of reconditioned and new-surplus electronic instruments installed in process plant hazardous (classified) locations.

Reconditioned instruments are also usually repainted to look like new. Further, these instruments usually have the original manufacturer nameplate left on, or it is reattached following reconditioning work. When this is done with an instrument having a nameplate with a NRTL Approved certification mark on it, and the reconditioned or new surplus instrument is sold by a supplier or service provider whose facility is not NRTL approved and audited, there is typically a misperception by both supplier and end user that the reconditioned instrument is still NRTL Approved.

But in this type of situation, a major NRTL *considers the instrument no longer compliant with the standards it originally certified the equipment to and thus would not be compliant to OSHA requirements for use in a hazardous (classified) location.*

And with the end user bearing the responsibility, once in use, for an instrument's continued compliance with applicable codes and standards, it is critical that no "changes" have unknowingly been made to the equipment after it has left an NRTL approved supplier and/or service-provider facility. [6]

Regulatory Background

The NRTL program is a part of OSHA's Directorate of Science, Technology and Medicine.[7] For purposes of meeting the NRTL product-approval requirements in OSHA standards, e.g., those under Subpart S of 29 CFR Part 1910, OSHA only accepts equipment or products approved by one of its listed NRTLs. One can access OSHA Web pages for information on each NRTL's scope of recognition, or contact OSHA or the NRTL for additional information at www.OSHA.gov.

OSHA's NRTL Program recognizes private sector organizations as NRTLs, and OSHA accreditation signifies that an organization has met the necessary qualifications specified in the regulations. The NRTL determines that specific equipment and materials ("products") meet consensus-based standards of safety to provide assurance, required by OSHA, that these products are safe for use in the U.S. workplace.

The hazardous locations standard, 29 CFR (OSHA)1910.307, regulates the use of electrical equipment and wiring in hazardous locations which are classified depending on the properties of the flammable vapors, liquids or gases or combustible dusts or fibers which may be present therein, and the likelihood that combustible concentrations or quantities are present.[3]

Pursuant to 29 CFR 1910.307(b): "Electrical installations: Equipment, wiring methods and installations of equipment in hazardous (classified) locations shall be"[3] either:

- "Intrinsically safe"
- "Approved for the hazardous location"
- "Safe for the hazardous location"

Using NRTL approved instrumentation is the most common and efficient method for an employer to reliably demonstrate that electrical instrumentation, intended for use in hazardous (classified) locations, meets the 1910.307 standard requirements. As stated in 29 CFR 1910.307(b)(i): "Equipment shall be approved not only for the class of location but also for the ignitable or combustible properties of the specific gas, vapor, dust or fiber that will be present."[3] This corresponds with the first two preceding bullets, with the NRTL performing the appropriate equipment testing and certification, audits and approves manufacturing and repair facilities, and then authorizes approved facilities to apply the NRTL's approval mark.

Electrical classification is also included within the PSM Standard[1] and RMP Rule[2], as noted in paragraphs 1910.119(d)(3)(i)(C) and 68.65(d)(1)(iii): "Electrical classification"

Published OSHA interpretations note that the employer must be able to demonstrate that the equipment will provide protection from the hazards arising from the combustibility and flammability of vapors, liquids, gases, dusts or fibers. The National Electrical Code, NFPA 70, contains guidelines for determining the type and design of equipment and installations which will meet this requirement. One method for an employer to demonstrate that equipment used in hazardous (classified) locations meets the requirements of the standard is to use equipment that is certified as intrinsically safe and approved for the hazardous (classified) location (29 CFR 1910.307(b).[8]

Distinguishing Between Compliant and Potentially Non-Compliant Instruments

As noted earlier, a purchaser may have an erroneous perception of whether a reconditioned or new-surplus instrument is still NRTL approved, i.e. is still compliant to the standards which the NRTL certified it to, and free from "changes" unknowingly made to the equipment after it has left an NRTL approved supplier and/or service-provider facility.

FM Approvals, an OSHA accredited NRTL, has recently communicated that such "change" can include equipment that is repaired, where repair also includes refurbished, remanufactured, reconditioned, salvaged, and new surplus. FM Approvals defines repair as "Work performed to the unit that would bring it back to its original condition approved by FM Approvals", i.e. reconfirming product compliance with appropriate standards as FM Approvals originally certified it to.[6]

Earlier, FM Approvals had specifically communicated its position regarding reconditioned and new-surplus instruments that were originally approved for use in hazardous (classified) locations at the time of OEM manufacture:

> "It is FM Approvals' position that only the original manufacturer of the Approved product or an FM Approved remanufacturer whose facilities are part of the FM Approvals follow-up audit program, can remanufacture a product and reissue the FM Approvals certification mark. Any suggestion, practice or inference to the contrary is wrong and must cease."

> Further: "Any salvaged, remanufactured or new surplus electrical instrument cannot be labeled or relabeled as FM Approved for use in a classified hazardous location unless the refurbishing/new surplus supplier entity is audited and approved by FM Approvals, LLC, for that specific type of instrument."
>
> "Absent the above being met, the device can carry the FM Approvals certification ONLY if the product has been resubmitted and approval granted by FM Approvals. Failure to follow these guidelines will invalidate the FM Approvals certifications. In such instances the FM Approvals certification mark shall be permanently removed from the product (including the nameplate)." [9]

The FM Approval mark is a statement of conformity that a product is in compliance with defined standards at the time the product leaves the manufacturing and/or repair facilities that have been approved and audited by FM Approvals. [6]

As noted, the certification mark and nameplate should be removed unless the supplier/service-provider is an FM Approved facility, i.e. has been approved and audited by FM Approvals. For third party and end user facilities, such approval and auditing is to FM Class Standard 3606 for specific product brands and models. [6]

Maintaining Safety and Regulatory Compliance

Safety and regulatory compliance may be jeopardized if end users cannot distinguish between compliant and potentially noncompliant devices. [6] Thus the recommendation that end users, who have responsibility for continued compliance with applicable codes and standards, need to fully qualify their product and repair service vendors as described later in this article.

The use of a reconditioned or new-surplus instrument purchased from a non-qualified vendor would require a MOC for installation and use in a hazardous (classified) location, since product compliance would not be reconfirmed with the appropriate standards FM Approvals originally certified it to. This raises a question about the probability that such a MOC process would be pursued for a noncompliant instrument. An alternate solution may be to have the process unit re-classified; however this may or may not be possible.

One may question the inclusion of new-surplus devices that were originally labeled by the OEM with the FM Approved certification mark. The issue is traceability since the typical origins of new-surplus instruments range from aged or obsolete unused inventory to instruments removed from units that may have been constructed but never started up. During the possible multiple transfers (resale) of new-surplus equipment there is limited, if any, knowledge and traceability of problems a device may have encountered that would unknowingly result in a "change", due to possible exposure to:

- Damage during prior installation and removal
- Overpowering during calibration and/or performance checks
- Inadequate re-packaging or damage during repacking
- Inadequate storage conditions
- Any other hidden handling or shipping damage

Importance

Currently, the presence of erroneously marked NRTL approved products in the marketplace is resulting in potentially significant numbers of noncompliant instruments being unknowingly installed, annually, into hazardous (classified) locations.

The issue appears to be exacerbated by a lack of industry awareness. As such, education, trade journal articles, published safety alerts, and symposium presentations are increasingly important.

Noncompliant instruments are potential ignition sources creating additional potential exposure to safety risks and regulatory citations. Even though there is no confirmation that incidents have directly resulted from the use of noncompliant instruments, the combination of situations required to cause an incident could very well occur. In a recent investigation, OSHA issued citations on the use of non-approved electrical equipment with each violation considered "willful" and accompanied by a fine and with the aggregate fines totaling several million dollars.

The risk of being a potential ignition source is one of the reasons there are OSHA requirements for the use of electrical and electronic equipment in hazardous locations. Although the subject of this article is not a treatment of how risks are realized, an important reminder is that a simultaneous combination of events can create an incident even though never experienced previously. At times, users cite their experience with salvaged instruments as having never given them a problem. However, safety experts typically concur that as the number of installed non-approved instruments increases, there is increased opportunity of such a combination of events occurring.

Industry is familiar with the "swiss cheese" or "light & disc" illustrations of this, where appropriate instrument electrical classification is an element of protection. Figures 1 and 2 illustrate situational risk by using discs with holes, representing layers of protection methods.[10] Any single disc (protection method) can keep the light source from penetrating through the entire box (Figure 1). If all of the holes in the discs become aligned, representing a combination of abnormal situations and failures occurring simultaneously, the light will shine through, representing an "incident".

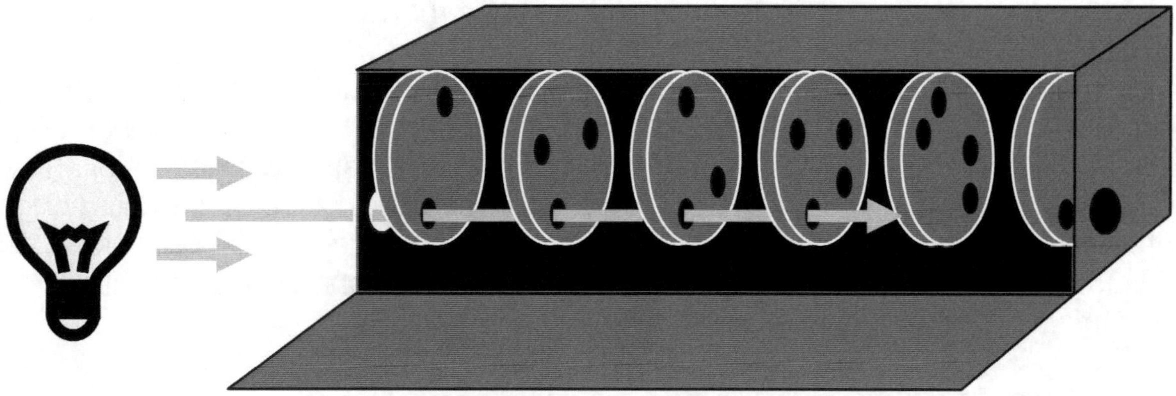

Figure 1.

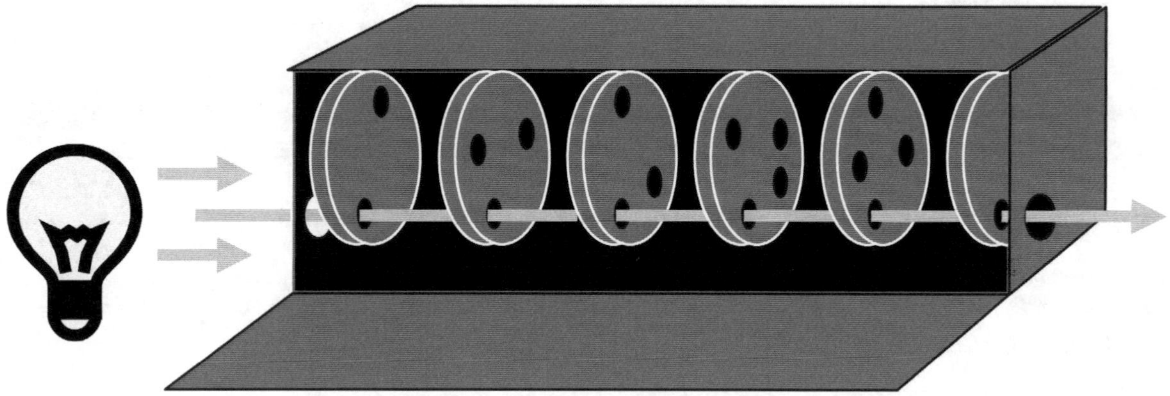

Figure 2.

Incident investigations have identified the occurrence of this type of situational concept, such as reports where prior start-ups had experienced similar abnormal situations and failures, but never the exact combination that resulted in an incident, until it finally occurred.

MITIGATING RISK

Again, the potential safety and regulatory compliance risks associated with the use of reconditioned equipment include:

- Loss of control valve containment (pressure integrity)
- Noncompliant instrumentation used in hazardous (classified) locations (not NRTL Approved)

So what can chemical processors do to help mitigate these?

Vendor qualification and technical awareness is critical, requiring initial education of all plant personnel associated with the specification, purchase, inspection or repair of reconditioned and new-surplus equipment. Further, ever-changing organizational structure and new personnel requires a sustained education program, including ongoing emphasis at safety meetings. End user issuance of specific corporate policy and guidance could be an effective method to appropriately emphasize and establish requirements for purchasing reconditioned equipment.

Vendor Qualification

Implementation and strict enforcement of three critical supplier qualification requirements can increase confidence in meeting safety and regulatory requirements when purchasing reconditioned control valves and instrumentation, or new-surplus instrumentation.

With appropriate documentation on file, such records can be valuable in discussions with insurance carriers or trade and regulatory organization safety audits.

Suggested Qualification Requirements

Note: Per prior definition, "reconditioned" includes salvaged, refurbished, and remanufactured.

1. For reconditioned or new surplus electrical/electronic instrumentation required to be compliant to OSHA 1910.307 per your plant's hazardous area classifications, require:
    - Signed NRTL authored documentation from your supplier and/or service providers of reconditioned and new-surplus instrumentation to include:
        - Certification that a supplier and/or service provider facility is approved and audited by the NRTL for the specific brands and models
        - Scheduled dates for the NRTL's follow-up audits of the facility
        - Results and status pursuant to the follow-up audits
    - That for any supplier/service provider whose facility is not approved and audited by the NRTL for the specific brands and models
        - Commitment, in writing, to remove any OEM nameplate containing a NRTL certification mark or at a minimum, completely remove the NRTL certification mark from the nameplate

2. For better evidence of control valve pressure integrity addressing plant safety and PSM and RMP Rule regulatory requirements, require:

- Signed documentation from your supplier and/or service providers of reconditioned or repaired control valves to include:
  - Certification of their ability to conform to all of an OEM's original specifications as designed to ASME B16.34, specifically including conformance to design requirements for additional metal thickness, as referenced by paragraph 6.1.7.
  - Certificate of Conformance documents for each control valve, upon request, certifying that a supplier's and/or service provider's reconditioned or repaired control valves meet all OEM specifications, with specific reference that body walls meet thickness specifications as designed to ASME B16.34 including paragraph 6.1.7 and are shell tested in accordance with section 7.1.

3. For any supplier and/or service provider incapable of complying with any of the above, require:
- Notation of noncompliance on all correspondence including:
  - Specification documents
  - Quotations submitted
  - Packing lists
  - Invoices

Note: Requiring noncompliance notations on correspondence increases site visibility to any anticipated application of non-compliant control valves or instrumentation, with an objective of triggering an appropriate MOC evaluation before installation of the equipment.

Beyond these first steps some manufacturers, in collaboration with end users, have developed programs addressing these pressure integrity and NRTL approval issues without significantly impacting an end user's personnel requirements.

Resulting work processes typically involve assessing and identifying existing, potentially non-compliant devices, developing appropriate abatement processes and sustaining compliance awareness through training. The following guidelines are presented in general terms without detailing specific execution steps since execution is often dependent upon site maintenance practices, equipment criticality and access, management appetite for risk, etc.

Suggested Work Processes

Work processes may include:
- Assessment and identification
  - Existence of potentially noncompliant instrumentation installed in hazardous (classified) locations
  - Existence of reconditioned control valves not having certified pressure integrity via body wall measurements

- Abatement
  - Instrument recertification performed by an NRTL approved and audited facility
  - Abatement executed via day-day or turnaround maintenance work processes
  - Control valve pressure integrity inspection and restoration (when required) that is compliant to all OEM specifications with individual control valve certification documentation via a Certificate of Compliance (COC).

- Sustaining
  - Implementation and enforcement of vendor qualification requirements to better prevent introduction or re-introduction of noncompliant equipment
  - Initial and ongoing communication and training of all potential safety and regulatory compliance issues and site prevention policies
  - Communication and training to include employees having any degree of involvement in the engineering, specification, purchase or inspection of reconditioned or new-surplus control valves and instrumentation.

OEM facilities that supply new and/or remanufactured instruments are typically audited and approved by the major NRTLs. Also, OEMs typically have authorized service centers with the technology, training, work processes, and access to intellectual property necessary to appropriately assess and certify compliance of reconditioned and repaired control valves to original OEM design specifications.

Summation

During the last few years there has been an increasing number of potentially noncompliant reconditioned control valves and instrumentation, as well as noncompliant new surplus instrumentation, being introduced into all types of chemical process industries and the refining industry. Depending upon variables such as equipment age, repair history, application risk, etc., this reconditioned or new surplus equipment may no longer be compliant with safety standards or to an OEM's original design specifications with respect to pressure retaining capability or safe use in hazardous locations.

Industry awareness of the technical and/or safety compliance issues associated with reconditioned equipment should provide the impetus for chemical processors and refiners to develop appropriate corporate policies and guidance directing inspection, engineering, operations, maintenance and procurement assessment of potential safety and regulatory issues.

Going forward, stringent supplier qualification can be a straightforward and efficient preventive solution. For suspect equipment that is already installed, identification and appropriate abatement processes may be needed.

Such actions assist in creating a safe workplace and can demonstrate a proactive safety culture by reducing the probability of deficient reconditioned or new surplus equipment being the focal point of a future, potentially significant incident.

References:

1. U.S. Department of Labor, Occupational Safety & Health Administration Regulations; Standard 29 CFR 1910.119: "Process safety management of highly hazardous chemicals"

2. U.S. Environmental Protection Agency, Code of Federal Regulations 40 CFR Part 68.73: *"Chemical Accident Prevention Provisions; Mechanical Integrity";*

3. U.S. Department of Labor, Occupational Safety & Health Administration Regulations; Standard 29 CFR1910.307: "Hazardous (classified) locations"

4. The American Society of Mechanical Engineers (2004) "*Valves – Flanged, Threaded, and Welding End*"

5. Hazan, Michael (2006), "Why is Mechanical Integrity So Tough?", AIChE 21st Annual CCPS International Conference

6. Gagliardi, Cheryl and Baker, Robert (2006), "Maintaining Certification Compliance of Equipment Used in Hazardous (Classified) Locations", 22nd Annual National VPPPA Conference

7. U.S. Department of Labor, Occupational Safety & Health Administration, Directive # CPL 01-00-003 – CPL 1-0.3, "NRTL Program Policies, Procedures, and Guidelines"

8. U.S. Department of Labor, Occupational Safety & Health Administration, OSHA Standard Interpretations: "9/21/05–Use of explosion proof certified equipment inside of pipes and manholes."

9. Martell, Robert (2004) Assistant Vice-President and Director of Approvals, FM Approvals,LLC

10. U.S. Department of Labor, Occupational Safety & Health Administration Regulations; Standard 29 CFR 1910.119: "Process safety management of highly hazardous chemicals"

11. U.S. Environmental Protection Agency, Code of Federal Regulations 40 CFR Part 68.73: *"Chemical Accident Prevention Provisions; Mechanical Integrity";*

12. U.S. Department of Labor, Occupational Safety & Health Administration Regulations; Standard 29 CFR1910.307: "Hazardous (classified) locations"

13. The American Society of Mechanical Engineers (2004) "*Valves – Flanged, Threaded, and Welding End*"

14. Hart, Christopher, Assistant Administrator for System Safety, Federal Aviation Administration (2005), "Lessons Learned from Unreported Incidents and Near-Misses", Mary K. O'Connor Process Safety Symposium

15. Rosenzweig, Mark (2005), "Take Safety Up A Notch", ChemicalProcessing.com

16. Wright, Joseph (2002), "Getting Closure on Compliance", Valve Magazine

17. Norvell, William C. Jr and Marrs, Scott D. (1998), "Counterfeiting - A Growing International Problem", Valve Magazine

18. Stoddard, Brooke (1993), "Valves in the Service of Safety", Valve Magazine

19. Frier, Len (2002), "What Electrical Inspectors Don't See", IAEI News website

20. ANSI/NFPA 70-1996,"National Electric Code"

The views and opinions expressed herein are those of the author himself. The author is not speaking and does not purport to be speaking on behalf of any other company or organization

Acoustic Modeling and Noise Plans: Implementation on Capital Projects

Leisa Porter & Stephen Johnson

Flour Corporation, Sugar Land, Texas

ABSTRACT

Incorporating Inherently Safer Design concepts regarding noise principles into a capital project can be a challenging task. Developing a noise plan early in the project can guide a project team to cost effective execution and equipment solutions. The purpose of a noise plan for capital projects is to indicate how the project intends to identify noise sources and the strategy for controlling noise at the facility.

Many engineers need to screen the acoustic impact of equipment to be installed on new construction and revamp projects. Powerful dedicated software tools exist to support such modeling, but many engineers have neither the access to such tools, nor the expertise to use them effectively. Further, such tools often represent gross overkill for the level of screening analysis required.

This paper will present the basics relevant to sound level screening as an introduction to noise. Simple web tools and spreadsheet-based solutions suitable to work at the conceptual through early engineering design levels will be discussed.

This paper will provide a guideline for developing a noise plan for a capital project; explore simple quantitative noise modeling tools that can be used to cost effectively screen preliminary designs; and discuss how these screening tools in turn allow for cost effective project execution and design solutions.

Acoustic Modeling and Noise Plans: Implementation on Capital Projects

Leisa Porter & Stephen Johnson

Flour Corporation, Sugar Land, Texas

INTRODUCTION

The purpose of a Noise Plan for capital projects is to indicate how the Project intends to identify noise sources and the strategy for controlling noise on the facility. Inherently safer design principles guide us to eliminate the hazard through substitution if feasible. Good safety engineering and industrial hygiene principles use the next hierarchy of a control/mitigation strategy to provide engineering controls. The option of using administrative controls, safe work practices or personal protective equipment such as hearing protection should be a last measure and not the first.

This hazard mitigation control strategy can be applied to capital project engineering designs by incorporating the concepts of specifying equipment noise levels to the vendors and the use of engineering controls such as insulation and equipment layout in the design. Administrative controls and safe work practices are typically addressed by the operating company through facility operating procedures.

In general, capital project design engineering noise plans should address how the following issues will be addressed on the project:

- Identification of potentially high noise equipment
- Noise control measures that may be required
- Estimation of area noise levels (based on vendor data)
- Designation of high noise areas and hearing protection requirements
- Estimation of off-site noise and potential impact to wildlife if in sensitive areas
- Computer modeling programs and noise calculations

DEFINITIONS

To provide a common basis of terminology as used in this paper, the following definitions for fundamental noise concepts are provided:

A-weighted - "A-weighted" means a specific weighting of the sound pressure level for the purpose of determining the human response to sound. The specific weighting characteristics and tolerances are those given in American National Standards Institute S1.4.

dB (A) – dB (A) means a unit of sound level expressed in decibels (dB) and A-weighted. The A-weighted sound level is also called the noise level. Sound level meters have an A-weighting network for measuring A-weighted sound level.

Ambient noise - The total of all noise in the environment other than the noise from the source of interest. This term is used interchangeably with background noise.

Attenuation - The reduction of sound by various means.

Broad-band noise - Noise with components over a wide range of frequencies.

Octave band level - The integrated sound pressure level of only those sine-wave components in a specified octave band.

Impulsive noise - "Impulsive noise" means either:

- a single sound pressure peak (with either a rise time less than 200 milliseconds or total duration less than 200 milliseconds) or,
- multiple single sound pressure peaks spaced at least 200 milliseconds apart.

Sound level - The weighted sound pressure level obtained by the use of a sound level meter and frequency weighting network, such as A, B, or C. Weighting shall be as specified in ANSI specifications for sound level meters (ANSI-S1.4). If the frequency weighting used is not indicated, the A-weighting is implied.

Sound Level Meter (Type I, Type II, Type III) - Three types of sound level meters differing in their performance requirements. The requirements are the most strict for the Type I or "Precision" Sound Level Meter, and progressively less strict for the Types II and III. Type I is commonly referred to as a "General Purpose" Sound Level Meter, and Type III a "Survey" Sound Level Meter. The Type I, or Precision instrument, is widely used in the laboratory where conditions can be carefully controlled and the accuracy of the instrument can be fully used. Because some Type I instruments are readily portable and easy to use, they are also widely used for field measurements.

Sound pressure level - "Sound pressure level," in decibels, means 20 times the logarithm to the base 10 of the ratio of the pressure to the reference pressure. The reference pressure shall be 20 micronewtons per square meter.

CAPITAL PROJECT NOISE PLANS

A capital project engineering plan to address noise should identify the requirements and the activities to be undertaken to meet the project noise design requirements. It should also describe the deliverables to be produced during the engineering design. In addition, the assumptions and basis for review of project noise data should be identified. The noise plan, when done early in the engineering planning stage, provides a tool to assist the project in achieving the project noise requirements. Typically capital projects will have noise requirements for both occupation limits and fenceline regulatory compliance that impacts the community.

Occupational noise level for personnel time-weighted average (TWA) will vary from project to project depending the owner/operators. Typically targets will be 82 dB (A) or less for a 12-hour working shift, with no instantaneous exposure over 110 dB (A), except for relief valves. 8 – hr shifts may be designated at a higher dB (A) level, dependant upon the operating phase hearing conversation program. In order to meet a 82 dB (A) personnel time-weighted average occupational exposure target, general equipment will typically be specified at 85 dB (A). Where the area noise level cannot be adequately controlled to an 85 dB (A) level, warning signs will be posted requiring the use of hearing protection.

Fenceline acceptable values can vary significantly. Construction fenceline monitoring is often required for reduction of noise pollution to neighbors during the construction phase. Operating plant fenceline limts are often determined by local or regional governmental requirements. Even remote locations in environmentally sensitive areas often require research into the affects of various noise exposures on wildlife populations. For these areas noise analysis is often required for submittal in an environmental impact assessment (EIA) with approval from a regulatory agency often required.

Acoustic induced vibration calculations for relief valves and flare lines are critical to some installations. Acoustic induced vibration studies are often misclassified as part of the noise study because of the pairing of the words acoustic and vibration. From a project execution standpoint and engineering discipline ownership on a project, the issue often becomes ignored or passed from one discipline to another. It is a difficult issue and typically nobody wants to own it. In most cases a specialist consultant should be engaged to undertake this kind of study and then managed by the HSE group, mechanical or process.

The design of foundations supporting major rotating equipment can become a significant issue in regards to resonance frequency/interference. Similar natural frequencies of a structure to that of a vibration emitted from a mechanical piece of equipment can result in significant structural loads with potential deformation and even failure. Structural and mechanical disciplines should be consulted to perform finite elements analysis for critical installations. Vendor requirements should be reviewed and recommendations made for vibration mounts, flexible connectors, etc., where appropriate.

CONSTRUCTION PHASE NOISE ISSUES

Construction noise plans are typically handled by field construction HSE specialists with nominal input from the engineering design team. The plans should be initiated during design stage so that planning, permit submittals or mitigation can be developed with minimal disruption to the project schedule once mobilizing to the field. The various stakeholders (owner, contractor, local authorities possibly non-government organizations) should have input to such plans and ensure that applicable regulations and requirements will be incorporated.

Site construction noise plans identifying noise limits near any island or water exclusion zones, neighborhoods, roadways or public areas will need to be considered depending upon the location of the facility. The construction execution plan developed for the project will generally identify the types and numbers of large equipment to be employed on the project, and a schedule of when the equipment is expected to be mobilized and demobilized to/from the facility. With these inputs, vendor information can be obtained and spread sheets developed to begin documenting the data for permit submittals. At this time any noise modeling that may be required can be initiated and any noise mitigation plans developed to prevent noise pollution to surrounding neighborhoods.

NOISE DESIGN BASIS ASSUMPTIONS

As is true of any design basis or analysis, the assumptions made in the design phase play a significant role in design development. The assumptions should be researched and proven to be generally workable on a large scale project an suitable for use at the beginning of design. Typical assumptions that need to be identified and agreed upon with the owner/operator as the starting point for a basis of design for project noise issues cross many engineering disciplines. The project goal for TWA for personnel as well as equipment impact and impulse noise limits need to be agreed upon, as well as the requirements for mechanical/rotating equipment, control valves and relief valves. Basic sound principles in regards to additive effects of noise and noise reduction due to distance are generally assumed to be as shown in Table I below and in Appendix I. The design phase noise plan needs to discuss how operational modes and sparing philosophy factors into the noise calculations and plots.

How to address piping is always controversial in regards to noise assumptions. Theoretically pipe thickness can provide some noise mitigation, but realistically is nominal and provides nominal value for most project applications. The same holds true for pipe vibration. Good engineering practices and piping project specifications/standards are adequate to reduce transfer of vibration to the structure thereby minimizing background noise.

Table 1: Noise Reduction Due To Distance

Distance	Decibel
X	Y
2X	Y-6 dBA
4X	Y-12 dBA

When the distance is doubled from a point source, the sound level drops 6 decibels (*Noise Control, A Guide For Workers and Employers, U.S. Department of Labor*).

CONSIDERATION OF PIPING AND VALVES IN DESIGN

The amount of noise emitted from a pipe is difficult to predict in the design stage of a project. Engineering design specifications for line sizing, pipe stanchions and hangers, etc., typically are such that noise emitted from pipe work remains background noise. Background noise from pipes typically does not create significant additive noise effects in the workplace. Exceptions may be areas where there are high pressure drops, pipe bends or expansion loops. These areas are typically not a concern unless they are in an area where employees spend a significant portion of their work day. Design factors that can impact the noise from pipe work include: velocity of the flow (operating or design), differential pressure across associated equipment, or pressure drop throughout the line itself. After startup, any vibration problems with rotating equipment will impact the noise emitted from associated downstream piping.

Acoustic insulation can be provided in areas where noise emitted from pipe may be a problem. Significant improvement (or dB reduction) from acoustic insulation is only gained if the pipe work provides a significant noise source to the total work area, or to an off-site receptor. The project PFDs and P&IDs can be reviewed to identify main piping connected to equipment specified at greater than 85 dB (A). Select lines can be reviewed to determine the extent of acoustic insulation required. Lines with high pressure drops or expansion loops should be reviewed in the design stage for the need for acoustic insulation.

Piping downstream of significant pieces of equipment should be evaluated for potential effects of radiant energy (noise). Recommendations can be made in the design stage for insulation in order to prevent additive effects to the general work area background noise levels. Areas typically recommended to be acoustically insulated include compressor suction and discharge pipe work to and from the nearest vessel or heat exchanger, and recycle pipe work downstream of recycle valves. In addition, good engineering practices will consider anti-surge valves and capacity control recycle valves as part of the piping system. These valves typically should be acoustically insulated as well.

To account for piping in noise contour maps, typical assumptions state that piping and valves which are not in high pressure service can be considered contributing to a background level of 3 dB (A). This is a conservative assumption that takes no credit for pipe thickness. The assumption is made that the noise emitted from the pipe is the same as the internal noise produced by the upstream equipment.

DESIGN ACTIVITIES

The following design activities are applicable to a facility project and can be undertaken early during a design:

- **Identification of facility existing sound level.** Prior to construction or field modifications, a sound level meter survey needs to be obtained to provide a baseline document. This can be obtained from the owner/operator, or by 3rd party specialists. It is critical that this data be obtained prior to overall noise design efforts to ensure the possibility of success in meeting facility noise requirements.
- **Identification of design noise sources.** Noise requirements for individual items of equipment should be identified on the equipment data sheets. Vendor requirements for attenuation and testing procedures should be identified as well as the method for noise measurements and requirements for submittal of the results are identified in the specification. Vendors can be requested to confirm equipment noise level guarantees in their proposals.

- **Estimating noise levels in Facility work areas.** Initially this will be a simple plot of the dB (A) values as identified on the data sheets. The noise levels for major pieces of equipment can then be overlaid on the approved for design general arrangement drawing. The operating/sparing philosophy for major pieces should be identified to indicate potential for additive effects of noise due to equipment layout.
- **Noise mitigation.** As noise control measures are identified by the vendors for equipment/skid packages and valves, the information will be maintained as project vendor data.
- **Estimating the noise levels in work areas after applying vendor noise control measures.** The noise plots will be updated to indicate the equipment dB (A) levels as submitted by the vendors. This will give a general indication of where potentially high noise areas may exist due to additive effects of noise.
- **Generate theoretical noise levels for the facility from the plot plan.**

NOISE CALCULATIONS AND PLOTS

For purposes of estimating the additive effects of noise in a general area, The Engineering Equipment and Materials Users Association (EEMUA) Publication No. 140 can be used as a guideline to develop a simple spreadsheet application. The following assumptions and EEMUA calculations can be applied through the application of a simple spreadsheet:

1. Noise sources are to be considered small sources and a directivity index is not required.
2. To account for ground reflection, hemispherical radiation calculations are to be used for sound propagation at ground level equipment.
3. Spherical radiation calculations are to be used for sound propagation of elevated sources.
4. The dB (A) readings provided by the vendor at 1 m out from the skid package/equipment are considered to be the sound power level (L_w).
5. It is assumed there are no sources provided with enclosures and reverberation effects are not applicable. Vendors will supply dB (A) readings 1 m out from enclosures provided with their equipment.
6. Additive effects from noise are considered as illustrated in Appendix I.

Quite often capital projects require sophisticated 3D computerized modeling to generate contoured noise plots. Usually the services of a third party are required to generate the theoretical additive effects of noise on the facility. Sophisticated modeling is often required when multi level noise sources are potentially contributing to the additive effects, or where piping is anticipated to potentially add significantly to the background level. If the facility is in an extremely sensitive environmental area and the fence line noise criteria is extremely low, 3D computerized modeling may be warranted. Many upstream oil and gas projects traverse through remote locations where disruption to the natural wildlife is required to be analyzed.

Another application for 3D modeling may be areas near tundra or over water simply because the effects and response to the surface may not follow standard sound principles. For environmental permitting submittal to regulatory authorities 3D computerized modeling by a third party is often the prudent execution strategy.

If need for a third party contract is agreed upon, a scope of work will need to be developed and the assumptions agreed upon. The scope of work document should be a formal project document. The scope of work should require that the programs inputs, outputs, and software algorithms be documented in the final report deliverable produced by the third party.

An alternative to 3D computerized modeling to evaluate fenceline values is the development of spreadsheets using standard noise calculations combined with free information/sources from the internet. This is adequate for most projects, especially in an early design stage. To develop a spreadsheet application a standard process can be used for the calculation of sound levels. Figure 1 illustrates this procedure. The steps involved in evaluating the sound levels across a plot are – in principal – fairly simple as described below.

The first step is to list the significant noise sources. For normal operations, these will typically be large rotating equipment such as turbines, diesels, large pumps, boiler fans, large air coolers, and the likes. Control valves can be very loud and quite important for work area levels, but will rarely drive fenceline values, because of their relatively small size. As mentioned above, vendor data is the most accurate, but is often not available when conducting screening studies prior to detailed engineering. Approximate values, however are usually fairly readily available in the literature.

Physical size of the primary sound source is important in the calculation as will be detailed in the continuity equation later – for practical reasons, sound measurements are usually taken at a minimum distance of 1m from the equipment being measured. In a typical screening calculation you will wish to model the source as an equivalent circular area of the same footprint.

Consideration needs to be given in the equation to sound intensity. Any given piece of equipment will have a specified sound intensity in dBA, measured as specified above. It is important to know whether this value is Sound Power or Sound Pressure. Sound Power is measured across a specified surface; pressure is at a given point. The authors generally prefer to work with sound Pressure, but under no circumstances allow the distinction to be lost.

The final key datum to possess about the source is its location. The coordinates of the individual pieces of equipment or noise sources can usually be obtained from the piping and layout disciplines on a project. Once the sources have been established, the global area to be evaluated needs to be considered. This will often be the entire plot for a given unit being examined. This can be simply accomplished by dividing the area up into a grid.

For the report to be meaningful, basal fenceline (or edge of platform, if offshore) values – without the unit / equipment being evaluated – will be needed for the work to have any practical value. This is often available in existing documents, but may need to be measured or estimated for onshore facilities. Offshore, the primary criterion is on plot values, rather than edge-of-platform, again for fairly obvious reasons.

Once the sources and area of interest are established, the next step is to calculate the sound pressure level at each point on the grid for each individual source. This will require the application of the continuity and attenuation equations, which are addressed separately below.

Finally, for each element on the grid, add up the individual sound pressures. Since decibels are a logarithmic measure, the authors prefer to work in actual sound pressures as Pa. Sound pressure normally being measured as a root mean square function, these can be directly added, and then back converted in dBA afterwards. Alternatively, one could work in dBA throughout; it's really a question of convenience in execution.

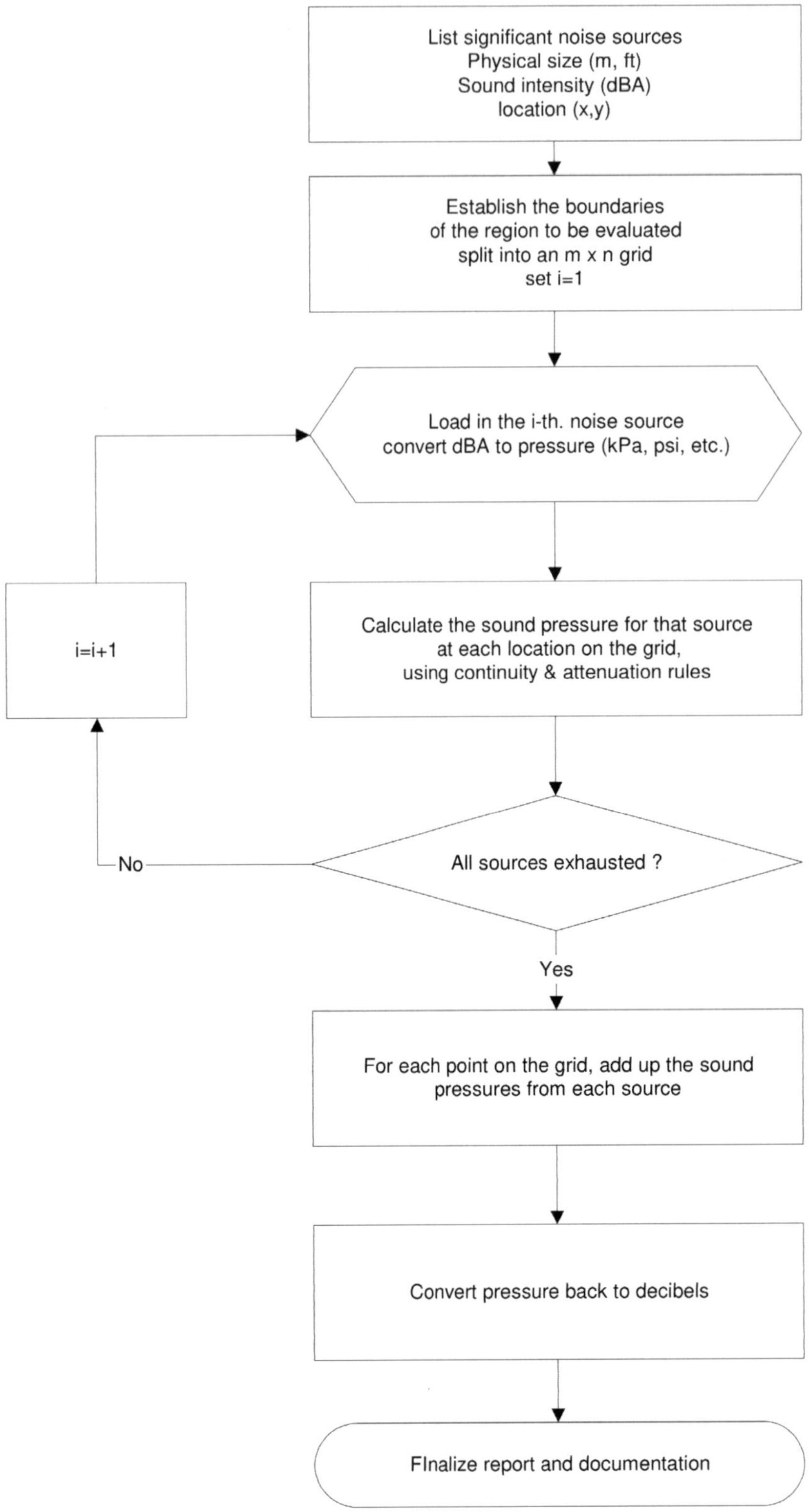

Figure 1 **Calculation of Sound Levels on a plot area**

For the true diehard engineers, you may wish to perform the calculations by hand. However this is a great application for a computer spreadsheet where the task can be automated. This is relatively easily accomplished, and makes it easy to check the sensitivity of your critical sound values to assumptions that you may have made going in.

Sound Attenuation in air is a fairly strong function of frequency. ISO 9613 Part 1 defines the calculation formula for sound attenuation in air, and an online calculator is available at http://www.npl.co.uk/acoustics/techguides/absorption/. Attenuation is influenced by frequency and relative humidity as shown in Table 2 below. Generally, however, the key feature to note is that attenuation requires rather exact knowledge of the properties of the atmosphere in question, and is highly sensitive to distributed frequencies - i.e. low frequencies are attenuated to a much lesser degree than higher ones. It is recommended that atmospheric attenuation factors – if used at all – be used with care.

Table 2 Sound attenuation dB/km at 0°C

Frequency	10% RH	50% RH	90% RH
1 kHz	14	4.7	5.3
8 kHz	180	110	63
16 kHz	230	360	90

kHz = kilohertz

RH = Relative Humidity

Extracted from NPL Kaye &Laby, Tables of Chemical & Physical Constants, http://www.kayelaby.npl.co.uk/general_physics/2_4/2_4_1.html

While attenuation factors may not offer a great deal of relief, energy conservation principles often do. For point or pseudo-point sources in a more-or-less free field, the conservation equation predicts that sound waves emanating out will make a hemispherical shell at any given distance, and that the pressure at any given radius R will be proportional to (R/Ro)^2, where Ro is the effective radius of the source. Most sound propagation will follow essentially free-field rules in onshore plots.

Linear sound reduction will occur in partially constrained environments, such as between decks in a platform, or a multi-level process plant. In such environments, the natural intensity reduction is linear with distance, i.e. proportional to R/Ro. This also applies in free fields with essentially linear noise sources, such as roads.

Finally, in some highly constrained circumstance, sound will not decay noticeably with distance except through attenuation. This would apply in an environment where the sound propagation path was essentially linear – this is the case, for instance, for sound transmission in a pipe or culvert, or between rows of highly congested equipment.

DESIGNATION OF HEARING PROTECTION REQUIRED AREAS

After the additive affects of noise and contour noise plots are generated, they should be reviewed for areas where mitigation may be needed. For personnel considerations, high noise areas should be identified to meet the requirements of the hearing conservation program that will be in effect during operation. This will provide a basis to review the anticipated employee work tasks and daily time exposure. The simple spreadsheet can be used as a sensitivity tool to determine if equipment spacing, inherently quieter equipment, or noise panels can feasibly provide reduction in an employee TWA. Some areas may require additional mitigation to reduce an employee TWA to below a required 82 dB (A) or 85 dB (A). If mitigation is not feasible, areas will need to be designated as high noise areas. High noise areas will require hearing protection, employee training and adherence to operating phase hearing conservation requirements for medical monitoring.

For fenceline mitigation efforts, the spreadsheet and contour plot can be used to determine which pieces of equipment should be mitigated and to what degree to meet requirements. Again, this can be ran as a sensitivity analysis to determine the most cost effective solution for individual pieces of equipment.

METHODS OF MITIGATION

The sensitivity analysis may show that noise mitigation plans are required where the theoretical noise plots indicate employee time-weighted average exposures are excessive or where neighborhoods or exclusion zones are determined to be adversely affected. Noise controls may be accomplished through measures at the source (orientation, vibration, vibration control, etc.), installation of acoustical shields, equipment layout, enclosures, or other barriers to interrupt the path of the sound.

Care must be taken to ensure noise mitigation does not interfere with or is detrimental to the safe, efficient operation and running maintenance of the equipment/item of concern. Nor should mitigation features in themselves constitute or create a hazard to safety in any way. This applies to mitigation methods for the facility and construction mitigation efforts.

The best means of controlling sound generation is to choose inherently “less noisy” or quiet machines or processes. Generally machinery is designed with performance in mind. To change internal components to reduce noise can reduce efficiency. However, one machine type or component design can often be substituted for another in the original design without sacrificing efficiency, or by altering the design to accommodate.

The further away one piece of equipment is from another the less overall increase in noise level from the additive effect of noise. During plot plan reviews the distance between equipment can be reviewed and increased where feasible. For offshore facility design or large revamps onshore, the luxury of real estate does not always exist. By recognizing that the space limitation is adding to the overall noise levels early in the design provides an opportunity to target strategic pieces of equipment, or trains of equipment, and work with the vendors to reduce the expected dB (A) levels from the equipment.

Vibration mounts, springs, dampeners, structural foundation materials, can be used to eliminate or reduce noise from vibration. When a large amount of noise reduction is required, an acoustic enclosure can be considered as a direct solution. The enclosure serves as a noise barrier that completely encloses a machine and will be designed such that its resonant frequency and the noise frequency of concern do not coincide. Acoustic enclosures may be acceptable as long as the enclosure does not present additional hazards. If acoustic enclosures are to be used, the following must be considered.

ENGINEERING DESIGN DELIVERABLES

Documentation for the noise efforts on the project are an important step in finalizing the design. Design documents include the original scope of work for the noise effort, equipment data sheets, vendor requirements and testing methods, accepted mitigation measures and any generated noise plots. A final noise report generated during the detail design execution typically identifies the following:

- Mechanical equipment packages and the associated noise level.
- Specific noise control measures submitted by vendors.
- List of process lines for which acoustic insulation has been recommended.
- Noise plots showing estimated noise levels in work areas.
- Identification of areas where the Project has agreed to waive the work area limits and has authorized designation as high noise areas with hearing protection required.

The final noise report should be a project document that is documented per the project document control procedures. This allows for the noise effort and design measures to be part of the standard project filing and retention process and retrievable if necessary after the project has been executed.

AFTER START-UP SOUND LEVEL METER SURVEY

After the plant has been in operation for a sufficient time period to allow for the equipment and process to "line out," field measurements of the actual noise level should be obtained. This will provide a quantitative description of general work area noise at the facility after startup. The survey shall be conducted after commissioning and startup of the plant facilities and conducted at full rated plant operations. Noise measurements should be made by an individual who is trained and skilled in the use of acoustic measuring instrumentation. For a project at an existing operating facility, this survey would typically be performed by the facility operator/owner for incorporation into their overall hearing conservation program noise monitoring. For a grassroots installation the survey may be conducted by a third party.

The results from the after start-up sound level meter survey should be compared to the theoretical maps. Areas where further field modifications can easily be made to address unidentified high noise areas on the theoretical maps can be addressed at this time. The after start-up sound level meter survey and any employee dosimeter TWA readings that are taken can be used in conjunction with the operating hearing conservation program to ensure hearing protection requirements are met.

SUMMARY

Incorporating Inherently Safer Design concepts regarding noise principles into a capital project can be a challenging task. Developing a noise plan early in the project can guide a project team to cost effective execution and equipment solutions. The purpose of a Noise Plan for capital projects is to indicate how the Project intends to identify noise sources and the strategy for controlling noise on the facility. A hazard mitigation control strategy can be applied to capital project engineering designs by incorporating the concepts of specifying equipment noise levels to the vendors and the use of engineering controls such as insulation and equipment layout in the design.

Capital project design engineering noise plans that address the issues and identify potentially high noise equipment provide the opportunity to use various modeling programs. Often simple quantitative noise modeling tools can be used to cost effectively screen preliminary designs. These screening tools in turn allow for cost effective project execution and design solutions in regards to noise control measures that may be required and provided for in a design.

REFERENCES

1. *American National Standards Institute S1.4*
2. *American National Standards Institute S1.11*
3. *OSHA 1910.95, Occupational Noise Standard*
4. *NOM-011-STPS, Health and Safety Conditions in Workplaces Where Noise is Generated*
5. *Noise Control, A Guide For Workers and Employers, U.S. Department of Labor*
6. *Noise Procedure Specification Publication No. 140, The Engineering Equipment and Materials Users Association (EEMUA)*
7. ISO 9613 Part 1
8. http://www.npl.co.uk/acoustics/techguides/absorption/
9. http://www.kayelaby.npl.co.uk/general_physics/2_4/2_4_1.html

APPENDIX 1
BASIC SOUND PRINCIPLES

Difference between levels in dB (A)	Number of dB (A) to be added to (or subtracted from) the higher level
0	3.0
1	2.5
2	2.1
3	1.8
4	1.5
5	1.2
6	1.0
7	0.8
8	0.6
9	0.5
10	0.4
12	0.3
14	0.2
16	0.1

Source: Noise Control, A Guide for Workers and Employers, U.S. Department of Labor, Occupational Safety and Health Administration

Right Sizing Fire Protection

Stephen Haines, Risk, Reliability and Safety Engineering, LLC, 6455 South Shore Blvd., Suite 400, League City, Texas 77573

ABSTRACT

How do you know if the level of fire protection at your facility is adequate or when fire preventive measures, such as equipment upgrades, would be more beneficial than installing fixed fire suppression? One approach is to utilize a Cost-Benefit Analysis.

In this paper, the author discusses two recent case histories where the cost-benefit analysis approach was utilized. In one case, this approach was utilized to identify an appropriate balance of fire prevention and suppression measures for a global corporation that processes metal ore using flammable and combustible liquids. In the second case, the cost-benefit approach was utilized to establish a corporate standard on fire suppression for a Liquefied Natural Gas (LNG) processing company.

INTRODUCTION TO COST-BENEFIT ANALYSIS FOR FIRE PROTECTION

Drawbacks of Traditional Fire Protection Decision Making Processes

The principal weakness within many companies is that the term "fire protection" has become synonymous with the term "fire control" only. As a result, fire prevention or fire frequency modification is often viewed as a separate field of study or expertise and is not regularly considered when making fire protection decisions.

This has been a huge mistake. The role of fire frequency modification cannot be understated and cannot be overlooked in the process industry. One need only look at the industry's fire and explosion history to come to the realization that most:

- Fire incidents start with a rather large flash fire or explosion, and then quickly recede back to a relatively localized fire at the source of the fuel. In fact, the majority of the damage incurred during some of the industry's largest and most notorious incidents occurred at their onset and not necessarily from incremental fire growth.
- Fire and explosion related fatalities and serious injuries are predominantly incurred in the first few seconds of an incident, as the result of the initial explosion or flash fire. It is physically and technologically impossible to use fire control systems, with the possible exception of explosion suppression systems, to intervene quickly enough to avoid these casualties.

This mind set has been so pervasive that many companies have divorced Process Safety Management from fire protection. As a result, the people worried about prevention and those worried about suppression have pretty much been working in their respective vacuums without realizing the impact they could have if a truly comprehensive approach were employed.

The other important weakness in past practice is that the fire protection strategies of many companies were not developed in a comprehensive manner that evaluated their individual risks and needs. Instead, the fire protection strategies of many companies have evolved as a result of:

- Placing too much emphasis on the lessons-learned during the suppression of fires, instead of looking at the entire event from ignition through extinguishment.
- Succumbing to the idea that: "If it's good for them, it must be good for us." As a result, the fire "suppression" standards of many companies include a lot of "copy-cat" engineering and do not necessarily address the company's individual risk profiles.
- Believing that because fire suppression helps to put fires out, it also improves life safety.

As a result of these traditional mind-sets and practices, there has been a great deal of thought put into how fires can be extinguished without considering the fact that it usually more effective to reduce the potential for the fire in the first place.

Advantages of Cost-Benefit Analysis

The primary advantage of utilizing CBA in the fire protection decision making process is that it allows the user to screen the effectiveness fire frequency modification options against that of fire control options. In this manner, the user is able to develop a fire protection strategy that employs a balanced combination of fire frequency modifiers and fire control features.

Other advantages include:

- The frequency of fire occurrence and potential consequences must be established in the process of performing the CBA. As a result, the user not only gets to evaluate the cost effectiveness of the various options on hand, but also gets to understand the risk profile of the fire scenario(s) under consideration. In the author's experience, the most cost-effective fire protection options generally have the greatest impact on risk.
- The company's fire protection strategy will be customized to its risk profile and unique financial position. As an example, during the design of a new process unit the design/build contractor had incorporated very little fire proofing compared to the company's usual practices. This decision had been made by the contractor because a global company that they had built a similar unit for had incorporated very little fire proofing in their unit (i.e., copy-cat engineering). When this situation was examined, it was determined that the proposed level of fireproofing was inappropriate for the regional company because:

- There were significant differences in the process equipment and instrumentation specifications employed by the two companies (affecting reliability, hence potential fire frequency).
- The global company had greater redundancies within its processing assets (units and plants) and greater cash reserves. As a result, they could take on a larger financial risk than the regional company could tolerate (or possibly survive if a significant loss ever occurred).
- The global company was self-insured and had significant risk engineering capabilities. As a result, they were able to closely manage their risk profiles with little input from external stakeholders. The regional company on the other hand, had limited risk engineering capabilities and had to contend with external stakeholders, such as insurers on a regular basis.

- Fire protection decisions are made with the mindset that they must support the company's bottom line to some degree and the continued employment of those who work for the company. As the old saying goes, "There's no use in being insurance rich, if it makes you cash poor".
- CBA allows the fire protection manager to speak with financial managers in terms and concepts that they can understand. This is important because at the end of the day most fire protection investments are not much different than other capital investments that the company makes, they all must show a reasonable probability of making or saving the company money. Most well chosen risk reduction investments will show a direct financial return from the reduced fire risk and possible indirect returns from improved personnel safety, process availability/reliability, or reduced environmental liabilities.

Fundamentals of Cost-Benefit Analysis

The application of CBA in the fire protection decision making process is essentially a marriage of traditional CBA methodologies and Quantitative Risk Assessment (QRA). There are numerous texts available on both of these topics. As such, this paper will focus on how the overall fire protection CBA process works and tips for applying it.

The following formula is the basis of the fire protection CBA process:

$$\textit{Annualized Risk (\$/yr)} = \textit{Frequency (events/yr)} \times \textit{Consequence (cost/event)} \quad [1]$$

In applying the CBA process, one first estimates the potential frequency of scenario occurrence and the potential financial consequences of the unmitigated or existing situation. The annualized risk, which is the product of these two factors, is in fact the annualized cost of doing business under the unmitigated or existing conditions.

If the annualized risk or cost of the scenario is deemed unacceptable one then identifies options for reducing it. Each option is evaluated to identify the potential frequency of occurrence and consequences that would remain following implementation of the option. These factors are then fed into Formula 1 to calculate the residual risk of the scenario, which is then applied to Formula 2. The annualized risk for the mitigated scenario is then calculated and input into the following formula:

$$\textit{Annualized Risk}_{\textit{unmitigated}} - \textit{Annualized Risk}_{\textit{mitigated}} = \textit{Risk Reduction (\$/yr)} \quad [2]$$

The risk reduction achieved is then taken as the annualized return on the investment. This value, along with the cost of implementation, can then be used to determine if the implementation of the option would meet the corporation's financial targets using its preferred means of evaluation (Pay Back Period, Return on Capital Employed, Internal Rate-of-Return, etc...).

In applying CBA in the fire protection decision making process, it is useful to establish a few guidelines:

- Some decisions should be exempted from having to meet CBA requirements. This may include:
  - Regulatory required protection.
  - Extreme risks, as identified by the applying the corporation's PSM Matrix.

- Basic fire protection features that act to reduce the frequency of significant fires (i.e., placement of fire extinguishers) or prevent uncontrolled fire spread through the entire facility (i.e., the fire water system). In many cases, these basic fire protection features will form the basis upon which the potential frequency and consequence estimates are developed.

- CBA can be used in evaluating life safety issues. However, this requires that one place a dollar figure on human lives (US government typically utilizes a value around $6.1 MM/life) in estimating the consequences of a scenario, which some companies are uncomfortable with.
- Life safety issues need to be addressed through fire frequency modification or by physically separating the humans from the hazard. Active fire control systems (water spray system, foam systems, etc...) tend to have very little life safety value in the process environment due to the size and rapidity of most events at their onset.
- Consequences should include potential equipment/structure damage, loss of stock and business interruption. Other concerns such as environmental impact and corporate image could also be included, if a value can be readily applied to them.
- Where feasible, the value of indirect benefits should be considered when evaluating the risk reduction achieved by implementing a fire protection option.
- The unmitigated scenario should always be compared against the corporation's PSM matrix to verify that its risks fall into a region of concern before moving further on in the process. If the situation is acceptable as-is according to the PSM matrix, no further evaluation is required.
- Don't spend a million dollars to answer a hundred thousand dollar question. The level of accuracy and effort (and thus cost) put into estimating potential fire frequencies and consequences should be commensurate with the risk on hand. In most cases, orders of magnitude accuracy can be tolerated and the needed data can be derived from:

  - Data from existing QRAs or other semi-quantitative risk analyses, such as Layers of Protection Analyses (LOPA).
  - Published industry data on:

    - Failure/fire frequencies.
    - Average losses per incident.

- The corporation or industry's historical experience.
- Basic consequence modeling using commercial software packages or accepted mathematical relationships.

Another important guideline that should be applied is that incremental improvements should have to stand on their own. In one example, it was determined that the annualized risk within a DCS building was too high and it was decided that gaseous automatic fire suppression should be considered. In this case, the fire detection system and the gaseous fire suppression system were looked at as incremental layers of protection because each system could significantly incrementally impact the risk profile of the building. As the study progressed it was found that:

- Smoke Detection - The frequency of reported fires in buildings protected with standard smoke detection was at least seventy percent lower than that of buildings without smoke detection according to reputable sources. It was further assumed that incipient or early detection systems would reduce the fire frequency even more. This data was then utilized to calculate the residual annualized risk that would remain following the installation of the smoke detection system. This residual risk was then compared against the risk of the unmitigated scenario to determine the value of the risk reduction achieved. In this case, the Rate of Return on Capital Employed (ROCE) methodology was utilized to evaluate the capital investments and it was found that the return was several times the corporation's minimum acceptable threshold.
- Gaseous Fire Suppression - The differences between average losses in protected and unprotected spaces computer spaces was identified and applied to the property and business values associated with the DCS building. This data was then utilized to calculate the residual annualized risk that would remain following the installation of the suppression system. This residual risk was then compared against the residual risk following installation of the smoke detection system to determine the value of the incremental risk reduction achieved. The Rate of Return on Capital Employed (ROCE) methodology was once again utilized to evaluate the capital investment. However, in this case the ROCE was found to be significantly lower that that of the corporation's minimum acceptable threshold.

As a result of this analysis, the corporation elected to install a smoke detection system within the DCS building and not to install fire suppression. Plant management was further sold on this choice based on the fact that incipient detection could sense some modes of impending computer failure (thermal breakdown of components), which would further improve the availability and reliability of the plant, in addition to providing a reduced risk from fire.

CASE HISTORY 1 - AN LNG PLANT

The LNG plant in question received natural gas via pipeline, refrigerated it or liquefied it, stored it in tanks and forwarded the product to market via ships. The fire suppression systems of the plant's first refrigeration train had primarily been laid out by the owner's design/build contractor.

The plant was interested in evaluating its fire suppression features, as they and their insurers had several concerns regarding the philosophies used in laying them out as well and about inconsistencies in coverage. The scope of this CBA project was to evaluate the need for improved monitor coverage and water spray protection.

At the start of the project, the plant was broken down into fire areas based upon its layout, process function and potential fire scenarios. The potential fire frequency for each area was developed from existing QRA data and loss estimates were developed using consequence modeling and industry accepted mathematical relationships. The consequence analysis took into consideration the properties of the process stream and both small and large leaks. The results for one area of the plant are shown below in Table 1.

Table 1 Fire Loss and Risk Estimates for One Area of Plant

Scenario	Status of Cooling Water Provisions	Worst-case Estimated Loss* (US$ MM)	Annual Risk** (US$ M/yr)
Area 1 (Liquid Phase)	Inadequate	291	16.3
	Adequate	10	0.6
	Reduction Achieved	281	15.7
Area 2 (Gas Phase)	Inadequate	391	237.8
	Adequate	24	14.6
	Reduction Achieved	367	223.2

* Worst-case for design fire scenarios only. Not related to QRA worst-case. This value includes estimated Property and Income Losses and is a weighted average (12mm leak is 8 times as likely as a 50 mm leak)

**Annual Risk = Estimated Frequency X Estimated Worst-case Loss

All values modified to protect client confidentiality

At this point of the analysis, it did not matter how the cooling water was applied, but that it was applied in a timely manner and in sufficient quantities to reduce the degree of damage sustained within the radiant heat affected zone of the fire. As it can be seen in Table 1, a significant reduction in annualized risk, thus a return on investment, could be achieved by providing adequate cooling water.

The residual risks in this case were on the order of:

- Area 1 (Liquid Phase)
  - Frequency: 10^{-3} fires/year
  - Consequence (Avg): US $2.0 MM (Property Damage Only)
- Area 2 (Gas Phase)

- Frequency: 10-4 fires/year

Consequence (Avg): US $ 106 M (Property Damage Only)

These frequencies and consequences would fall into the "Acceptable As-Is" or "As Low as Reasonably Practical (ALARP)" region of the most PSM risk matrices. Therefore, no further risk reduction was needed beyond providing adequate cooling in these areas.

The facility in question had a reasonable degree of fire detection, manning and emergency response capabilities. All of these combined to assure that the manual application of cooling streams could be initiated in this area in time to avoid significant damage, thus alleviating the need for automatic water spray systems. A CBA was then completed to identify the maximum dollar expenditure that could be sustained in providing the additional monitor coverage:

- US $1.4 MM based on an acceptable pay back period
- US $2.8 MM based on an acceptable ROCE

In either case, the cost of installing the needed monitors would be well below the maximum sustainable values and was deemed cost-effective.

CASE HISTORY 2 - ORE EXTRACTION PROCESS IN THE MINING INDUSTRY

The process in question received a weak water/acid mixture that had leached through a pile of ore and become copper rich. This weak acid solution was passed through a mixing chamber where it was combined with a hydrocarbon liquid (a kerosene - diesel weight material) that has a strong affinity for copper and then into an open top settling tank where the acid-water mixture and hydrocarbon would separate. This process was repeated several times to assure that as much of the copper was recovered from the weak acid solution before it was recycled back into the ore pile.

At the other end of the process, the now copper rich hydrocarbon was mixed with a strong acid solution that had a stronger affinity for the copper than the hydrocarbon. Once again, several passes through mixing chambers and open top settling tanks would be undertaken to assure that as much copper as possible was extracted from the hydrocarbon. At this point the hydrocarbon was returned to the initial settler circuit and the copper rich acid solution was forwarded to an electrical process through which the elemental copper would finally be recovered.

The client in this case had become concerned about their facilities because the industry had experienced several large loss fires involving this process over the years. This concern was further compounded by insurer pressure for improvements and the fact that the degree of fire protection at their sites varied greatly, ranging from almost nothing to fully automated foam fire suppression systems. As a result, the client elected to have a fire protection assessment conducted with an emphasis being placed on the use of CBA to assure that the protection at each site was commensurate with its criticality.

The first step in addressing this project was to better understand the fire hazards associated with the process, its fire history and to develop fire frequency estimates. This required a significant degree of research using trade group literature and several references and studies that the client possessed. As a result of this research it was found that:

- The fire frequency at each site fell into the range of 10^{-4} to 10^{-5} fires/year, very low by hydrocarbon industry standards. In most cases, the hydrocarbons in use at these facilities are handled at temperatures that are well bellow their flash points.
- The majority of the industry's large loss fires had originated in and propagated through the piping trench systems that interconnected the settling tanks. This was an interesting fact, as most fire suppression, where it was provided, tended to focus on the extinguishment of settling tank fires.

- The primary contributing factor to the origination and propagation of fires within the pipe trenches was the presence of hydrocarbon liquids and residues on the floor, walls, piping and electrical equipment within the trenches. In many cases, these liquids and residues extended out of the trenches and up the sides of the settlers, thus providing a continuous path of combustibles between the trenches and the liquid surface of the settlers.

The second step in addressing this project was to develop a standardized method to estimate potential losses for each facility. In doing so, event trees were developed and rules for estimating the probability of fire spread between individual settlers and trains of settlers were developed based upon:

- Siting and layout
- Drainage and pipe trench configuration
- Availability of onsite and mutual aid emergency responders
- Fixed fire water and fire suppression systems
- Actual and projected copper production rates, based on the corporation's 30-year business plans

A blended loss estimate, considering minor fires and catastrophic fires, was developed for each of the sites surveyed. These loss estimates, along with the estimated fire frequencies were then utilized to estimate the annualized risk from fire of each facility. These estimated risks were found to be on the order of US $1,000/yr at each facility. This figure is very low compared to that tolerated within the hydrocarbon processing industry were it is not uncommon to see annualized risks of US $ 3,000 - 30,000/yr accepted at the ALARP level.

As indicated above, the risks imposed by fire at these facilities is very low and would in fact receive little, if any attention in the hydrocarbon industry. However, the objective of this project was not only to identify the level of risk, but was also to determine the level of fire protection warranted based on a CBA.

Several fire preventive and fire suppression options were considered for each site. In order to simply the CBA process it was decided that it would be more efficient to calculate the maximum installation costs that could be sustained while meeting selected financial thresholds. In doing so, the matrix labeled as Table 2 was developed.

Tables 2 and 3 are arranged so that incremental measures can be individually evaluated. Using Table 2 as an example, this table would be interpreted to convey the following:

- Case 1/Condition B - Implementation of the fire preventive Option B would be cost effective if it cost no more than US $12,500.
- Case 3/Condition A - Implementation of Advanced Fire Suppression Option 1 would be cost effective if it cost no more then US $4,900. The effectiveness of this option assumes that:

  - The Basic Fire Suppression Upgrade of Case 2 was implemented and that this implementation did not cost more than US $2,000.

- Case 3/Condition B - Implementation of Advanced Fire Suppression Option 1 would be cost effective if it cost no more then US $180. The effectiveness of this option assumes that:

  - Fire Preventive Option B was implemented and that this implementation did not cost more than US $ 12,500, and
  - The Basic Fire Suppression Upgrade of Case 2 was implemented and this implementation did not cost more US $ 2,500.

Table 2 Maximum Expenditure Based on Net Positive Value Criteria

Case No.	Suppression Options	Max. Expenditure Returning an Acceptable NPV		As Compared to Case ...
		A	B	
1	Existing Fire Protection	N/A	12,500	N/A
2	Basic Fire Suppression Upgrade	2,000	2,500	1
3	Advanced Fire Suppression Option 1 plus Basic Upgrade of Case 2	4,900	180	2
4	Advanced Fire Suppression Option 2 plus Basic Upgrade of Case 2	10,600	7,100	2
5	Advanced Fire Suppression Option 3 plus Basic Upgrade of Case 2	4,100	6,000	2

Modifying Conditions (Fire Preventive Options)

A = Existing Conditions

B = With Potential Improvements

All values and NPV requirements modified to protect client confidentiality

The conclusions drawn from this analysis and tables were that:

- The greatest return on investment would come from the implementation of Fire Preventive Option B, which essentially called for improved housekeeping of hydrocarbon spills and residues. This option could not only be employed on a one-time basis (cost not exceeding that stipulated in Table 2), but could also be implemented on an annual basis provided the cost did not exceed the annual risk reduction it would provide.

- Implementing the fire suppression options of Cases 2 through 5 would not be cost effective, as the cost of doing so would be many times the maximum limits called for in Tables 2 through 3.

Towards the end of this project, the client requested that a comprehensive fire protection guideline be written for the corporation so that they could implement the findings of the CBA across all of their sites. However, they expressed concern that they were not comfortable with the limited level of fire protection that the CBA called for at a few of their more important sites due to regional political reasons and financial concerns outside the realm of the CBA.

This concern was addressed by providing basic fire protection options for sites of unusual importance within the comprehensive guideline. A one-size fits all fire protection scheme was not called for at these sites due to the international nature of the company, which meant that the lowest cost option at one site may not necessarily be lowest cost option at another. Instead, a performance specification was written that described the minimum level of suppression capability to be provided while at the same time placing limits on these capabilities. In this manner:

- The site could select a fire suppression option that would meet the minimum performance specification for the least cost.
- The corporation was assured that the sites would not expand their fire suppression capabilities beyond that called for. Therefore, preventing excessive costs.

CONCLUSION

The use of CBA in the fire protection decision making process can help companies to identify inadequately protected risks and to avoid fire protection options that are inappropriate, or too costly, for the risks to be mitigated. The CBA process further helps to bring the fire protection decision making process from one based on emotions or guessing to one that is based upon reasonable estimates of risk.

In most cases, the CBA process can be completed using existing propriety and/or published data making it very cost-effective to employ, particularly when its cost is compared to that of installing unnecessary fire protection.

AUTHOR'S SHORT BIOGRAPHY

Stephen W. Haines, P.E. is a principal with RRS Engineering, where he provides fire protection and process safety consulting services in the oil, chemical, LNG, power generation, mining and other heavy industries. Prior to joining RRS Engineering, Mr. Haines was the senior corporate fire protection engineer for Sunoco, Inc. In this capacity the presenter was responsible for all fire protection matters in corporation's refining, chemical, pipeline and terminal operations. Prior to joining Sunoco, Mr. Haines worked as a career municipal fire chief and spent an additional six years as a consultant in the refining and chemical industries.

How Can Simplification Enhance Process Safety?

C. P. Ganatra
W. L. Gore & Associates, Inc.
Elkton, MD 21921
cganatra@wlgore.com

Abstract:

This paper addresses the question of how simplification through material standardization in the area of gasket material can enhance process safety. The paper describes the process used by W. L. Gore to assess whether standardization of process piping gaskets increases safety. Testing for critical performance factors is central to that assessment process. After establishing where gaskets are the "weak link" in a process piping system, it explains the relationship between reliability and process safety. We first focused the investigation on manufacturing and maintenance/reliability groups, where simplification is an important stated goal. Often, the piping gaskets, which serve as an important link in the reliability/safety effort, get a lesser amount of attention. Next, the paper describes the testing work done to objectively determine impact to process reliability. Going further, the paper describes the gasket properties that are required to achieve process reliability and safety. Learnings on key properties indicated the difficulties of meeting the varied challenges of sealing across different pipe classes – where pipe classes have distinctly different requirements. The paper also explains how findings were confirmed in the primary process industries. The paper provides results of third party tests on several of the most commonly used gasket materials. This testing reveals how the inherent properties of those materials compare. The types of tests discussed are those which measure resistance to blow out, torque retention (for maintaining seal tightness), as well as basic strength tests. Included among the materials examined were PTFE and ePTFE based products that offer the best balance of properties for consideration as simplification candidates. A broad and thorough investigation of comparative properties was conducted, and is contained in the paper. Finally, and in support of the off-line testing, real-world case studies are provided to connect reliability and process safety with the test findings.

9th Plant Safety Symposium
AIChE Spring Meeting
Houston, TX
April 22-26, 2007

Analyzing Hazards of Sour Water Spills

J. L. Woodward, Ph.D.
Baker Engineering and Risk Consultants, Inc.
3330 Oakwell Court, Suite 100
San Antonio, TX 78218
(210) 824-5960

Abstract

This paper illustrates evaluating the risk of a large spill of sour water by modeling the plume of hydrogen sulfide and ammonia to toxic limits. This modeling uses chemical equilibrium constants and rigorous thermodynamic formulas, coupled with dispersion modeling to show how the plume contours of H_2S and of NH_3 vary with initial concentrations and pH.

Introduction

Sour water is developed in petroleum refining as a byproduct in hydrogenation plants that remove sulfur from petroleum products. Sour water in petrochemical plants can contain a significant concentration of hydrogen sulfide and ammonia. Design of sour water storage tanks, including location and dike area, should account for the hazards of a large leak. The hazard of a large spill of sour water can be evaluated by applying dispersion analysis to predict the contours of both H_2S and of NH_3 plumes out to ppm concentrations. The driving force for vaporization is the vapor pressure of the toxic gases. Fortunately, the vapor pressure of H_2S and of NH_3 is substantially reduced by the ionization of the molecular species to the ionic form (hydrosulfide and ammonium ions). The equilibrium concentration of these ions is a strong function of pH since a hydrogen ion is lost in forming the hydrosulfide ion and is gained in forming the ammonium ion. This suggests that pH adjustment can be an effective mitigation measure for a spill event.

Example of Sour Water Storage Tanks

A typical large sour water storage tank is a low-pressure cone-roof cylindrical tank with a diameter of 40 to 60 feet, cylinder height 40 to 50 feet. Sour water can come from the process warmed to around 100 °F (37.8 °C). A range of concentrations of H_2S and NH_3 are evaluated here as listed in Table 1 labeled as a base case and minimum and maximum H_2S cases.

To emphasize the potential hazard posed by sour water storage tanks, a sensitive area is assumed to be located less than 50 ft from the sour water dike, as illustrated later.

Table 1. Range of Sour Water Concentrations Evaluated

Component	Minimum H_2S Case, mole %	Base Case Mole %	Maximum H_2S\ Case, mole %
NH_3	0.70	0.60	0.35
H_2S	0.05	0.15	0.40
Water	99.25	99.25	99.25

Vapor/Liquid Equilibrium

Edwards, Maurer, Newman, and Prausnitz [1] provided the definitive work on the vapor/liquid equilibrium for dilute aqueous solutions containing both a weak acid and a weak base. The defining relationships are given here, and the coefficients are tabulated in Appendix A.

The vapor/liquid equilibrium relationship, Eq. 1, gives the vapor concentration of H_2S and NH_3 in terms of the mole fraction of a component, C, in the vapor phase, y_C, the vapor-phase activity coefficient, φ_C, the total pressure, P, the molal concentration in the liquid phase, m_C, the liquid phase activity coefficient, γ_C, and the Henry's Law coefficient $H_C(T,P)$ as a function of liquid temperature, T, and system pressure, P. The other terms in the exponent argument are the partial molal volume at infinite dilution, $v_{C\infty}$, in cm^3/gmole and the gas constant, R in gmole units (8.3143 J/(gmole ^oK) and the saturation pressure of water, P_{Wsat}, (6555.7 Pa at 100 ^oF) The relationship is:

$$y_C \varphi_C P = m_C \gamma_C H_C(T, P_{Wsat}) \exp\left[\frac{v_{C\infty}(P - P_{Wsat})}{RT}\right] \quad (1)$$

The molality, m_C is defined as the moles of C per kg of water as listed in Table 2. The vapor pressure of saturated water at the initial temperature is 6555.7 Pa. The activity coefficient in the liquid phase is given in terms of the self-interaction coefficients, β_C^0

$$\ln \gamma_C = 2\beta_C^0 m_C \quad (2)$$

Table 2. Sour Water Cases in Molality Units

Component	Minimum H_2S Case, mole/kg water	Normal Case Mole/kg water	Maximum H_2S\ Case, mole/kg water
NH_3	0.3915	0.3356	0.2517
H_2S	0.0280	0.0839	0.1678

The coefficients correlating liquid activity, β_C^0, are correlated as a function of temperature by:

$$\beta_C^0 = E + \frac{F}{T} \quad (3)$$

with coefficient values in Table A.1 in Appendix A by Edwards et al (ibid).

The partial molar volume, Henry's Law coefficients and the other thermodynamic values needed for H_2S and NH_3 in dilute solution are given by Edwards et al (ibid). Table A.2 in Appendix A lists the partial molar volume as a function of temperature.

Henry's Law constants are given by the semi empirical equation:

$$\ln H = \frac{B_1}{T} + B_2 \ln T + B_3 T + B_4 \qquad (4)$$

with the coordinate values given in Table A.3 of Appendix A.

At the operating temperature, substituting the molal concentration of undissociated dissolved components from Table 2 into Eq. 1 gives an upper limit to the vapor concentration of H_2S and NH_3. This is shown in Table 3. Of the terms in Eq. 1, the liquid-phase activity, φ_c, the vapor phase activity, γ_c, and the exponential term are near unity. Since Henry's law coefficient for H_2S, H, is a high value, the product m_C H exceeds unity for the base case and the maximum H_2S case. This can be interpreted as predicting a partial pressure of H_2S exceeding atmospheric pressure, just as the vapor pressure of pure H_2S is larger than atmospheric at 37.8 °C. The practical result is a prediction of nearly 100 mole % H_2S in the vapor for these two cases. This result is modified substantially by considering dissociation in the liquid phase.

Table 3. Theoretical Vapor Concentrations Using Henry's Law With Undissociated Dissolved Gases

Case	Component	Molal Conc. m_i	H_0 Atmos/gmole	Undissociated Partial P, atm.
Minimum H_2S	NH_3	0.3915	0.0291	11.4E-3
	H_2S	0.0280	12.83	0.359
Base Case	NH_3	0.3356	0.0291	9.77E-3
	H_2S	0.0939	12.83	>0.99
Maximum H_2S	NH_3	0.2517	0.0291	7.33E-3
	H_2S	0.1678	12.83	>0.99

Dissociation into ionic moieties reduces the concentration of molecular H_2S and NH_3 in the liquid phase and hence the vapor-phase concentrations as well. Eq. 1 applies also after dissociation is taken into account, so the calculations leading to Table 3 are repeated later with decreased liquid concentrations.

Dissociated Vapor-Liquid Equilibrium

The concentration of H_2S and of NH_3 in the vapor phase is strongly decreased by ionization in the liquid phase. This process effectively reduces the concentration of molecular species in solution. The pH of the aqueous phase is the controlling factor for the concentrations of H_2S and NH_3. The treatment of dissociation continues following Edwards, et. al. (ibid) and also Zemaitis et al[2]. The following ionization equations occur:

$$NH_3 + H_2O = NH_4^+ + OH^- \tag{5}$$

$$H_2S = H^+ + HS^- \tag{6}$$

$$H_2O = H^+ + OH^- \tag{7}$$

$$HS^- = H^+ + S^= \tag{8}$$

The corresponding equilibrium constants are defined in terms of the activity coefficients of the species:

$$K_1 = K_{NH3} = \frac{a_{NH4^+} a_{OH^-}}{a_{NH3} a_{H2O}} \tag{9}$$

$$K_2 = K_{H2S} = \frac{a_{H^+} a_{HS^-}}{a_{H2S}} \tag{10}$$

$$K_3 = K_W = \frac{a_{H^+} a_{OH^-}}{a_{H2O}} \tag{11}$$

$$K_4 = K_{HS} = \frac{a_{H^+} a_{S^=}}{a_{HS^-}} \tag{12}$$

There are other equations that can be written, such as:

$$NH_3 + H^+ = NH_4^+ \tag{13}$$

$$NH_3 + HS^- = NH_4^+ + S^= \tag{14}$$

but these can be shown to be redundant since they can be obtained by adding two or more of the equations 5-8. For example:

$$NH_3 + H_2O = NH_4^+ + OH^-$$

plus $H^+ + OH^- = H_2O$

gives Eq. 13 and the equilibrium constant is K_1/K_3. This recognition significantly simplifies the algebraic solution.

The equilibrium constants are given as a function of temperature by:

$$\ln K = \frac{A_1}{T} + A_2 \ln T + A_3 T + A_4 \tag{15}$$

Table A.4 in Appendix A provides the coefficients for Eq 15, along with the value of K at 100 °F.

The constant K_4 is given (ibid) by an addition of equations to give:

$$K_4 = K_{HS^-} = 0.018K_3 \quad (16)$$

Mass balance provides three equations:

1) Total of NH_3: $m^0_{NH3} = m_{NH3} + m_{NH4^+}$ (17)

2) Total of H_2S: $m^0_{H2S} = m_{H2S} + m_{HS^-} + m_{S^=}$ (18)

3) Total of H_2O: $m^0_{H2O} = m_{H2O} + m_{OH^-} + m_{H^+}$ (19)

Electrical neutrality requires:

$$m_{NH4^+} + m_{H^+} = m_{HS^-} + 2m_{S^=} + m_{OH^-} \quad (20)$$

For dilute solution, the activity coefficients are related to the molalities by:

$$\ln a_i = 2\beta^0_i m_i \quad (21)$$

where the index applies to NH_3, NH_4^+, H_2S, HS^-, $S^=$, NH_4S^-, H^+, and OH^-. For water with dilute solutions:

$$\ln a_{H2O} = -(\beta^0_{H2O} m^2_{H2O} + m_{H2O})M_W \quad (22)$$

A simplification that is justified in the present case is to replace activities, a_i, by molarities, m_i' in gmole/liter. Hereafter, dropping the prime, understand m_i to be molar units. There are eight variables, the molecular species in the aqueous phase, m_{NH3}, m_{H2S}, m_{H2O}, and the ionic species, m_{NH4+}, m_{HS-}, m_{S-}, m_{H+}, m_{OH-}. There are four equilibrium relationships and Eqs. 17-20, and a solution is found as follows:

Solve for four ionic species from the equilibrium constants. Leave m_{H^+} to be specified. This is done by specifying pH which is defined as:

$$pH = -\log_{10} m_{H^+} \quad (23)$$

The remaining ion molarities are:

$$m_{OH^-} = \frac{K_3 m_{H2O}}{m_{H^+}} \quad (24)$$

$$m_{NH_4^+} = \frac{K_1 m_{NH3} m_{H2O}}{m_{OH^-}} = \frac{K_1 m_{NH3} m_{H^+}}{K_3} \quad (25)$$

$$m_{HS^-} = \frac{K_2 m_{H2S}}{m_{H^+}} \quad (26)$$

$$m_{S^=} = \frac{K_4 m_{HS^-}}{m_{H^+}} = \frac{K_4 K_2 m_{H2S}}{(m_{H^+})^2} \quad (27)$$

Substituting Eqs. 24-27 into Eqs. 17 and 18, gives the mass balance equations for ammonia and hydrogen sulfide as:

$$m_{NH3} = \frac{m^0_{NH3}}{1 + \frac{K_1}{K_3} m_{H^+}} \quad (28)$$

$$m_{H2S} = \frac{m^0_{H2S}}{1 + \frac{K_2}{m_{H^+}}(1 + \frac{K_4}{m_{H^+}})} \quad (29)$$

Binary interaction coefficients could also be applied for slightly greater accuracy, but the improvement is minor. Eqs. 24-29 are evaluated and plotted as a function of pH in Figure 1 and Figure 2. Figure shows that the molecular H_2S in solution remains essentially at the initial, undissociated value for acidic pH (pH <6) as would be expected for a weak acid. Similarly, since ammonia is a weak base, it remains at its initial, undissociated value in alkaline solutions (pH > 9). The HS^- ion is even more basic, and drops off in concentration below pH of 7.

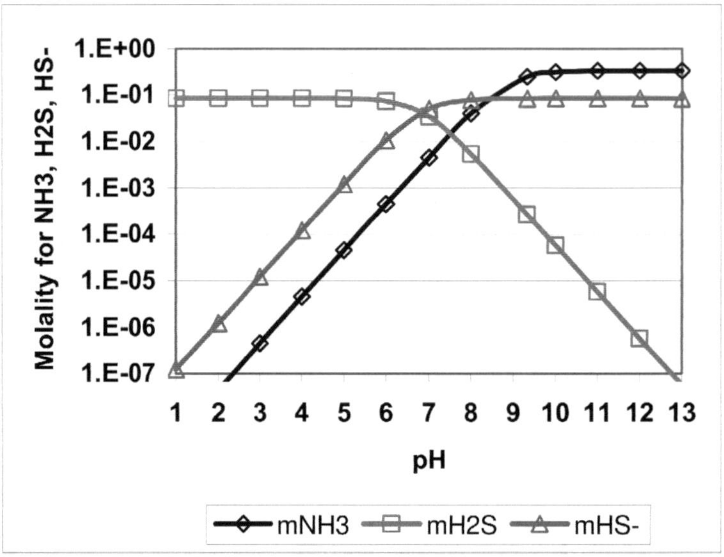

Figure 1. Concentration of Molecular NH_3 and H_2S and Ionic HS^- in Solution as a Function of pH at 100 ^oF for Base Case

For completeness Figure shows the concentration of the $S^=$ ion which ranges in value down to 10^{-22} at a pH of 1. The hydroxyl ion (OH^-) is linear on the log pH scale.

The electrical neutrality equation, Eq. 20, provides the final information to solve the set of dissociation equations. The left-hand-side (LHS) of Eq. 20 gives the concentration of positive ions, and the right-hand-side (RHS) adds the concentration of negative ions. These are plotted in Figures 3 to 5 against pH for the three cases. The point of intersection defines the solution pH and the concentration in the dissociated solution. For these example cases, the LHS curve changes little from case to case but the RHS curve moves decidedly to the left as the NH_3 concentration is reduced. The concentrations of the primary species at the point of equilibrium are listed in Table 4.

Using Eq. 1 with dissociated aqueous concentrations gives the equilibrium vapor mole fractions of NH_3 and H_2S in Table 5. The ammonia vapor-phase concentration is reduced by dissociation between 10 and 66% of the undissociated value. The hydrogen sulfide molecular concentration is reduced between 96.3 and 99.9%.

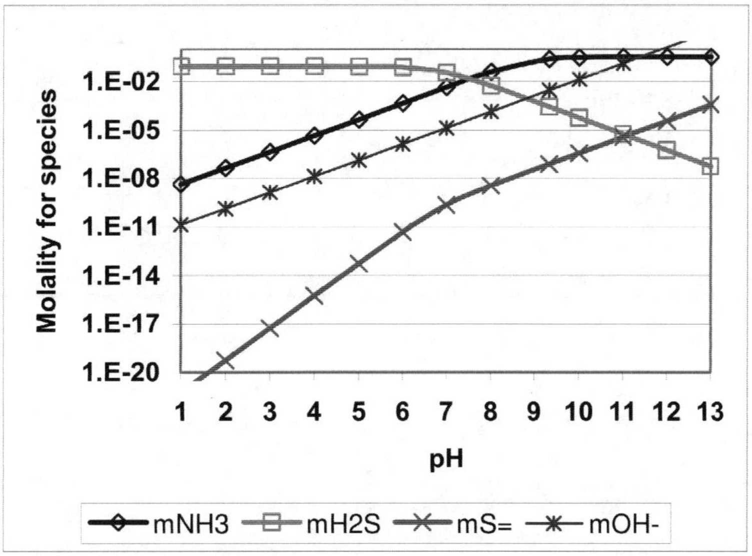

Figure 2. Concentration of Molecular and Two Other Negative Ionic Species in Solution at 100 °F as a Function of pH for Base Case

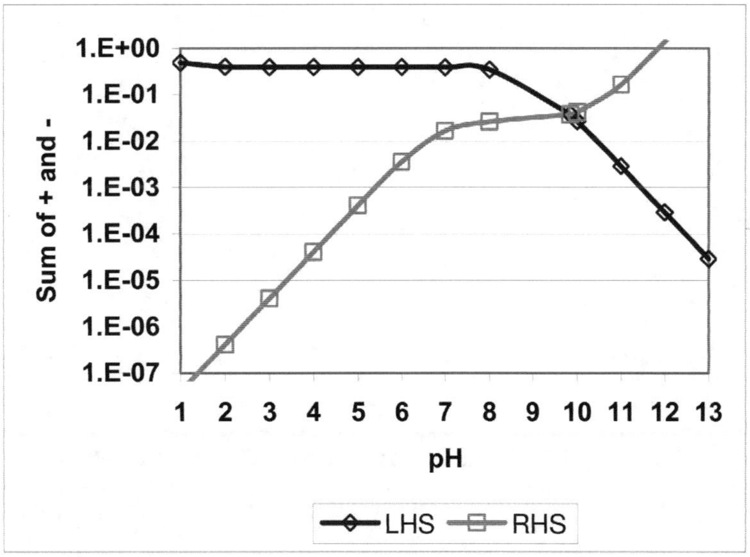

Figure 3. Finding pH by Point of Electrical Neutrality, Minimum H_2S Case

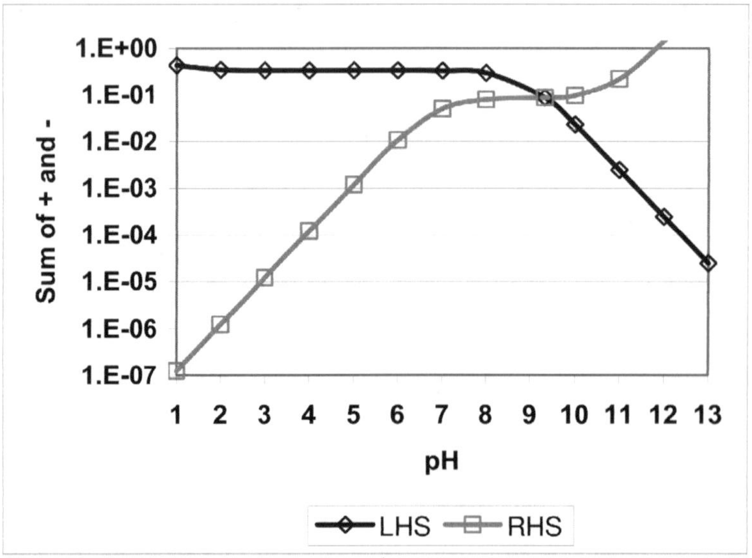

Figure 4. Finding pH by Point of Electrical Neutrality, Base Case

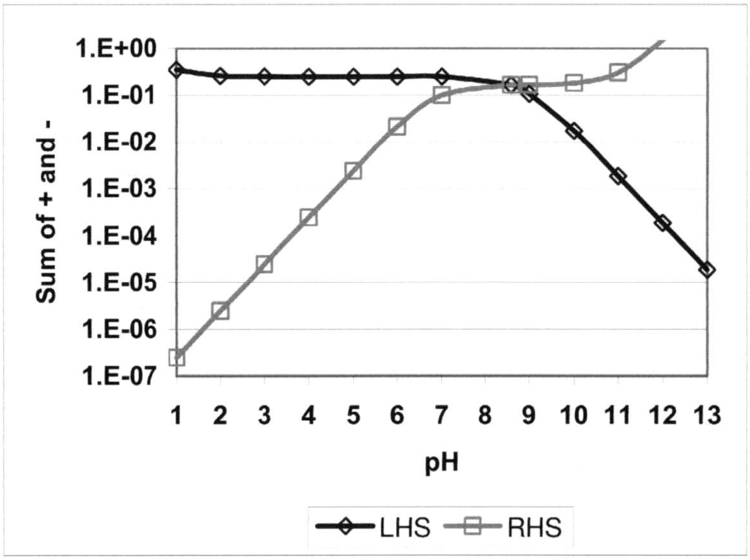

Figure 5. Finding pH by Point of Electrical Neutrality, Maximum H_2S Case

Table 4. Dissociated Aqueous Concentrations

Case	pH	Molal Concentration, gmole/liter			
		m_{NH3}	m_{H2S}	m_{NH4+}	m_{HS-}
Minimum H_2S	9.84	0.3539	2.747E-5	0.03760	0.0279
Base Case	9.33	0.2490	2.692E-4	0.0866	0.0836
Maximum H_2S	8.59	0.08632	2.923E-3	0.1654	0.1649

Table 5. Comparing Equilibrium Vapor Concentrations for Dissociated and Undissociated Solution Concentrations

Case	Component	Undissociated y_C mole fr	Dissociated y_C mole fr	Dissociated/ Undissociated %
Minimum H_2S	NH_3	0.01136	0.01026	10
	H_2S	0.353	0.00347	99.9
Base Case	NH_3	0.00972	0.00719	26
	H_2S	0.99	0.00340	99.7
Maximum H_2S	NH_3	0.00727	0.00248	66
	H_2S	0.99	0.00369	96.3

Toxic Limits

Published toxic limits for ammonia and hydrogen sulfide vapor are given in Table 6, along with the odor threshold. The Emergency Response Planning Guidelines (ERPG) are used here for contour plots. The ERPG 1 value is essentially a threshold of irritation or "mild transient health effects," ERPG2 is a threshold for injury or "irreversible, serious health effects," and ERPG3 is a threshold for "life-threatening health effects." These are defined for 60-minute exposures. Other values are the IDLH (Immediately Dangerous to Life and Health) defined by NIOSH (National Institute for Occupational Safety and Health) for 30-minute exposure, and STEL (Short Term Exposure Limit) defined for 15 minute exposure.

Table 6. Toxic and Odor Thresholds for Main Sour Water Components

Component	ERPG1 ppm 60 min.	ERPG2 ppm 60 min.	ERPG3 ppm 60 min.	IDLH ppm 30 min.	STEL ppm 15 min	Odor ppm
NH_3	25	200	1000	300	35	5-20
H_2S	0.1	30	100	100	15	0.003-0.02

Dispersion of Sour Water Plume

A large leak scenario is evaluated that drains the entire contents of a 44 ft diameter upright cylinder storage tank. The leak conditions, discharge rate, and duration are summarized in Table 7. Base case aqueous concentrations are used with the weather conditions in Table 8. The vapor concentrations listed in Table 5 are the initial concentrations in the vapor over the pool.

The plumes developed using dispersion modeling apply soon after a spill develops and while fresh sour water is spilling. The plumes will shrink as the contaminants are depleted by evaporation. For the example case, the inventory spilled is 1, 207,600 kg. The evaporation rate for the first hour averages 1.37 kg/s, allowing for evaporative cooling. This is a depletion rate of 0.408 mass %/hr, so the concentration of contaminants in the bulk liquid would be depleted only 0.408 %. If turbulence in the pool is very limited, a thin surface film would be depleted by a larger proportion, but the initial plume is likely to be representative over an exposure period of 30 to 60 minutes.

Predictions are made with the SafeSite$_{3G}$^{TM} model developed by BakerRisk[3]. A 30 minute averaging time is applied to the dispersion contours. Plan view contours are evaluated up to the breathing height of 2 m at an elevation of maximum downwind distance. The discharge fills the dike area except for the area occupied by the non-damaged sour water tank. Dispersion plumes are plotted with H_2S contours for the Base Case water concentrations in Figures 6 through 8. At ground level, the ERPG3 and ERPG2 contours extend about 30 m from the leak point. This distance is enough to engulf nearby buildings if they are located near the sour water storage tanks and dikes as illustrated in Figure 6. The very low concentration ERPG1 contours extend much farther as seen in the side-view plot, Figure 7. Figure 7 illustrates that the plume centerline becomes essentially level, indicating neutral buoyancy. A near-field side-view plot in Figure 8 illustrates that the plume spills over the dike and then lifts off slightly until the lower edge of the plume grows faster than the plume centerline rises. This explains a shallow arc on the bottom edge of the side-view plume in Figure 8. The shallow arc leaves a gap in the plan-view plots that is bridged by a straight line in Figure 6.

Table 7. Discharge Conditions for Near Rupture of Sour Water Tank

Temp, °F	Head P, psig	Level, ft	Hole Diam, ft (m)	Bottom of Hole, ft	Ave.Dischg. Rate, kg/s	Blow Down Duration, s
100	0	30	6.0(1.83)	0	12,540	96

Table 8. Dispersion Modeling Conditions

Time	Wind Speed, mph (m/s)	T_{amb} °F	Rel. Hum. %	Stability Class	Solar Rad., W/m^2
Summer Day	7.5 (3.35)	90	70	D	500

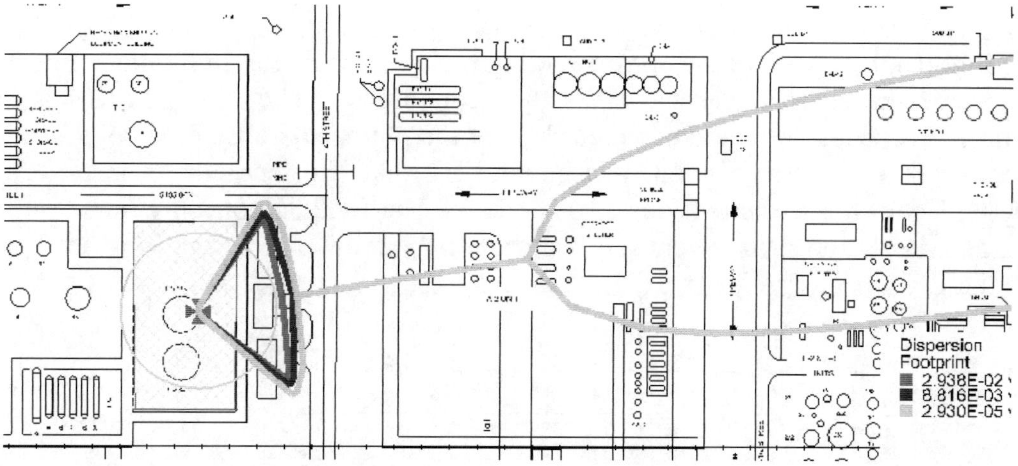

Figure 6. Plan-View Ground Level H_2S Contours for Summer Day Weather, Base Case [*Red= 100 ppm, Blue= 30 ppm, Green=0.1 ppm]*

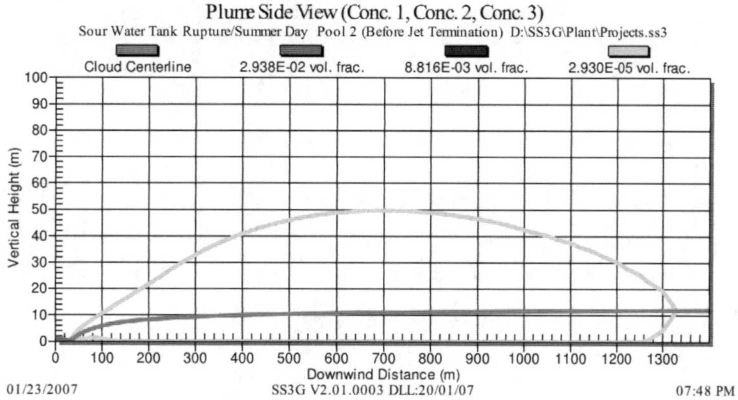

Figure 7. Side-View Plot of H_2S Contours for Summer Day Weather, Base Case

Red= 100 ppm, Blue= 30 ppm, Green= 0.1 ppm

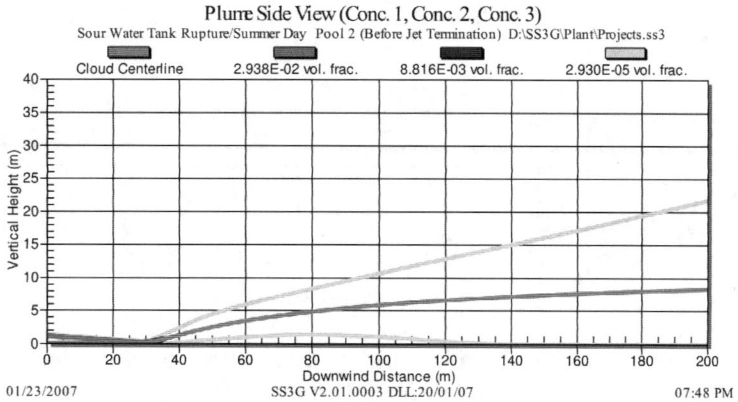

Figure 8. Near Side-View Plot of H_2S Contours for Summer Day Weather, Base Case [*Red= 100 ppm, Blue= 30 ppm, Green= 0.1 ppm]*

Figure Figures 9 and 10 are the corresponding plan view plot for ammonia concentration contours for the same release and weather conditions. The high ERPG3 ammonia concentrations occur only near the centerline of the plume. The ERPG2 contour extends 32 meters from the release point, just beyond the dike, but enough to pose a threat if nearby buildings are sufficiently open or porous. The ERPG1 concentration contour for ammonia extends about 50 meters and is centered about the plume centerline in Figure 10 just as the ERPG1 profile for H_2S is in Figure 8.

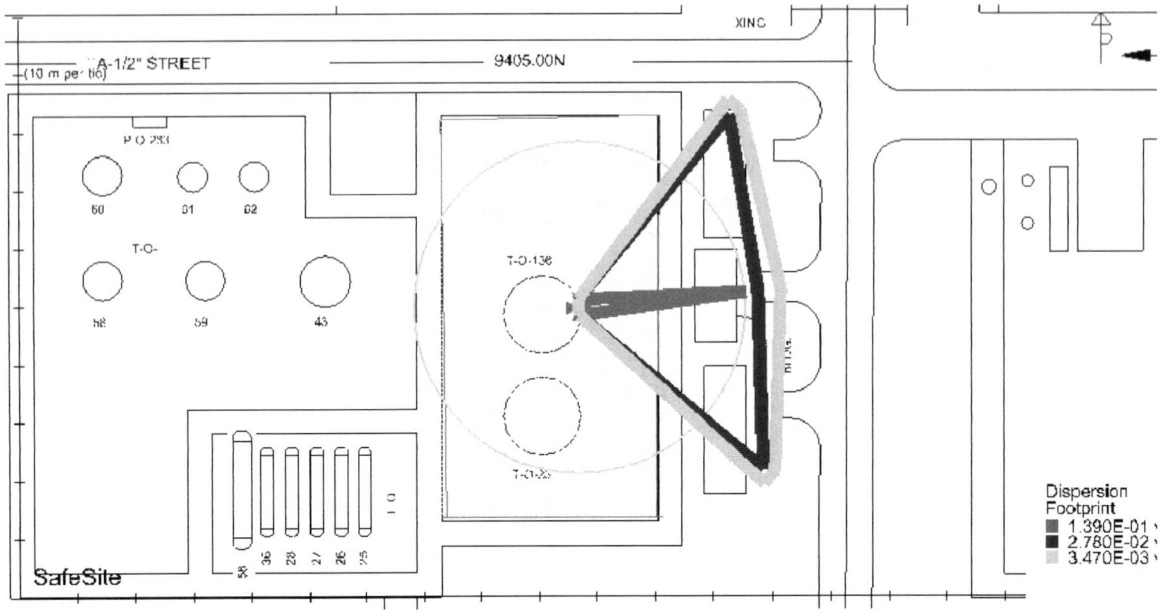

Figure 9. Plan-View of 2-m High NH_3 Contours for Summer Day Weather, Base Case [*Red 1000 ppm, Blue=200 pp), Green=25 pp]*)

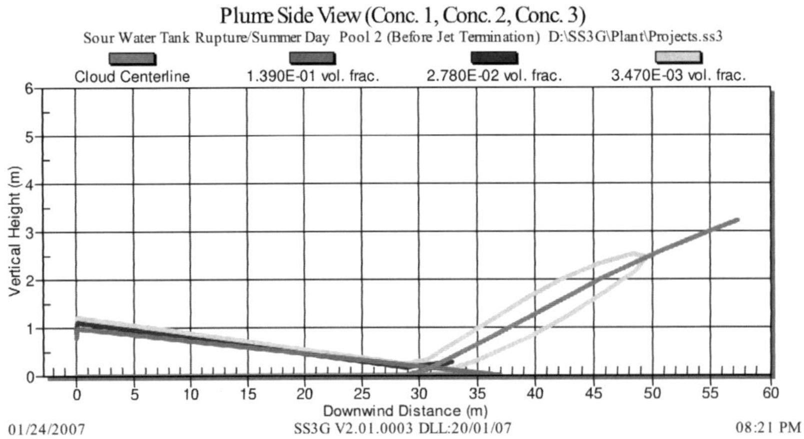

Figure 10. Side-View Plot of NH_3 Contours for Summer Day Weather, Base Case

Red= 1000 ppm, Blue= 200 ppm, Green= 25 ppm

For the maximum H_2S water concentration case, the H_2S contours at 2 m elevation extend slightly longer than for the base case (up to 50 m to the ERPG2 concentration). This is shown in Figure 11 at the same weather conditions as before. The outer green line in Figure 11 pinches in to form a narrow "neck" because of the same trade-off between a rising centerline and a growing plume width (lower edge) that produced a "gap" in Figure 6. Elevated platforms and building air intakes at elevations below about18 meters could experience H_2S concentrations above the ERGP2 concentration up to 200 m away as seen in the side-view plot, Figure 12.

The ammonia plume contours for the maximum H_2S case are shorter than for the base case because the initial vapor concentration is considerably weaker. Figures 13 and 14 illustrate this for both side-view and plan-view plots.

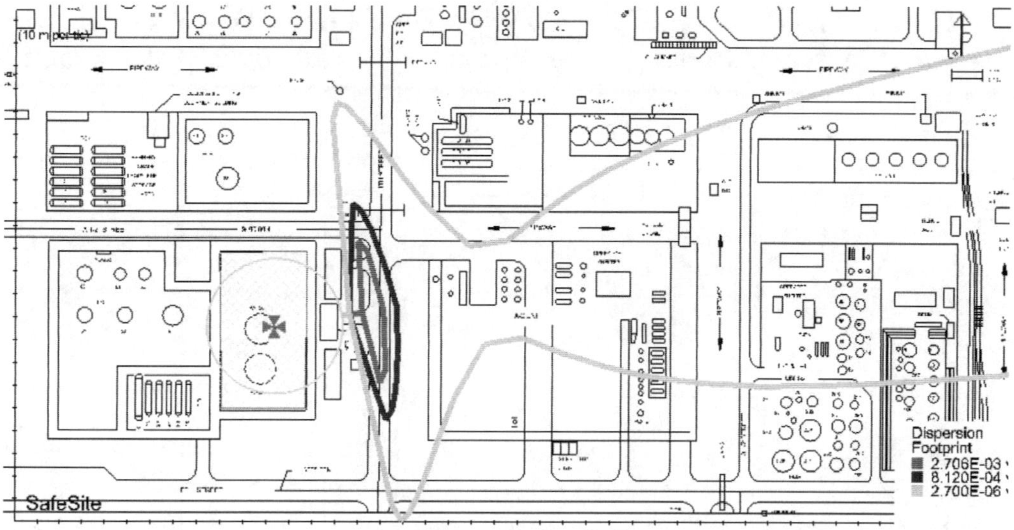

Figure 11. Plan View of 2-m High H_2S Contours for Summer Day Weather, Maximum H_2S Water Conc. [*Red= 100 ppm, Blue= 30 ppm, Green= 0.1 ppm]*

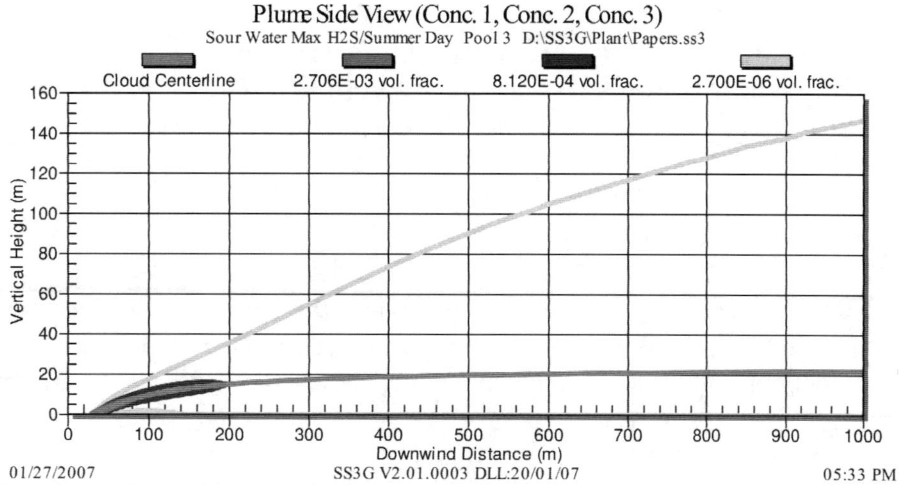

Figure 12. Side View o High H_2S Contours for Summer Day Weather, Maximum H_2S Water Conc. [*Red= 100 ppm, Blue= 30 ppm, Green= 0.1 ppm]*

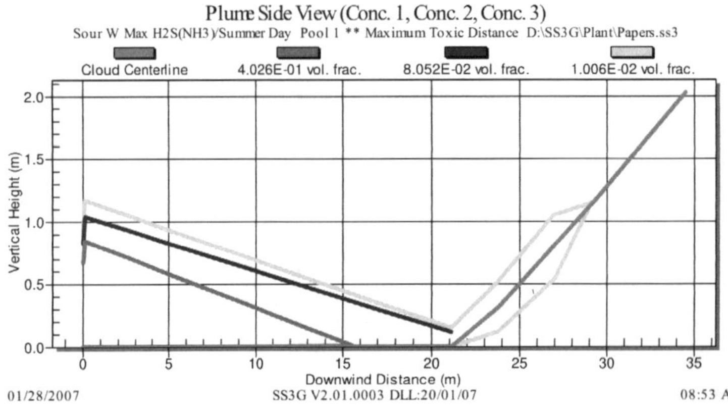

Figure 13. Side View of NH_3 Contours for Summer Day Weather, Maximum H_2S Water Conc. [*Red= 1000 ppm, Blue= 200 ppm, Green= 25 ppm]*

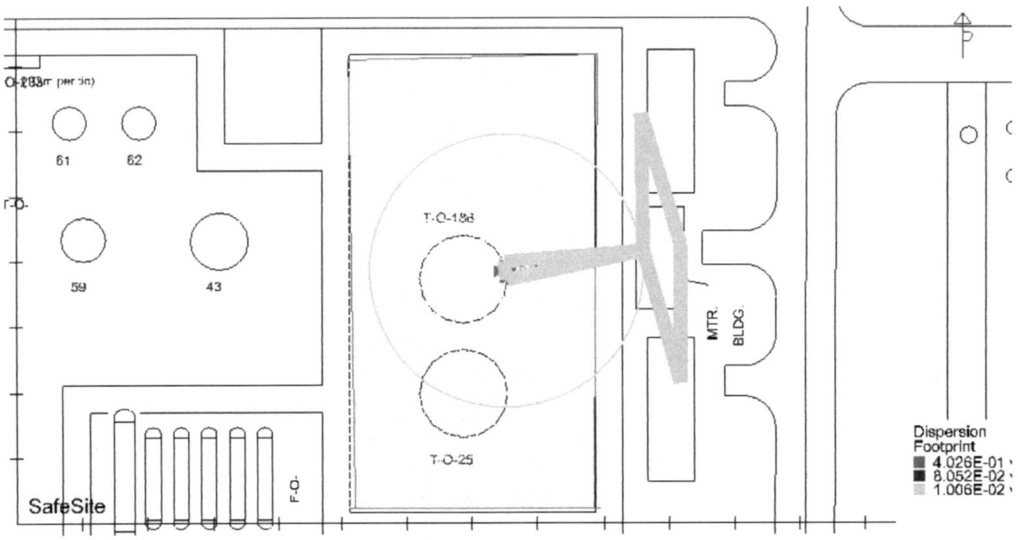

Figure 14. Plan View of NH_3 Contours for Summer Day Weather, Maximum H_2S Water Conc. At 0.5 m Elevation [*Red= 1000 ppm, Blue= 200 ppm, Green= 25 ppm]*

Conclusions

A large release from one of the sour water tanks could develop concentrations of H_2S and NH_3 above toxic limits over the area of the dike confinement. For the range of aqueous concentrations considered here, the plume contours to the ERPG2 concentration are found to be approximately the same size for both H_2S and NH_3. Both plumes extend to ERPG2 values for about 30 meters from the source for base-case water concentrations and up to 50 m for H_2S with the maximum H_2S case. This finding justifies a buffer distance between sour water storage tanks and occupied buildings.

Considering volatility alone (Henry's law constants) the vapor/liquid concentration for H_2S is considerably higher than that for ammonia. That is, hydrogen sulfide can have a lower concentration than ammonia in the liquid phase and yet have a much higher concentration in the vapor phase. Fortunately, dissociation has a strong influence in reducing the liquid-phase concentrations, and consequently, the vapor-phase concentrations for both H_2S and NH_3. This is even more pronounced when these two components are present together. In the present case, comparing the theoretical vapor concentration for undissociated H_2S with that for dissociated H_2S finds that ionization is responsible for a reduction of 96 to 99.9%. Vapor concentrations of ammonia, in the examples considered here, are reduced between 10% and 66%.

A pH between 8.6 and 9.8 is predicted for the example aqueous concentrations considered here. Such pH values, on the basic side, reduce the vapor concentration of the acidic, H_2S, more so than the basic, NH_3. Since H_2S is more toxic than NH_3, it is desirable to ensure that the pH remains alkaline. Adding bicarbonate or caustic solution to a sour water spill is, therefore, a prudent mitigation measure.

References

1. T. J. Edwards, G. Mauer, J. Newman, and J. M. Prausnitz, "Vapor-Liquid equilibria in multicomponent aqueous solutions of volatile weak electrolytes", AIChE J, 24 (6) 966-976, 1978.

2. J. F. Zemaitis Jr., D. M. Clark, M. Rafal, N. C. Scrivner, "Handbook of aqueous electrolyte thermodynamics", Design Institute of Physical Property Data (DIPPR), AIChE, NY, NY, 606-647, 1986.

3. Baker Eng. and Risk Consultants, Inc., "SafeSite$_{3G}$™ Theory Manual" Version 1.04, Apr. 28, 2005.

4. S. W. Brelvi and J. P. O'Connell, "Corresponding states correlations for liquid compressibility and partial molal volumes of gases at infinite dilution in liquids", AIChE J., 18, 1239, 1972.

Appendix A Tabulated Coefficients for Vapor/Liquid Relationships from Edwards et. al.

The coefficients used in the thermodynamics correlations for sour water dissociation are tabulated below.

Table A.1. Correlation Coefficients for Liquid Activity, β_C^0

Component	E	F
H_2S	-0.0260	12.29
NH_3	-0.2106	61.56

Table A. 2. Partial Molar Volumes in Dilute Aqueous Solution, $v_{C\infty}$ in cm^3/gmole(Edwards et al calculated from correlation of Brelvi and O'Connell[4])

Temp, ° C	NH_3	H_2S
0	28.7	34.8
50	30.0	36.5
100	33.9	41.1
150	40.4	48.4

Table A.3. Coefficients for Henry's Law Constants for Weak Water Solutions

Component	B_1	B_2	B_3	B_4	Range of Validity, °C
H_2S	-13268.8	-55.0551	0.0595651	342.595	0-150
NH_3	-157.552	28.1001	-0.049227	-149.006	0-150

Table A.4 Regression Coefficients for Ionic Equilibrium Constants

Component	A_1	A_2	A_3	A_4	K Value at 100 °F
NH_3, K_1	-3335.7	1.4971	-0.0370566	2.76	1.85056E-05
H_2S, K_2	-12995.4	-33.5471	0	218.599	1.45642E-07
H_2O, K_3	-13445.9	-22.4773	0	140.932	2.49444E-14

Resilient Engineered Systems: The Development and Demonstration of an Inherent System Property

Susan Mitchell, Mary Kay O'Connor Process Safety Center, Texas A&M University Chemical Engineering, 3122 TAMU, College Station, TX, smitchell@tamu.edu

M. Sam Mannan, Mary Kay O'Connor Process Safety Center, Texas A&M University Chemical Engineering, 3122 TAMU, College Station, TX, mannan@tamu.edu

ABSTRACT

Engineered systems today must be protected from a variety of hazards including engineering failures, terrorism effects, and natural disasters. With the development of increasingly complicated engineered systems has come a rise in the difficultly of determining all possible failure modes of these systems. Therefore, it is desirable to design and develop systems that not only have well-defined failure modes but also can withstand or endure some unanticipated failures without catastrophic loss of life or property. To aid engineers in approaching this challenge, the concept of a "resilient" system has been developed.

The concept of system resilience has been developed by taking inspiration from material resilience from the materials science field. A definition has been proposed that defines system resilience as the amount of energy a system can absorb before that system reaches its yield stress. System properties such as input energy, input exergy, and exergy destroyed, have been used to develop specific variables for characterizing system resilience. System characterization curves (similar to material stress-strain curves) have been developed to aid in visualizing the concept of resilience for a variety of different systems and system changes. These curves also help with identifying specific yield stress points for each system.

This paper will present the background information which inspired this research, show the process used to develop both the system resilience definition and the specific variables used to characterize system resilience, as well as show examples of system characterization curves for specific process systems including a steam pipe and a water pump. Additionally, future opportunities, directions, and questions for this research will be explored.

INTRODUCTION

As engineers design and operate more complex systems, the ability to predict and protect systems from all possible failures becomes more daunting. Within chemical engineering, we have sophisticated tools to help assess the probability of system failure, however most of these tools rely on the ability to predict all possible failure modes of the system. Due to the complexity and interconnectivity of modern systems, there are limits to our ability to understand their operation. With the added threat of terrorism, it becomes more desirable to design systems that can not only operate safely under expected conditions, but also can tolerate variable fluctuations and other changes to the system state without experiencing catastrophic failure. Therefore, this research seeks to assist with this challenge by defining the concept of system resilience, developing quantitative methods to allow resilience to be assessed and compared, and then incorporating these methods into the design process.

RESILIENCE

Resilience was chosen due to its incorporation of both the idea of systems being strong, or robust, against failure but also exhibiting flexibility so that if the system does fail it can in a "graceful" manner – with warning and in a gradual, safe manner. Resilience can be semantically defined as "the capability of a strained body to recover its size and shape after deformation caused especially by compressive stress[1]"

The attention paid to resilience in research settings has increased dramatically in recent years. Much of this research is in its nascent stages but there are a few examples of note. Within the discipline of materials science, self-healing plastics have been developed that can resist or reverse damage through different mechanisms. The plastic of White et al[2] has imbedded polymer capsules which disperse polymer via capillary action if they are intercepted by a crack. The polymer then reacts with dispersed catalyst pellets, forming new plastic. Thermally cross-linked polymers "heal" by undergoing a thermally reversible reaction[3]. If the cracked plastic is heated above a certain threshold temperature, bonds will reform across the cracks healing the plastic.

Examples of resilient systems include power grids and naval ships. Power grids have been developed which can anticipate changes in electricity demand and ramp up or scale back production in response[4] while modular naval ships have been developed which allow operation and electricity to be rerouted around damaged areas allowing the ship to continue operation[5]. However, with both of these examples, tools developed are of limited use outside of that specific example – resilience was studied for that example, not systems in general.

An area where resilience has received considerable attention as a desired end product is complex systems research. Complex systems are composite systems containing numerous components and interconnections where the sum of the parts does not give all information about the whole[6]. Complex systems research seeks to describe and understand the behavior and properties of a wide variety of these systems using similar tools and methods. However, systems are generally examined on the whole – all aspects including human interactions, financial stresses, as well as physical aspects are studied holistically. Systems can be classified as having physical (material and equipment), financial (money as well as financial markets), and behavioral (human interactions and decisions) components. When approaching system resilience from this holistic view, resilience can be defined as the ability of system to withstand perturbations, where perturbations could be anything from different human decisions to physical forces. This research will focus solely on physical systems – no human decisions, economic factors or the like will be included.

While complex systems research often focuses on resilience as a desirable end product, this research will study resilience as an inherent property of physical systems. A potential starting point determining a specific physical system resilience definition is the examples of resilience within academic disciplines. The concept of resilience can be found within ecology, psychology, and materials science.

Material Resilience

Materials science will be examined in more detail due to the fact that it alone defines resilience as an easily quantifiable, inherent property. Material resilience is the amount of energy that can be applied to a material before the material displays yielding behavior. It can be expressed in terms of the modulus of resilience (U_R) given below, which is the area under the stress-strain curve from the point of zero stress to the yield stress (σ_{yield})[7]. The yield strain is denoted by ε_{yield}.

$$U_R = 0.5\sigma_{yield}\varepsilon_{yield}$$

Some important properties of a material's resilient range is that within this range the energy added to the material is almost entirely recoverable, the system undergoes essentially reversible deformation, and there is minimal dissipation of stored energy. For linearly elastic materials, the resilient range can be characterized by linear slope of the stress strain curve. However, not all materials are linearly elastic and thus the resilient range can be characterized by different types of behavior for different materials.

Material resilience will be used for inspiration in defining system resilience. Since material resilience is the amount of energy a material can withstand before yielding, system resilience will similarly be defined as the amount of energy a system can withstand without exhibiting instability or failure[8]. Since resilience has been defined in terms of energy, concepts such as system thermodynamics will be important in developing variables for system resilience.

With resilience defined, how it is to be manifested in systems must be established. Material resilience can be easily viewed and assessed using a material stress-strain curve. It would be desirable to develop an equivalent system curve. However, if such a curve is to be created, equivalent system stress and strain variables must be defined. Therefore, the properties of energy within physical systems will be explored in more detail.

Since this research aims at developing the resilience concept to aid in improving safety, it is important to explore the connections between energy and safety. Most safety incidents can be termed "loss of containment" events, meaning some process flow escaped its designed bounds. All systems are designed to withstand a certain range of energy levels. If the energy applied to the system is outside that range, failure can occur and result in loss of containment. Understanding how energy loads affect systems may assist in better defining appropriate energy ranges. But while energy analysis yields a wealth of information, it has some limitations – energy cannot be destroyed, so the energy will balance no matter the change as well as that energy balances do not give a great deal of information on the energy's potential or density.

ENERGY AND EXERGY

While material resilience can be viewed as a reversible process with minimal dissipation of input energy, this analogy is not directly applicable to systems because no system is fully reversible. All real process systems are irreversible – while they do not destroy energy, they irreversibly dissipate energy through various modes including friction, unrestrained expansion, and spontaneous chemical reaction. The more irreversibility within the system, the less efficient the system is.

A concept that captures information about these irreversible energy processes is exergy. Exergy can be defined as "the amount of work obtainable when a system is brought to equilibrium with its surroundings[9]." Exergy is not conserved and depends on a reference state that is generally set to atmospheric temperature and pressure. While reference states can be determined for chemical composition, this is only needed for systems involving reaction – no reactions will be studied during this research. This research will apply exergy for steady-state material streams with no change in composition, thus the exergy for these streams can be calculated using the difference between the process enthalpy (H) and reference enthalpy (H_o) and process entropy (S) and reference entropy (S_o)[9]. The equation is shown below with process mass flow rate as m_f and reference temperature as T_o.

$$Exergy = m_f[(H - H_o) - T_o(S - S_o)]$$

SYSTEM STRESS AND STRAIN

Now that the concepts of energy and exergy have been explored, the stress and strain variables from materials science will be revisited for guidance in defining system stress and strain. Material stress has dimensions of force per unit area and is a measure of a load applied to a cross-sectional area. The energy applied to a system (E_{in}) can be viewed as its load while the system's volume (V_{system}) is the space the load is distributed over. The system stress variable (Ss) will thus be defined as shown below:

$$Ss = E_{in} / V_{system}$$

Material strain is non-dimensional and measures the percent elongation of the system. Elongation is used as a measure of the material's deformation. Therefore, system strain will also be non-dimensional. Since systems deform in a wide variety of ways, it would not be reasonable to pick one variable. Therefore, the source of the deformation will be used to measure the system's "elongation." The applied energy load causes deformation, however not all that energy load has the potential to deform the system – only the portion of the energy load which has potential to do work on the system or its surroundings. Therefore, system strain (Sn) will be defined as the ratio of the exergy destroyed by the system ($Ex_{destroyed}$) over the exergy into the system (Ex_{in}) as shown below.

$$Sn = Ex_{destroyed} / Ex_{in}$$

The system stress and strain variables will be tested using simple process systems. A steam pipe and a water pipe will be studied as well as a water pump and a steam condenser. The system stress and strain variables will be used to create system response curves for different applied energy loads to each of the test systems. Different loads were applied by changing process variables such as mass flow rate, temperature and pressure. Each of the test systems used water as a process fluid since physical properties can be determined from steam tables. All calculations were performed in Excel, so the Excel add-in Water 97_v13 version 1.3 was used for steam properties[10].

TEST SYSTEMS

The steam and water pipe test cases both used a 50-foot, uninsulated, carbon steel pipe with assumed roughness of light rust. For simplicity, the pipe was assumed to not contain fittings. A centrifugal, adiabatic water pump was chosen with a pump curve chosen to relate the pump's volumetric flow rate, head, and efficiency of liquid compression. The motor efficiency for the pump driver was assumed to be 0.65. The heat exchanger used was a horizontal, tubular steam condenser. The condenser was assumed to have standard geometry with one-inch tubes at a triangular pitch. The hot fluid, steam was placed on the shell side, while the cold fluid, water, flowed through the tubes.

Input data was set for each system. Energy balances were performed on each system to determine outlet conditions. Then, the flowing stream exergy equation shown earlier was used to calculate the exergy of each inlet and outlet stream. This allows the exergy destroyed by each system to be calculated via an exergy balance. Once the energy and exergy of all streams were calculated, system stress and strain variables (and thus characteristic system curves) were determined by calculating the system's stress and strain values at the initial base conditions and then calculating stress and strain again each time system parameters were incrementally increased.

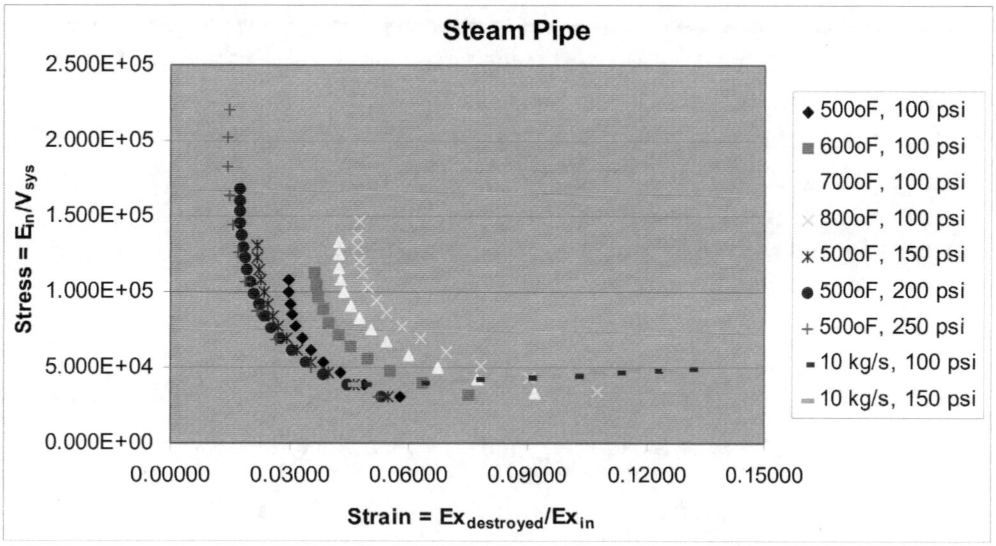

Figure 1. Steam Pipe Characteristic System Curve

Figure 1 shown above is for the steam pipe system. These curve were created by increasing the pipe's mass flow rate, temperature, and pressure from an initial base case of 500°F, 100 psi, and 8 kg/s. The variables were changing individually – the first six curves shown the mass flow rate increasing from 8 kg/s in 2 kg/s increments with the temperature and pressure constant while the last two curve show the temperature increasing as the mass flow rate and pressure were held constant. Multiple curves are shown on this graph – the legend shows two of the variables for each curve, while the variable not indicated is the one changing. This combination of the data allows trends for different variables and ranges to be observed.

As can be seen, the changing mass flow rate curves shown an inverse relationship between stress and strain with higher temperature curves positioned at higher stress and strain rates and higher pressure curves positioned at slightly lower stress and strain values. As the mass flow rate increases, the friction factor (f) increases, but the increase tapers off toward the end of the mass flow rate range. Each higher flow rate point destroys more exergy by transforming mechanical flow energy into thermal energy, but the system destroys proportionally less exergy due to the smaller percentage increase in the friction factor. The system thus operates at higher efficiency levels as the mass flow rate increases, but if the system has been designed to operate at lower efficiencies, the higher efficiency may results in more exergy than the system has been designed to withstand. At higher flow rates, the system has (percentage-wise) more potential to do work on its surroundings than originally anticipated.

The two curves for changing temperature show a direct relationship between stress and strain with the stress increasing much slower than the strain rate. The strain for the temperature curve increases as the stress increases and while this trend remains consistent throughout the investigated temperature range, the strain values begin to increase at a slower rate for higher temperatures resulting in an increase in the curve slope. This behavior can be explained by observing that while the pipe does dissipate more energy via heat loss to the atmosphere, the amount of energy added to the system increases at a faster rate. Thus, while the percentage of exergy lost increases for higher temperatures, it grows at a slower rate. Thus, the amount of exergy contained within the system is less than what would be predicted by the initial displayed trend.

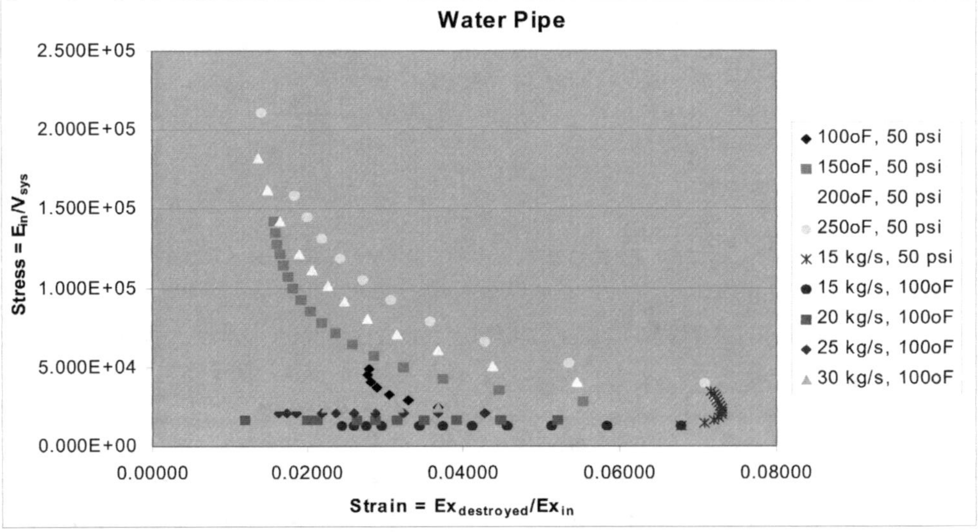

Figure 2. Water Pipe Characteristic System Curve

The water pipe curve base case was 100°F, 50 psia, and 15 kg/s. As explained in the legend, the first four curves show the mass flow rate increasing from the base case of 15 kg/s with the temperature and pressure held constant. The fifth curve increased the temperature from 100 to 280°F while the mass flow rate and pressure were held constant and the last four curves increased the pressure from 50 to 550 psia while the temperature and mass flow rate were steady. The water pipe curves in Figure 2 show similar trends, with changing temperature curves displaying an inverse relationship and higher temperature curves having higher initial stress values. However, the strain relationship is less clear. While the initial strain for the 150°F curve is higher than the strain for the 100°F curve, the initial strain value decreases for the 200 and 250°F curves. Examining the changing temperature curve yields more information – as the stress increases, this curve reaches a maximum strain value around T=150°F. This inflection point is due to the fact that at lower temperatures, operation is close to the reference temperature making process exergy minimal. As the temperature increases, exergy increases at a faster rate than the temperature. Exergy destroyed also increases due to more heat lost to the atmopsphere. The initial direct relationship between stress and strain is due to the exergy loss increases having more of an effect on the strain than the overall process exergy increases.

The changing pressure curves show very little change in stress with increasing strain with higher mass flow rate curves positioned at slightly higher stresses. The curve shows stress values increasing at higher pressures as strain decreases. The small stress ranges of the changing pressure curves is due to the fact that pressure changes cause very small changes in the energy of process streams as compared with mass flow or even temperature changes.

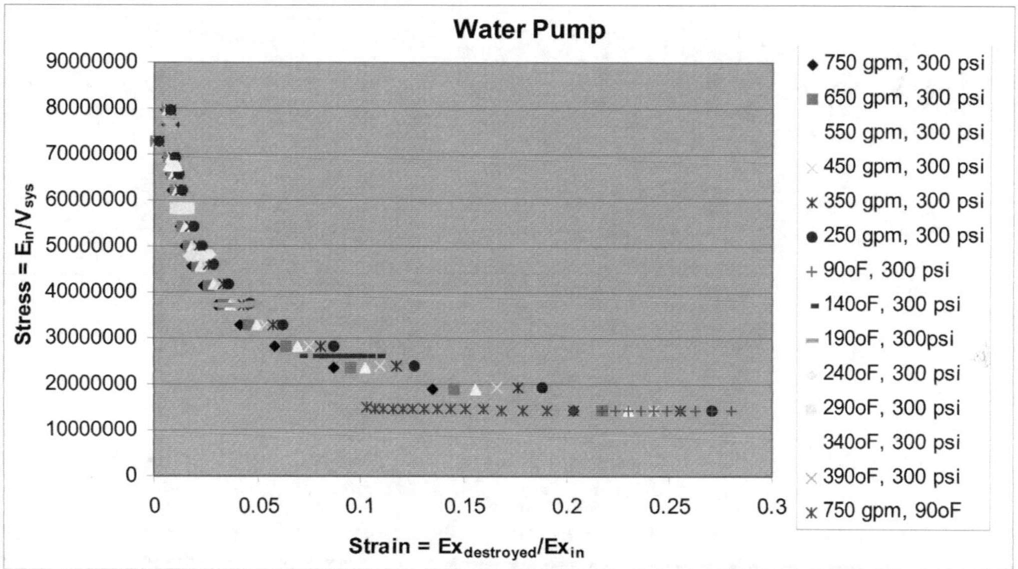

Figure 3. Water Pump Characteristic System Curve

The water pump base case was 90°F, 300 psia inlet pressure, and 750 gallons per minute flow rate (gpm). The first five curves varied the temperature between 90°F and 410°F with the flow rate and inlet pressure constant. The next seven curves varied the flow rate between 750 and 200 gpm with the inlet pressure and temperature constant while the last curve varied the inlet pressure with the temperature and flow rate constant. The water pump graph (Figure 3) shows again show an inverse relationship between stress and strain for the changing temperature curves with decreasing flow rate curves positioned at higher strain and slightly higher stress values. The curves show markedly similar shape. This behavior is due in part to the way the pump volume is defined – volume equals the volumetric flow rate divided by the pump's specific speed (constant for this example). Volume decreases for decreasing mass flow rates resulting in a slight increase in the pump's stress even though the total energy into the pump is decreasing. Reasons for the gradually decreasing slope include a decrease in pump efficiency with lower mass flow rates. The pump begins to destroy a higher percentage of the input exergy at lower mass flow rates.

As the temperature is increased, the data points significantly scatter – the reason for the scatter is the onset of a phase change and cavitation – the onset of cavitation represents an unsafe operating point, so it is encouraging that the behavior of the curve indicates significant deviations from the previously displayed trend.

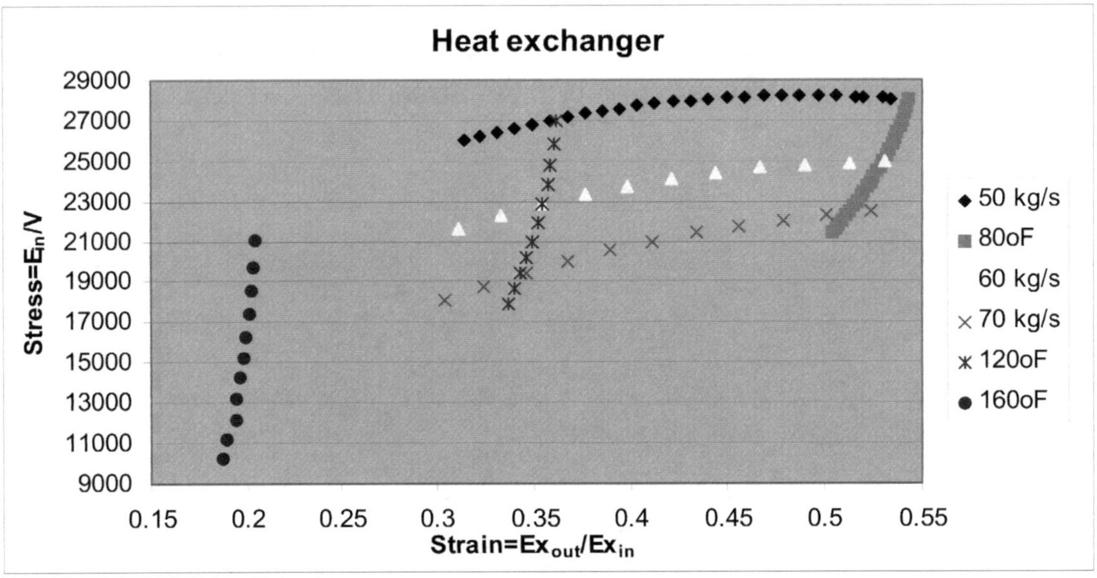

Error!

Figure 4: Heat Exchanger Characteristic System Curve

For the heat exchanger base case, the inlet steam mass flow rate was set at 50 kg/s while the inlet cooling water temperature was set at 80°F. The first, third, and four curves varied the inlet cooling water temperature while the steam mass flow rate was constant while the second, fifth, and sixth curves held the cooling water temperature constant while the inlet mass flow rate was increased. For the heat exchanger graph (Figure 4), changing steam mass flow rate curves show a direct relationship with higher cooling water temperature curves positioned at higher initial stress and strain values. The slope of these higher temperature curves also decreases. The changing cooling water temperature curves display a direct relationship, however the curves appear to reach or approach a maximum stress value – higher steam mass flow rate curves are positioned at lower initial stress and strain values.

CURVE SIGNIFICANCE

A few identifiable trends can be gleaned from these graphs – first, similar trends were observed for different data ranges and curves displayed the same general shape for similar variable changes. However, there are significant differences between the different systems and even between different variable changes to the same system. Other than the heat exchanger, all systems displayed both inverse and direct relationships. While qualitative comparisons and observations are beneficial, it would be desirable to develop quantitative methods by which resilience could be assessed by determining where the system transitions from a resilient to a non-resilience behavior region. All curves show changing trends between stress and strain, so there is likely some predictive ability to determine the expected behavior of the system, such as a specified functional relationship between stress and strain in the resilient region (for example, exponential, polynomial, linear, etc). The onset of non-resilience would then be when the curve deviates from that functional relationship.

Within the predicted relationship, the relative storage and dissipation of exergy is expected to be proportional to the added or decreased energy "concentration" within the system. If the slope increases more than expected, the relative storage and dissipation of exergy is not proportional to the stress, or energy concentration. If the added energy concentration increases faster than the expected relative exergy destruction by the system, the system could be operating more efficiently. Since exergy destruction is essentially wasted energy potential, systems that conserve exergy are more sustainable as well as more cost effective and efficient. Another explanation is that the exergy destruction modes of the system could be overwhelmed. System can shed excess energy load through exergy destruction and if the capacity of these modes is reached, the system must store the added energy load. If the system reaches the upper bound of its energy storage capacity failure could occur.

For inversely related stress and strain variables with a higher than expected increases in slope, exergy destruction is not decreasing as expected. The system may be dissipating excess energy load by destroying more exergy. The system could be dissipating more exergy to stay within its energy capacity range. At some point the system may not be able to dissipate or store the addition energy load and then a failure may occur.

RESILIENCE QUANTIFICATION

If the resilience of a system under certain conditions is to be quantified, the point where the system deviates from resilience must be established. As stated earlier, for linearly elastic materials, the resilient region is characterized by linear slope and where resilience ends can be determined by observing deviations from that linear slope. However, not all materials are linearly elastic and display this linear relationship – different materials are characterized by different stress-strain relationships.

This analogy can be extended to systems – as observed in the previous graphs, many trends and relationships exist for systems. However, it would still be useful to determine a functional relationship between stress and strain such that departure from resilience could be determine by assessing when behavior deviates from what is expected from and predicted by the functional relationship.

A power law relationship of the form y equals A times x to the B power was chosen to fit the data due to its flexibility of fit. Power laws can be used to fit inversely related data (with B less than zero), linear data (B=1), directly related data with a decreasing slope (positive B less than one), and directly related data with increasing slope (B is greater than one). There is precedent to using power law relationships in complex system analysis to predict system behavior. Power laws trends have been observed in data for the probability of a certain size forest fire and the distribution of file download sizes as well as data for species extinction, traffic jams, and financial market volatility[6]. Power laws can also be found to describe stress behavior of certain fluids[11].

If the data follows the power-law trend, the use of a trend-line to fit the data will introduce little error and thus the coefficient of determination, R^2, will be close to one. If R^2 begins to decrease, this indicates the data is deviating from the expected power law trend. Therefore, the resilient region will be assumed to end when the R^2 value falls below 0.99. The system is assumed to be designed to safely withstand the initial conditions at which the system operates, therefore, the range predicted will vary for different systems, different types of variable changes, and even different initial conditions. While trend lines can be fit in Excel, power law regression was performed so the R^2 value could be determined as each subsequent point was added to the range.

Data for the changing mass flow rate steam pipe system with T=500°F and P=100 psia is shown in Table 1. The R^2 value falls below 0.99 when the mass flow rate is increased above 22 kg/s. It can be observed that while the R^2 value decreases slowly from 8 to 22 kg/s, it decreases more rapidly after it falls below 0.99, indicating increasingly large deviations.

Table 1: Curve Fit Data for the Steam Pipe Case, T = 500°F and P = 100 psia

Mass Flow Rate (kg/s)	System Stress	System Strain	A	B	R^2
8	3.083E+04	0.05762			
10	3.853E+04	0.04883	657.8	-1.348	1.000
12	4.624E+04	0.04276	638.0	-1.359	1.000
14	5.395E+04	0.03843	601.3	-1.379	0.9999
16	6.165E+04	0.03531	551.1	-1.408	0.9995
18	6.936E+04	0.03306	491.2	-1.446	0.9985
20	7.707E+04	0.03149	425.9	-1.493	0.9965
22	8.477E+04	0.03045	359.1	-1.548	0.9927
24	9.248E+04	0.02986	294.9	-1.612	0.9862
26	1.002E+05	0.02964	236.5	-1.684	0.9758
28	1.079E+05	0.02975	186.3	-1.761	0.9597

Table 2 shows a summary of the points just above and just below R^2 equals 0.99 for the steam pipe system. Identifiable trends include that the allowable mass flow rate increases as the initial temperature increases. One possible reason for this is that higher temperature systems have been designed to withstand a larger base energy load and thus may tolerate larger energy fluctuations. The allowable range for the changing temperature runs appear to vary significantly less – when the initial pressure was changed from 100 to 150 psia, no change was observed in the allowable temperature range. An explanation for this may be that that increasing the pressure within a system does not cause as large of a change in the energy of the process stream as the change caused by changing temperature or mass flow rate. Similar analysis was performed for all four systems, however for brevity sake those results will not be presented.

Table 2. Curve Fit Data for the Steam Test Case

Curve #	m_f (kg/s) value / range	T (°F) value / range	P (psia) value / range	System Stress	System Strain	A	B	R^2
1	8 - 22	500	100	8.477E+04	0.03045	359.1	-1.548	0.9927
	8 - 24	500	100	9.248E+04	0.02986	294.9	-1.612	0.9862
2	8 - 20	600	100	8.009E+04	0.03968	794.9	-1.421	0.9980
	8 - 22	600	100	8.810E+04	0.04060	622.3	-1.509	0.9790
3	8 - 26	700	100	1.080E+05	0.04263	904.6	-1.491	0.9905
	8 - 28	700	100	1.163E+05	0.04209	789.3	-1.542	0.9840
4	8 - 26	800	100	1.120E+05	0.04831	1346	-1.437	0.9931
	8 - 28	800	100	1.206E+05	0.04746	1205	-1.481	0.9883
5	8 - 30	500	150	1.151E+05	0.02211	471.9	-1.422	0.9930
	8 - 32	500	150	1.228E+05	0.02178	421.6	-1.457	0.9891
6	8 - 38	500	200	1.453E+05	0.01754	531.6	-1.364	0.9926
	8 - 40	500	200	1.529E+05	0.01735	490.2	-1.388	0.9896
7	8 - 43	500	250	1.637E+05	0.01489	625.3	-1.305	0.9950
	8 - 48	500	250	1.827E+05	0.01452	542.6	-1.346	0.9896
8	10	500 - 900	100	4.460E+04	0.1021	6.915 E+04	0.196 0	0.9901
	10	500 - 1000	100	4.616E+04	0.1127	7.245 E+04	0.212 8	0.9841
9	10	500 - 900	150	4.455E+04	0.09686	7.004 E+04	0.197 3	0.9914
	10	500 - 1000	150	4.611E+04	0.1073	7.329 E+04	0.213 4	0.9859

Overall System Simulation

While the results from the individual system analysis are promising, it is important that the method be scalable – i.e. an overall system composed of smaller units give results compatible with the individual system results. To demonstrate this, all four test systems will be combined into an overall system as shown below. Stresses were applied to this system using the process simulator ASPEN. The process layout is shown below in Figure 5. Water flow through a pipe, pump, and then into the tube side of the heat exchanger. Steam flows through a pipe and then into the shell side of the exchanger.

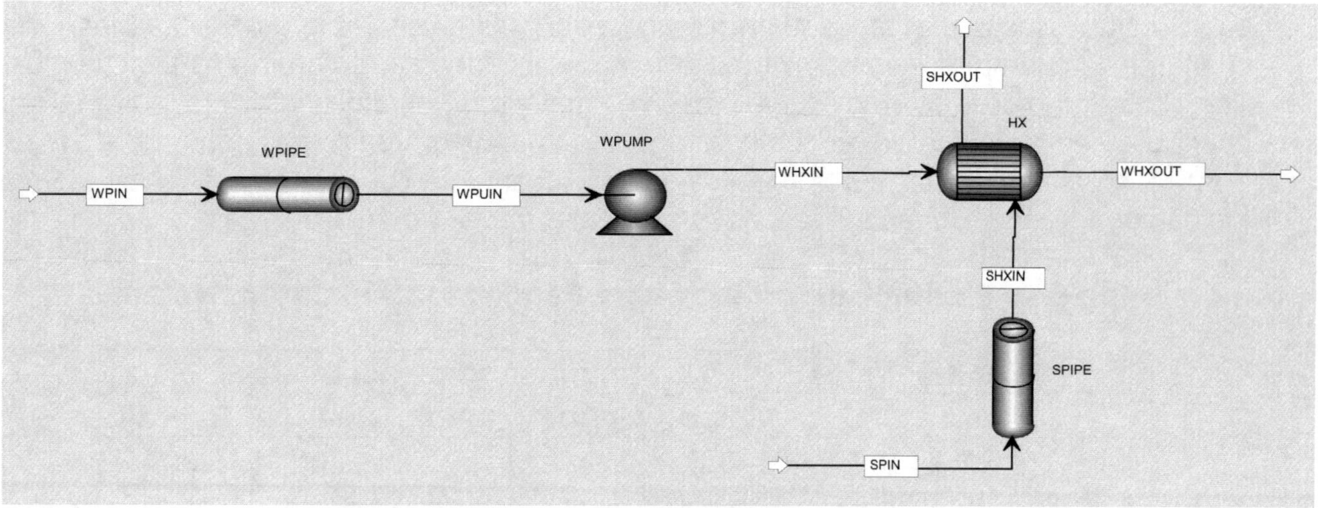

Figure 5. ASPEN Simulation Layout

The overall system curve can be constructed using the material streams into the water and steam pipes, the electricity stream into the water pump, and the steams out of the shell and tube sides of the exchanger. The curve below in Figure 5 shows different variable changes to the input streams. The base conditions for the inlet streams were set as 100°F, 50 psia, and 674 kg/s for the water stream and 500°F, 100 psia, and 30 kg/s for the steam stream.

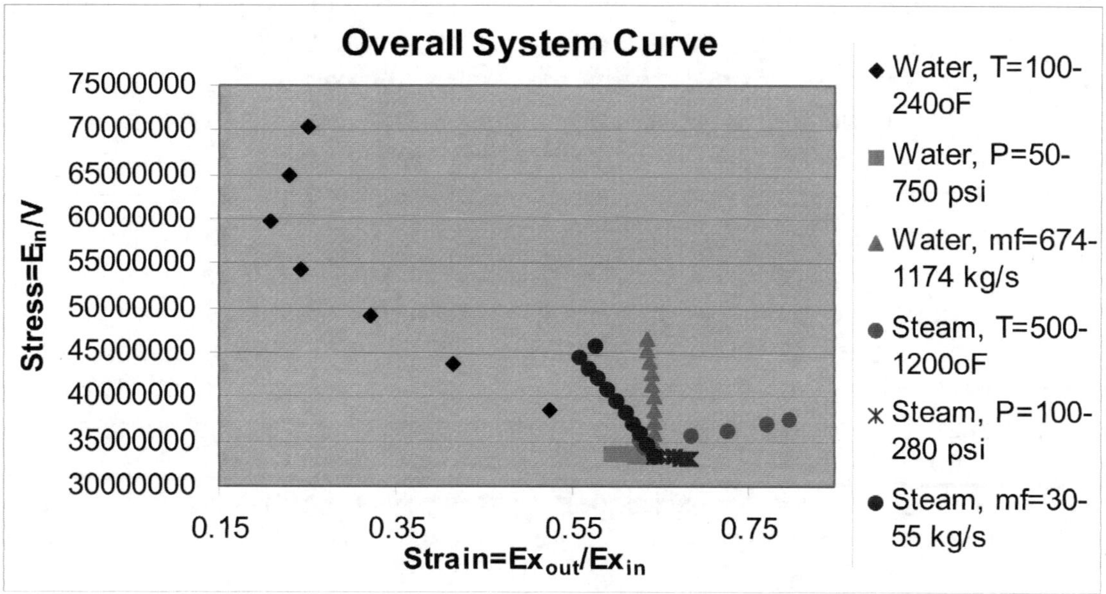

Figure 5. Overall Simulated System Curve

As can be seen, while all curves begin at a common stress and strain point, they trend differently for different variables. There are commonalities – both changing pressure curve involve miniscule stress changes and small but observable strain changes. The changing temperature curves involve large strain changes and varying stress changes. The changing mass flow rate curves involve small strain changes with moderate stress changes. The changing water temperature and changing steam mass flow rate curves both reach minimum strain points followed by a change in curve behavior. These points indicate where the exchanger ceases to condense the entire steam stream.

Table 3. Comparison of Allowable Ranges for Simulated System

Curve #	Initial Value	Maximum Allowable Value				
		Wpipe	Wpump	Spipe	HeatEx	Overall
1	W T=100°F	140°F	all ok	n/c	220°F	200°F
2	W P=50 psi	500 psi	all ok	n/c	all ok	all ok
3	W m_f=674 kg/s	all ok	1074 kg/s	n/c	824 kg/s	774 kg/s
4	S T=500°F	n/c	n/c	900°F	900°F	900°F
5	S P=100 psi	n/c	n/c	200 psi	180 psi	180 psi
6	S m_f=30 kg/s	n/c	n/c	all ok	45 kg/s	47.5 kg/s

Table 3 shows a summary of the allowable data ranges for each unit and the overall system for each variable change series. The notation "n/c" indicates the system properties did not change for the variable series while the notation "all-ok" indicates the R^2 value did not fall below 0.99 for the entire tested range. As can be seen, maximum allowable value is less than or equal to the max value allowed by the individual units for all but a few exceptions. The water pipe unit is the main exception – all series maximum values are lower than the corresponding overall value. The main reason for this is the order of magnitude of the variables – the water pipe strain values are ten, sometimes one hundred times smaller than the strains seen by other units. The water pipe operates extremely close to ideal – thus, very small changes can cause large percentage changes in the unit strain values but not affect the overall system a great deal. The other exception is the heat exchanger for curve 6 – its allowable mass flow rate is one increment lower than the overall system. However, this may be due to increment choice or small rounding errors or assumption errors affecting the system.

Specific design choices such as exchanger type (shell and tube, plate and frame, or concentric tube) could be compared by finding resilient variable ranges for each design and then determining which exchanger type yields the widest allowable ranges. Comparison of ranges could also be used to answer questions such as optimal starting point, equipment order, as well as equipment size. However, these design choices and other questions will not be addressed in this paper due to the necessity of additional simulations.

DISCUSSION AND ANALYSIS

While these test systems were set-up to include equipment and process limitations, because of the simplicity of test systems, some aspects of the system's response to detailed limitations may not be present. For example, detail properties of the pipe such as additional vibrations that may occur at higher mass flow rates were not included. It is recognized that certain systems may have limitations such as required partial pressure for reaction or minimum concentration for sensor detection. Possible approaches for using this methodology in light of these limitations include not allowing those parameters to vary (vary other system parameters) or ensuring the system's energetic response is detailed enough to reflect departure from those conditions. For example, if studying a reaction that requires a certain concentration, the system's energetic response (in the form of displayed stress-strain trend) would change if the reaction did not occur (base case set to where reaction was occurring). This change would likely cause the allowable range to end due to deviations for the stress-strain trend established for the system when the reaction was occurring.

The concept of exergy has been used extensively in sustainability research because exergy measures the potential to do work and thus destroying exergy destroys the potential to do work and does not imply sustainability. However, the primary goal of this work is not to design more efficient or environmentally friendly processes – the resilience concept was explored with the end goal of creating safer systems. Therefore, while processes with low strain values indicate less exergy destroyed and thus more sustainable processes, the actual value of the strain variable is not as concerning as the degree it changes within the process. It has been assumed that all systems are designed to operate safety at their base conditions, thus departure from those initial conditions is the primary concern. While the exact stress-strain values may yield valuable information in the future, this research focused on the change in those variables. All systems have been designed to withstand a range of energy values or initial energy loads and if a load outside that designed range is applied, failure could occur. For example, if a reactor is designed to withstand a certain pressure and a higher pressure is applied, the relief value will lift allow process fluids to vent to a safe place. By assuming the initial conditions are the start of that allowable range, departure from the expected behavior (in the form of stress-strain behavior) can be used as critieria to determine when the system's behavior changes significantly for a specific variable – as the behavior changes, the system is more likely to undergo failure (an extreme example of changing or unexpected system behavior). Thus, the resilience of the process is highest close to the initial conditions while resilience decreases as the system conditions move further from the initial stress-strain point.

One reason for the difference in accuracy of the ranges may be the type of system characteristic by which the range is limited. The range may be limited due to changes caused by fluid properties or changes from equipment limitations. For example, if the allowable pipe operating range is limited due to fluid property limitations, the range will end well short of any equipment limits and thus an action like increasing the pipe's design pressure will have no effect. However, if the range limits are due to equipment limitations, changes to equipment specifications will cause changes in the energy behavior of the system as it approaches those limitations and thus an action like increasing the design pressure of the pipe will increase the allowable range (the pipe will cease to dissipate energy through modes such as material deformation). The physical properties of the process fluids (density, internal energy, etc) many not show predictable or consistent behavior, thus these property changes may be sufficient to cause the energetic performance of the system to deviate from the power-law trend. While deviations caused by fluid properties will probably not cause catastrophic failure (like equipment failures may), they make the system more difficult to control and operate as well as making it harder to understand system behavior and interactions. The different types of system deviations that cause the power-law trend to end can be viewed as examples of the dual aspects of strength and flexibility. The strength aspect of resilience is well embodied by equipment limitations while the fluid properties and changes contribute in part to flexibility aspects. While these dual properties are manifested differently in systems and the consequences of overstepping their related system bounds differ, both contribute to resilience and must thus be considered when understanding system behavior.

While departure from a power-law trend has been used to signify changing system behavior for this research, it is possible that for certain systems this may not be sufficient or appropriate. However, the ability to have one, unified method for determining appropriate ranges makes the power-law method very desirable. As with any other methodology, it is possible that the ranges predicted from power-law analysis may be more accurate for certain applications. For example, the pump ranges allowed operation to within approximately 10 to 20°F of a known failure point, the onset of pump cavitation, whereas the water pipe ranges cut-off well before any known pipe failure points. Thus, the ranges predicted for the water pipe seem overly conservative. While the inclusion of more detailed process parameters and limitations may improve the accuracy of predicted limits, it is certainly more desirable that ranges be overly conservative rather than the opposite.

METHODOLOGY STRENGTHS AND WEAKNESSES

While the resilience method developed has shown promise, there are some notable weaknesses that provide opportunities for future improvement. The effect of noise on the methodology has not be assessed – since the methodology relies on the accuracy of the data, it has not be determined if the presence of noise would introduce enough scatter to cause problems. Also, the sensitivity of the method to individual units within an overall system is not as robust as would be desired – this was demonstrated with the results for the water pipe within the overall simulation. The methodology also relies on being able to definitively determine boundaries for analysis. While there were only a few material and energy streams present for the test systems, actual systems have numerous streams. Finally, the methodology is not applicable to non-physical system components – while any energy changes caused by a human decision would be included, the actual decision and its associated factors cannot be represented.

Some notable strengths of the methodology include its flexibility of reference states – since the exergy reference temperature and pressure can be set to ambient conditions, the reference state can be set to different values for processes in warm or cold climates. Also, the use of a general "energy in" stream allows multiple energy sources to be represented by the same variable.

FUTURE OPPORTUNITIES AND CONCLUSION

There are a number of opportunities for future development and expansion of this research. Connections to process control analysis can be probed – since both control and safety include determination of operating ranges, this methodology might be of use in the control arena. Also, results from this methodology could be compared with results from optimization techniques – for example, if determining an optimal pipe diameter, the results from an optimization could be compared with diameters which allow the largest variable ranges. The methodology could be augmented by the addition of other exergy destruction modes like chemical reaction or mixing. Other energy sources such as elevation changes could also be included.

One extremely promising application is integration with sustainable approaches. Sustainability uses exergy to assess the "green footprint" or environmental impact of various processes. If analysis from safety could be applied to environmental applications or vice versa, this would save both monetary and time resources. Finally, the application of this method to real data would allow verification of the usefulness of this work.

In conclusion, the background and development of the concept of system resilience has been presented. A general framework for visualizing the resilient behavior was developed as well as a method allowing acceptable ranges to be computed for different systems, different stress types, and different initial conditions.

ACKNOWLEDGEMENT

This research was performed while on appointment as a U.S. Department of Homeland Security (DHS) Fellow under the DHS Scholarship and Fellowship Program, a program administered by the Oak Ridge Institute for Science and Education (ORISE) for DHS through an interagency agreement with the U.S Department of Energy (DOE). ORISE is managed by Oak Ridge Associated Universities under DOE contract number DE-AC05-

00OR22750. All opinions expressed in this paper are the author's and do not necessarily reflect the policies and views of DHS, DOE, or ORISE.

REFERENCES

1. The Merriam-Webster Dictionary Online. Accessed 6/7/05, http://www.m-w.com/cgi-bin/dictionary?book=Dictionary&va=resilience

2. White, S. R., et al (2001), "Autonomic Healing of Polymer Composites," *Nature*, 409, pp. 794–797.

3. Chen, X. et al (2002), "A Thermally Re-mendable Cross-Linked Polymeric Material." *Science*, *295*, pp. 1698-1702.

4. Venere, E. (2001), "Engineers Will Demonstrate New System to Prevent Power Failures," available at www.globaltechnoscan.com/7thMarch-13thMarch01/power.htm.

5. Venere, E.(2000), "Future U.S. Warships Will be Automated," *Purdue News,* available at http://news.uns.pudue.edu/UNS/html4ever/000107.Sudhoff.futureships.html.

6. Carlson, J. M. and John Doyle (2002), "Complexity and robustness." *Proceedings of the National Academy of Science*, 99, pp. 2538-2545.

7. "Resilience" (1983). *Van Nostrand's Scientific Encyclopedia, Sixth Edition*, Van Nostrand Reinhold Company: New York, pp. 2433.

8. Mitchell, S.M. and M.S. Mannan (2006), "Designing Resilient Engineered Systems." *Chemical Engineering Progress,* 104, April, pp. 39-45.

9. Szargut, J (1980), "International Progress in Second Law Analysis." *Energy*, 5, pp. 709-718.

10. Spang, Bernhard (2002), *Water97_v13.xla – Excel Add-In for Properties of Water and Steam in SI-Units*, Accessed http://www.cheresources.com/iapwsif97.shtml.

11. Barnes, Howard A (1999), "The yield stress – a review or 'παντα ρει' – everything flows?" *Journal of Non-Newtonian Fluid Mechanics*, 81, pp. 133-178.

Safety Contributions of Ramcap Plus Risk Management Software

Jay Robinson
Alion Science and Technology
Alexandria, VA 22311
jarobinson@alionscience.com

and

Jake Stenzler
ASME Innovative Technologies Institute
Washington, DC 20036
jstenzler@asme-iti.org

Abstract:

The ASME Innovative Technologies Institute, LLC (ASME-ITI) developed the Risk Analysis and Management for Critical Asset Protection (RAMCAP), a risk management methodology being deployed by the U.S. Department of Homeland Security (DHS) to address terrorism threats to the seventeen Critical Infrastructures and Key Resources as identified in Homeland Security Presidential Directive 7. Recognizing the needs of industry to manage both safety and security risks, ASME-ITI and Alion Science and Technology have established a collaborative effort that uses the CounterMeasures,,· risk analysis software to automate a more powerful methodology named RAMCAP Plus. RAMCAP Plus goes beyond RAMCAP by analyzing natural hazards, unintentional hazards, business continuity risks, and additional terrorist threats, as well as incorporating a robust cost-benefit tool. The software provides a systematic method for identifying, analyzing, and managing all risks associated with critical assets and system components. RAMCAP Plus is a 7-step process that allows for repeatable and consistent results; yet, it is inherently flexible to accommodate user needs and unique facility specifications. The current version is tailored to the chemical industry and is the only software available to meet DHS-recommended criteria. Future versions will address the needs of other sectors. The software assists users in the calculation of consequences, vulnerabilities, and threat/hazard likelihood to compute absolute risk. A library of mitigation systems and countermeasures is supplied to fully support the risk management process. RAMCAP Plus generates data and information that is necessary to make cost-effective and logical safety and security enhancements. Ultimately, the use of RAMCAP Plus will reduce risk, increase resilience and improve the safety of the workplace and surrounding communities.

Ignitability of Sprayed Liquid due to an Electrostatic Spark (II)

Kwang-Seok CHOI, Mizuki YAMAGUMA, and Atsushi OHSAWA

National Institute of Occupational Safety and Health
Address: 1-4-6 Umezono Kiyose Tokyo 204-0024 JAPAN
**Corresponding Author: choiks@s.jniosh.go.jp*

ABSTRACT

In the current paper, we deal with Minimum Ignition Energy (MIE) due to an electrostatic spark of a sprayed liquid relative to the percentage of nitrogen (N_2), including compression in an air cylinder. The charge amount while spraying and electrostatic sparks in a sprayed region were also observed. Two types of hand-held spray guns and a supply of air pressure in the range of 0.1 to 1 MPa were used in this study. With regard to the materials, kerosene was selected for ignition tests while spraying, and water, including some sodium chloride, was used to investigate the charge amount of the sprayed liquid. The following noteworthy results were obtained: (1) the ignitability of sprayed liquid was dramatically reduced by atomizing with an air changing the percentage of N_2 instead of pressurized pure air, and its efficiency increased with increasing the supply of air pressure. (2) the charge amount, -1.72 to -2.62 μC/kg, was obtained using a spray gun A (-0.82 to -3.75 μC/kg with spray gun B); the values obtained in this study are far from safe in the painting industry, and (3) several electrostatic sparks in the spraying region were observed while spraying. Experimental details and a discussion are presented in this paper.

Keywords: Electrostatic spark; liquid, Ignitability; Minimum Ignition Energy.

1. INTRODUCTION

Explosions and/or fires involving liquid while spraying frequently occur due to electrostatic sparks in various chemical processes and while cleaning with the use of water jets and solvents. These dangers are especially great when liquids spew through small holes or cracks that develop unexpectedly in pipes, resulting in explosions and/or fires that are ignited by electrostatic sparks. Accordingly, we conducted Minimum Ignition Energy (MIE) tests for four types of sprayed liquid related to safety engineering and the prevention of accidents involving sprayed liquids [1]. The results obtained were interesting and significant: (1) in a normal temperature range (0 °C to 30 °C), any sprayed liquid can be ignited by a spark with discharge energy below several mJ irrespective of its flash point, and (2) the optimum region for the ignition of a sprayed liquid was observed, and it depended on the velocity, concentration, and particle size of the liquid.

However, more quantitative research into the ignitability of liquids will be necessary to assess hazards in industrial operations under various conditions and to propose effective countermeasures. Therefore, the experiments in this current study showed the degree of ignitability of a sprayed liquid relative to the percentage of nitrogen (N_2), including the compression in an air cylinder, and the charge amount while spraying.

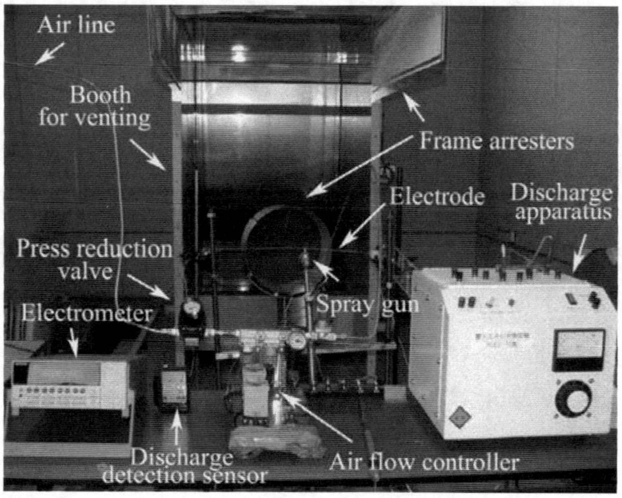

Figure. 1 MIE measurement system for sprayed liquid.

2. EXPERMENTAL

2.1. Apparatus and Method

An improved version of the MIE measurement system used in previous research was used in the current study. The experimental MIE measurement system for sprayed liquid is shown in Fig. 1. The system consists of a hand-held spray gun with a 1-mm internal-diameter orifice (Anest Iwata, W-100), a metal container for storage liquid, a super-mist separator (SMC, AME250), an air flow controller, a press reduction value, an air compressor (Hitachi, 0.40P-7S), and an open-fronted booth for venting (Anest Iwata, WD-10D).

The electrostatic discharge unit and measuring instruments included a power source (Amano, MIES-10) equipped with a high-voltage dc generator and a set of capacitors ranging from 10 pF to 2500 pF, a digital storage oscilloscope (Textronix TDS430A), a current amplifier (Textronix AM503), a personal computer (PC) for controlling and data processing, and other auxiliary devices.

A liquid was supplied to the spray tube (liquid input line) with air from the pressure supply line. The liquid was automatically sprayed for 15 s to measure the ignitability. A supply air pressure, P, in the range of 0.1 to 1 MPa was used. The discharging (sparking) electrodes had sharp tips made of stainless steel rods with a diameter of 2 mm. The electrode spacing was in the range of 2 to 4 mm for this study. The position, S, of the sparking electrodes in the spray liquid could be changed through the rails.

The net energy, W [J] for the discharge spark was calculated using the following formula (1):

$$W = 1/2\ C\,(V_1^2 - V_2^2) \quad \text{--}(1),$$

where C [F] is the capacitance of the capacitor and V_1 [V] and V_2 [V] are the voltage of the power source before and after sparking, respectively.

The maximum discharge energy was 320 mJ in this study. In this experiment, ignition took place when the sprayed liquid caught fire by an electrostatic spark within 20 times of the same energy, as shown in Fig. 2.

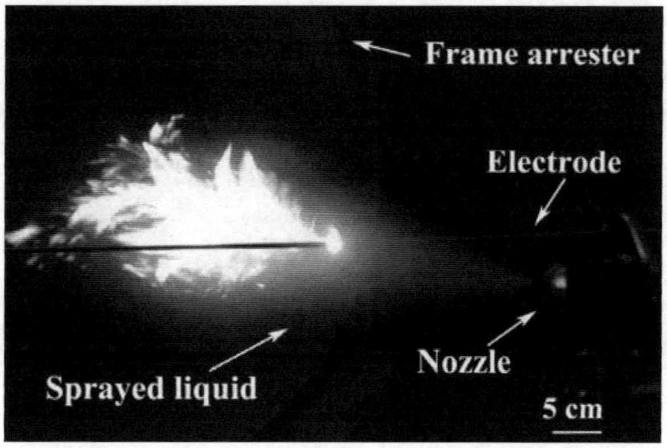

Figure. 2 Ignition of a sprayed liquid due to an electrostatic spark.

The spray gun was isolated with insulating material (PTFE) for evaluating the electric current of the liquid while spraying. The electric current, I [A], from the spray gun to the ground was measured by means of a very sensitive instrument measuring the current (Keithley-6514). The charge to mass ratio, Q [C/kg], of sprayed liquid was obtained with formula (2) below:

$$q = -\frac{1}{m}\int_0^t i dt \quad \text{--(2)},$$

where m [kg] is the mass of sprayed liquid, i [A] is the electric current I, and t [s] is the sprayed time.

In order to prevent environmental pollution, a venting booth that was 2 m wide and 2 m high was set up. A water stream was set up in the front of the booth to avoid a fire and/or explosion during the tests. The maximum volumetric flow was 90 m^3/min. In addition, a cylinder-type flame arrester was also set up in front of the nozzle at a 1.5-m distance to quench and extinguish the flames.

2.2. Samples

Kerosene (flash point, *f*: 50 °C; conductivity, σ: 3.50×10^{-14} S/m) in this study was selected for comparison with the data from our previous research[1]. Kerosene was cleaned with a clay filter (Wako Industrial, 1F1AJ, 10 *l*/min) to remove impurities prior to the test. Water (σ: 0.1 S/m), including sodium chloride (0.5wt% NaCl), was used as a test specimen to investigate the charge amount of the sprayed liquid. The conditions for all tests were 30 ± 3 °C and 60 ± 5 % RH.

3. RESULTS AND DISCUSSION

3.1. Influence of the Rate of Inert Gas on the Ignitability of a Sprayed Liquid

An inert gas is non-flammable and contains little or no oxygen, such as, for example, carbon dioxide (CO_2), nitrogen (N_2), and boiler flue gas [2]. The addition of such a gas to a flammable atmosphere can reduce the oxygen concentration to such a low level that ignition is impossible at any concentration of flammable material. It is well known that the ignition hazard of ethanol mist was reduced by atomizing with CO_2 instead of pressurized air [3]. However, the use of CO_2 will be restrained to prevent any contribution to the greenhouse effect. Accordingly, N_2 as an inert gas was selected for our tests because the benefits would be environmentally friendly and useful.

The main objective of this test was to evaluate quantitatively and systematically the relationship between the concentration of N_2 including air and the ignitability of liquid sprayed at various pressures in order to obtain new information for researchers working on the prevention of liquid explosions. Four different mixtures, A (O:20%, N_2:80%), B (O:15%, N_2:85%), C (O:10%, N_2:90%), D (O:5%, N_2:95%), and *P*, in a range of 0.1 to 0.5 MPa were used.

The following noteworthy experimental results are shown in Tables 1 to 4. First, under all conditions, kerosene was ignited with 7.2 mJ of spark energy irrespective of its flash point. The optimum regions for the ignition of sprayed liquids affected by factors such as the velocity, concentration, and particle size of the liquid are from 200 mm to 500 mm in the *X*-axis. These were almost identical results obtained with our previous research [1]. Secondly, the ignitability of sprayed liquid was drastically reduced by atomizing with an air increasing the composition ratio of N_2 instead of pressurized pure air. The increasing efficiency, η [times], of MIE increased with increasing P in the tests. For example, when mixture B instead of mixture A (MIE:7.2 mJ) was used, η at the 0.5 MPa had the largest value, 14 times (125 mJ), among the four conditions, while the value decreased to 5.2 times (38 mJ) at 0.3 MPa, 2.9 times (21 mJ) at 0.2 MPa, and 1.7 times (12.8 mJ) at 0.1 MPa, as shown in Fig. 3. This could be because the lower the P, the more easily oxygen can be introduced into the surrounding spray area.

This fact suggests that both conditions, a 90% rate (or over) of N_2 in air and 0.3 MPa (or over) of P, are necessary to prevent effectively explosions and fires that are the result of electrostatic sparks in a practical industry.

Table 1. Ignitability of a sprayed liquid as a function of the composition ratio of air under 0.1 MPa.

*X** / Substance**	Minimum Ignition Energy [mJ]								
	*X*50	*X*100	*X*200	*X*300	*X*400	*X*500	*X*600	*X*700	*X*800
A(O:20%, N_2:80%)	**180**	**29.4**	**7.20**	**9.80**	**21.6**	**38.4**	**38.4**	**128**	**NI**
B(O:15%, N_2:85%)	**320**	**50**	**12.8**	**12.8**	**29.4**	**50**	**72**	**NI**	**NI**
C(O:10%, N_2:90%)	**NI**	**50**	**12.8**	**12.8**	**29.4**	**50**	**72**	**NI**	**NI**
D(O:5%, N_2:95%)	**NI**	**72**	**21.6**	**12.8**	**38.4**	**128**	**NI**	**NI**	**NI**

*: X-axis (horizontal distance from the nozzle tip, mm). **: Composition ratio of air.

Table 2. Ignitability of a sprayed liquid as a function of the composition ratio of air under 0.2 MPa.

*X** / Substance**	Minimum Ignition Energy [mJ]								
	*X*50	*X*100	*X*200	*X*300	*X*400	*X*500	*X*600	*X*700	*X*800
A(O:20%, N_2:80%)	**NI**	**50.0**	**9.80**	**7.20**	**21.6**	**29.4**	**38.4**	**72**	**128**
B(O:15%, N_2:85%)	**NI**	**72**	**29.4**	**21**	**38.4**	**72**	**128**	**NI**	**NI**
C(O:10%, N_2:90%)	**NI**	**72**	**29.4**	**21**	**50**	**72**	**320**	**NI**	**NI**
D(O:5%, N_2:95%)	**NI**	**128**	**38.4**	**50**	**50**	**NI**	**NI**	**NI**	**NI**

*: X-axis (horizontal distance from the nozzle tip, mm). **: Composition ratio of air.

Table 3. Ignitability of a sprayed liquid as a function of the composition ratio of air under 0.3 MPa.

Substance** \ X*	Minimum Ignition Energy [mJ]								
	X50	X100	X200	X300	X400	X500	X600	X700	X800
A(O:20%, N_2:80%)	**NI**	**180**	**12.8**	**7.20**	**12.8**	**29.4**	**29.4**	**72.0**	**180**
B(O:15%, N_2:85%)	**NI**	**180**	**38.4**	**50**	**320**	**320**	**NI**	**NI**	**NI**
C(O:10%, N_2:90%)	**NI**	**180**	**50**	**98**	**NI**	**NI**	**NI**	**NI**	**NI**
D(O:5%, N_2:95%)	**NI**	**180**	**128**	**NI**	**NI**	**NI**	**NI**	**NI**	**NI**

*: X-axis (horizontal distance from the nozzle tip, mm). **: Composition ratio of air.

Table 4. Ignitability of a sprayed liquid as a function of the composition ratio of air under 0.5 MPa.

Substance** \ X*	Minimum Ignition Energy [mJ]								
	X*50	X100	X200	X300	X400	X500	X600	X700	X800
A(O:20%, N_2:80%)	**NI**	**180**	**29.4**	**7.20**	**7.20**	**7.20**	**9.8**	**21.6**	**21.6**
B(O:15%, N_2:85%)	**NI**	**NI**	**180**	**125**	**320**	**NI**	**NI**	**NI**	**NI**
C(O:10%, N_2:90%)	**NI**	**NI**	**NI**	**NI**	**NI**	**NI**	**NI**	**NI**	**NI**
D(O:5%, N_2:95%)	**NI**	**NI**	**NI**	**NI**	**NI**	**NI**	**NI**	**NI**	**NI**

*: X-axis (horizontal distance from the nozzle tip, mm). **: Composition ratio of air.

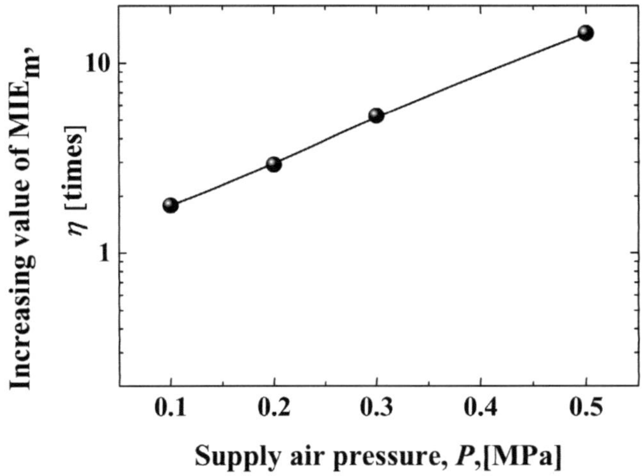

Figure. 3 η with a B cylinder as a function of air pressure supplied to the nozzle.

3.2. Charge Amount while spraying liquid

The *I* leaking from the nozzle to the ground was measured when the liquid was automatically sprayed for 15 s. Two types of spray guns were used. One has a 1-mm internal-diameter orifice (spray gun A, Fig. 4 (a)), and the other has a 3-mm internal-diameter orifice (spray gun B, Fig. 4 (b)), which was made especially for this test. The capacitor values, C [F], of spray guns A and B measured with an LCR Hi-tester (Yokogawa: 4332A) were 65 pF and 84 pF, respectively.

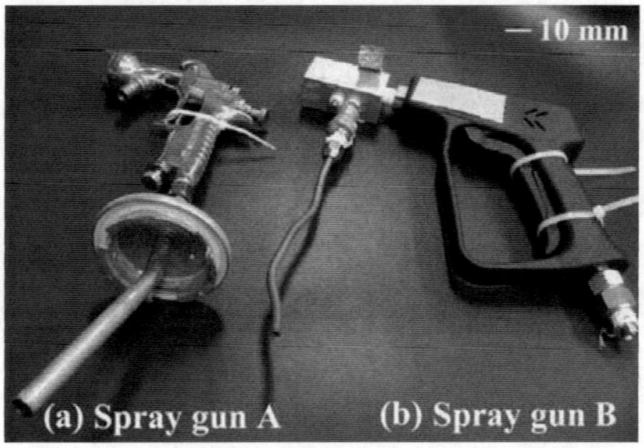

Figure. 4 Photo of spray guns used in this study.

The experiment results are shown in Figs. 5 and 6. First, it was found that a large part of I occurs while spraying liquid. In addition, when using spray gun B, the values were relatively large (from +5 nA to 60 nA) compared to the values obtained using spray gun A (from +1 nA to 2.7 nA). In general, if the spray guns were made from the same material, the I would increase with a reduction in the size of the spray-gun orifice [4]. It was concluded that the geometrical shape of the spray gun would be responsible for the amount of I. A pipe with 11 mm in internal diameter and 50 mm in length was attached to spray gun B. As a result, without the pipe, the I hardly occurred. The I had a large effect on the P; that is, the I increased clearly with an increase in P, as shown in Figs. 5 and 6. This result can be explained as follows: when a liquid passes through a certain section at a high velocity, the intensity of friction increases, and the exchange of charges is promoted.

Table 5 shows m and q in various P. With spray gun A, the q related to m, -1.72 to -2.62 μC/kg, was similar to the level obtained using spray gun B, -0.82 to -3.75 μC/kg. In the paint industry, 60 to 70% of a solvent changes into gases and vapors while spraying [5]. Solvents, such as toluene, acetylene, and xylene, are used in this industry. Their MIEs are about 0.2 mJ [6]. The electrostatic spray equipment included automated electrostatic spray and hand-held equipment. The accidents with the automated equipment decreased since an interlock safe system was used. However, the hand-held equipment can be moved easily from one point to another. Thus, the conductive part of the equipment is often not grounded, and, therefore, it could become charged and produce electrostatic sparks. A large number of accidents occurred with the hand-held equipment, and it was attributed to an electrostatic spark. The degree of risk about the electrostatic charge with the hand-held equipment remains unknown. This is discussed in detail in the following. When an operator works with hand-held equipment, the capacitor of the entire arrangement for the worker (100 pF) and spray gun (84 pF) is 184 pF. The value of 0.2mJ reported above equals the electrostatic energy when an operator grabs a spray gun charged at 1.5kV. The charge q related to the capacitor C [F] and potential V [V] was 0.27 μC. The results from Table 5 show that the values obtained with spray guns A and B in 0.1 MPa are -1.76 μC/kg and -0.82 μC/kg, respectively, even when the minimum p is used in this test. Thus, the values obtained in our test are high enough to result in the ignition of flammable solvents.

In order to evaluate the risk of electrostatic sparks in greater detail, the electrostatic sparks surrounding the nozzle while spraying were observed with a camera with an image intensifier (Hamamatsu Photonics, Ltd., Night Viewer C3100, Gain 60,000), as shown in Fig. 6. The brightness in Fig. 6 (b) was controlled so that the electrostatic sparks could be easily observed. Spray gun A and 0.5 MPa of P were used for this test. As a result, we confirmed that there were several electrostatic sparks while spraying, even though the incendiary sparks had not been detected in our tests. These sparks were particularly conspicuous at the nozzle. Concerning the relationship between the q and P (Table 5), these suggest that the spraying of liquids under high pressure of several tens MPa as water jets or airless paint spray must be carefully managed since it is possible that incendiary electrostatic spark could be occurred. In December 1969, three very large crude carriers experienced cargo tank explosions while being washed with water jets. This may be attributed to an electrostatic spark [7].

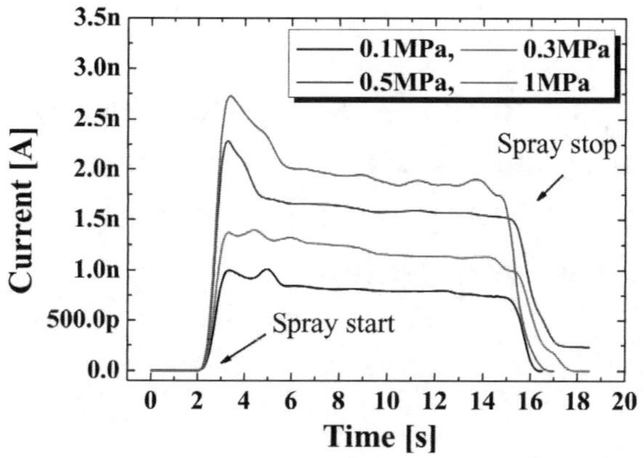

(a) spray gun A

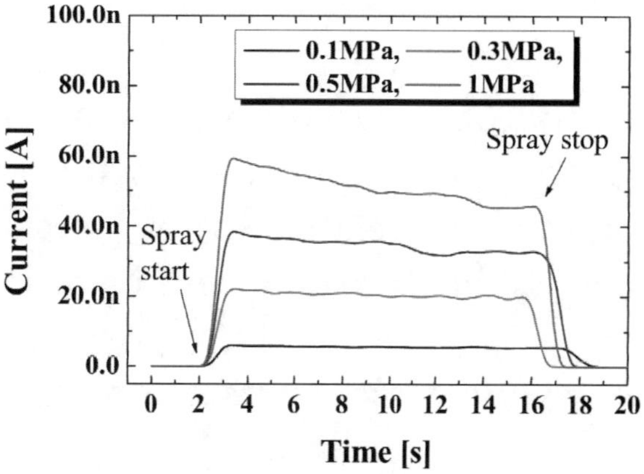

(b) spray gun B

Figure. 5 Change in current while spraying using spray guns A and B.

Table 5. Flow rate and charge density of liquid with various air pressures supplied to a nozzle.

Substance	0.1 MPa*		0.3 MPa		0.5 MPa		1 MPa	
	m^{**} [g/s]	q^{***} [μC/kg]	m [g/s]	q [μC/kg]	m [g/s]	q [μC/kg]	m [g/s]	q [μC/kg]
Water (spray gun A)	0.60	-1.76	0.72	-2.19	0.87	-2.51	0.98	-2.62
Water (spray gun B)	8.50	-0.82	15.0	-1.78	15.7	-2.81	17.5	-3.75

*: air pressure supplied to nozzle, m^{**}: flow rate. q^{***}: charge to mass ratio.

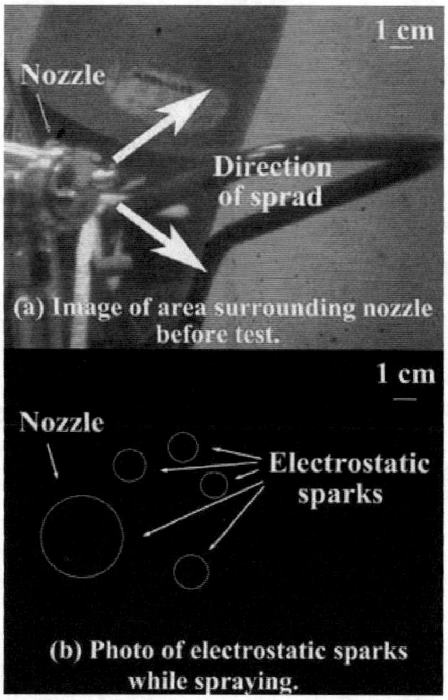

Figure. 6 Observation of the electrostatic sparks while spraying.

4. CONCLUSIONS

With the objective of improving the understanding of the ignitability of sprayed liquid due to electrostatic spark, the degree of ignitability of a sprayed liquid relative to the percentage of N_2 in an air, and the charge amount while spraying were investigated experimentally.

The experimental results can be summarized as follows;

(1) The ignitability of sprayed liquid was reduced by atomizing with an air increasing the percentage of N_2 instead of pressurized pure air. Its efficiency increased with increasing the supply of air pressure. Specially, we found that both conditions, a 90% rate (or over) of N_2 in air and 0.3 MPa (or over) of a supply air pressure, are necessary to prevent effectively explosions and fires that are the result of electrostatic sparks in a practical industry.
(2) The charge amount, -1.72 to -2.62 μC/kg, was obtained using a spray gun A (-0.82 to -3.75 μC/kg with spray gun B). The values obtained in this study are far from safe in the painting industry.
(3) Several electrostatic sparks while spraying were observed. These sparks were particularly conspicuous at the nozzle.

Consequently, these suggest that a sprayed liquid in all conditions involving the painting process, spew from small holes or cracks unexpectedly due to problems, must be carefully managed since it is always possible that electrostatic spark energy could exceed the values established for safety in the industry.

Acknowledgements

We are grateful to Mr. Shunji Nakajima, for his assistance and participation in discussions concerning the experiments.

Refereences

1. K.S.Choi, M. Yamaguma, and A. Ohsawa(2007): Ignitability of spayed liquids due to an electrostatic spark, "Advanced Powder Technol.", Vol. 18, No. 1, pp.105-115

2. NIIS(1988), "Recommended practice for protection against hazards arising out of static electricity in general industries", RIIS-TR-87-1, p. 38-39.

3. H. Matsui, and T. Ohtsuka(1999), "Measurement of minimum ignition energy for flammable mists", NIIS-SRR-No. 17 pp. 13-18 (in Japanese)

4. M. Yamaguma(2004), "Electrostatic hazards associated with spraying, featuring spray cans used for nondestructive examination", Vol. 43. No.4 pp.229-237 (in Japanese)

5. CEMA(2000), "New handbook for coating", p.129 (in Japanese)

6. NIIS(1988), "Recommended practice for protection against hazards arising out of static electricity in general industries", RIIS-TR-87-1, p. 132-151.

7. S. Masuda(1974), "Recent electrostatic engineering", KHK-E 006, p. 64 (in Japanese)

Why Qra for Emergency Planning?

Farshad Nourai
Rastar Farayand Consulting Engineers
Tehran, Iran
f.nourai@rastarco.com

Abstract:

An Emergency Management Plan (EMP) is an important ingredient of any successful process safety management system. On the other hand, in a typical chemical plant there are several possibilities for emergencies. One must thoroughly analyze any possibility in order to reach a set of representative scenarios on which the emergency response plan can be built.

In this study, Quantitative Risk Assessment (QRA) is used for emergency planning in a real chemical processing facility. Release scenarios were identified using a detailed HAZOP study, followed by a qualitative risk assessment for screening credible accident scenarios. Final quantitative risk assessment of these scenarios helped screen 9 representative scenarios that were used for preparation of the EMP.

Briefly reviewing the plan itself, the current paper mainly discusses three crucial questions concentrated on the practicality of application of QRA for emergency planning based on the experience of this project, namely: (1) Is QRA necessary? (2) Is QRA economical? (3) Is QRA advantageous? Each question has been analyzed in terms of the standard content of modern emergency management plans, as classified into four major categories: mitigation, preparedness, response, and recovery.

The paper also gives remarks on a number of issues that need to be addressed if use of this technique is going to be extended. A part of the paper is dedicated to a real-time emergency awareness and response system that was specifically developed for this project.

Simplified Pressure Relief Requirements for Indirectly Heated Equipment Containing a Homogeneous Fluid

Robert N. D'Alessandro
Degussa Corporation
Theodore, AL 36590-0606
Robert.Dalessandro@degussa.com

and

Christoph Hertel
Otto-von-Guericke University Magdeburg
Magdeburg, Germany
robert.dalessandro@degussa.com

Abstract:

The motivation for this paper comes from the consideration of vertical shell and tube vaporizers where the shell side fluid receives uncontrolled heat input from the hot tube side fluid. The even distribution of tubes in this type of exchanger can result in volumetric boiling, and therefore, significant level swell similar to the situation that occurs during a bulk uncontrolled exothermic reaction in a vessel. Therefore, the pressure relief device on this type of equipment must consider the possibility of two-phase flow. Given this, new simplified equations are derived for the required relieving rate assuming the contents act as a boiling vapor- liquid homogeneous fluid during emergency venting. Unlike previous work in this area, the expressions presented here account for a temperature dependent heat source. Since the homogeneous vessel assumption is employed, the equations are applicable to any indirectly heated equipment whose contained fluid is expected to exhibit homogenous character due to equipment geometry, equipment internals, or simply the fluid properties.

Ethylene Decomposition Relief Sizing High Pressure Polyethylene Industry

David I Shannon
E. I. DuPont de Nemours & Co., Inc.
2695 Old Bloomington Hwy (P.O. Box 526 – 77902)
Victoria, Texas 77905
Phone: 361-572-1813 E-Mail: david-irvin.shannon@usa.dupont.com

ABSTRACT:

This paper discusses revisions to a method previously used for sizing relief devices (rupture disks) on vessels in high pressure Polyethylene processes. The Re-evaluation was forced by an unexpectedly high peak pressure in a product separator (High Pressure Separator - HPS) when an ethylene decomposition occurred. A method, based on revised equations for calculating the rate of pressure rise, is presented for determining relief requirements for an ethylene decomposition in a confined space. The method applies to both HPSs and Polyethylene Autoclave Reactors. Causes of decompositions and effect of comonomers are also discussed.

INTRODUCTION:

As a result of an incident at a DuPont facility, ethylene decomposition relief device sizing was revisited. In the process of doing this, other incidents were identified at licensee facilities that also had unexpected results. After of the reanalysis of ethylene decomposition data, a revised method of sizing was developed. This information has been published in the past at the June 1992 ASME Pressure Vessel and Piping Conference in New Orleans, La., and at a Society of Polymer Industry safety Conference in February 1995. In recent years it has been observed that many High Pressure Polyethylene Facilities have not made changes needed to address the risk identified by these previous presentations. Additional discussion has been added about causes and associated changes in decomposition severity, and effect of comonomers (vinyl acetate).

INCIDENTS:

In January 1987, a DuPont facility experienced an incident where the HPS pressure exceeded 5000 psig (extrapolated to >6000 psig, instrument maximum scale was 5000). The relief sizing for this vessel indicated that the rupture disks should have been large enough to keep the event pressure from rising after the disk burst (3300psig). Technically, even this event would not be a problem for this vessel if it had a high enough design pressure, but the design pressure of 4500 psig only allowed a peak pressure of about 5735 psig based on 85% of the hydrostatic pressure. The hydro-static test pressure was 6750 psig (designed based on ASME Section VIII, Div. 1). The piping in the off-gas system of this vessel was only designed for 2700 psig. This piping opened at the joints during this incident. Relief protection for the piping was sized for process upsets only; the HPS disks are there to protect from a decomposition. The event itself actually started in the line exit the bottom of the HPS. The operator involved was unable to establish flow through this 550 psig steam jacketed line. To break through the perceived plug, pressure was increased and decreased 3 times in a short period of time, not only increasing the pressure in the line, but also increasing the temperature in an already hot line (250°C). This resulting temperature exceeded the ethylene decomposition temperature (see reference 1, 2 and 3 for this temperature(s)) due to adiabatic compression. The temperature increase is substantial due to poor mixing in the line, essentially acting as a piston compressing the hot gas at the end of the line which happened to be a broken guard valve.

Two similar events occurred in 1994 at another facility. These resulted from a repair of the steam condensate drain of the HPS bottom exit (tails) line and a poor but accepted operating procedure; the line (25 ft long) was pressured up, with steam on the jacket, against an automatic valve. The poor operating procedure was masked by a condensate drain that had required flow out the top of the jacket. When flow out the bottom of the jacket was restored, the poor procedure was exposed by the 2 decomposition events. Each of these incidents resulted from one pressurization of this line adiabatically compressing 250°C ethylene from less than 10 psig to 2400 psig.

SIZING BASIS:

The original work on ethylene decomposition sizing was done, within DuPont, in 1952 through 1954 to assist in the design of a new High Pressure LDPE Autoclave process. Based on that work the Reactor and High Pressure Separator (HPS) relief protection was designed. This design was based strictly on the operating conditions expect while making polymer. It did not consider start-up or shutdown conditions that might resulted in different relief protection.

Reactor relief design has resulted in no events on a DuPont Reactor where all of the disks ruptured, indicating that the available relief area was high enough to handle the pressure increase generated by the decomposition.

The original HPS rupture design was for a decomposition that started at normal operating conditions, not low temperature conditions of start-up or shutdown. It was assumed that decompositions only result from propagation from Reaction or from external heat input. Based on experience this assumption was invalid.

DECOMPOSITION CAUSES:

Ethylene decompositions result from ethylene breaking down. The chemical reactions that are normally proposed for this reaction are below:

$$C_2H_4 \rightarrow C + CH_4 \quad \text{and} \quad C_2H_4 \rightarrow 2C + 2H_2$$

In reality the result of an ethylene decomposition event is methane, hydrogen, but also many different oligomers, and possibly some polymer as reported in reference 3, 4, 5 and 6. So, with indefinite complex mixtures, that change with operating conditions and compositions, an empirical calculation is necessary to get calculable results.

In a Polyethylene Reactor decompositions can be initiated by too much reaction initiator (oxygen, peroxides and other catalysts if used). They can be initiated by not removing all of the oxygen from the process during equipment purging. And they can be catalyzed by impurities such as metal compounds from packing, particular copper.

In the High Pressure Separator (HPS) in a Low Density Polyethylene plant, decompositions have multiple potential causes. The first one expected is propagation of a decomposition from reaction. Another is heat input from equipment or pipe jackets. Equipment purging or plug removal can adiabatically compress gas in heated downstream piping; e.g., using HPS high pressure gas to quickly pressure up a heated bottoms line that can be plugged with polymer or may have multiple valves, the downstream one being closed. Or there can be flow regime problems in the HPS bottoms line that can adiabatically compress gas to very high temperatures; flow is in the slug flow regime due to the wrong line pipe diameter for the length of line required. Decompositions that start in heated lines can, and do, propagate back into the HPS

Original Rupture Disk Sizing Basis:

Work in some unpublished laboratory work done in the 1950's showed development of equation (1) for the rate of pressure rise during a decomposition. This was developed from tests run by initiating decomposition in a number of different sized static bombs and stirred autoclaves. The decompositions were initiated by hot wire, squibs, external heat, oxygen and peroxide initiators. The equation was developed by fitting the pressure rise versus time data. It should be recognized that the equations which follow do not represent a rigorous kinetic model. They are approximations intended strictly for sizing rupture disks for ethylene decompositions.

$$(1)\quad \frac{dP}{dt} = 2.3 * K' * P, \text{ where}$$

K' = Violence coefficient
P = Pressure, psia.
T = time, sec.

The violence coefficient K' for the HPS involved in the 1987 incident was estimated based on the laboratory data at similar pressures. Violence coefficients for other vessels, such as the reactor, were estimated the same way.

Equation (2) was also developed for disk venting rate as a function of the pressure, the vessel volume, and the total rupture disk flow area. This was done by fitting the rate of pressure decrease after the decomposition reaction was complete.

$$(2)\quad \frac{dP}{dt} = - 0.147 * P^{1.5} * S, \text{ where}$$

S = A/V
A = rupture disk area, in^2
V = vessel volume, ft^3

Note that the available relief area includes the effective capacity of any relief devices attached to the process lines and the flow through any line that leads to another system with significant capacity, such as off-gas line with relief valves.

Process vessels are not closed bombs, they have rupture disks so equations (1) and (2) can be combined to yield a single relationship, equation (3). This equation can only be used once the vessel pressure actually opens the rupture disks, until then only equation (1) can be used.

$$(3)\quad \frac{dP}{dt} = 2.3 * K' * P - 0.147 * P^{1.5} * S$$

Ideally, the value of dP/dt in equation (3) is negative. So, when a rupture disk bursts, the pressure immediately starts back down. Based on the three incidents mentioned above and a number of incidents on other facilities known to DuPont, this is not normally the case. The design case for the vessel involved in the 1987 incident used a K' of 0.2.

With a vessel volume of 355 ft^3, rupture disk area of 31.7 in^2 and a burst pressure of 33 psig, the value of equation (3) at burst is negative (-550 psi/sec). Since the dP/dt was greater than zero the data was revisited resulting in the changes discussed below.

Effect of Initial Density

The actual peak pressure experienced in the 1987 incident indicated that something was not being considered. All DuPont laboratory data and plant incident data were reviewed to look for variables not accounted for in these equations. This revealed that the closed bomb pressure (P_{max}, psia) and Violence Coefficient (K') are functions of the mass of ethylene in a vessel at the start of the decomposition as measured by the initial density (Den_0, lb/ft^3). Equation (4) shows the closed bomb pressure relationship and equation (5) the K' relationship, as reported in the 1992 ASME Paper (8).

(4) $Ln(P_{max}) = 1.3423 * Ln(Den_0) - 0.4097$

(5) $Ln(K') = 2.663 * Ln(Den_0) - 6.896$

The data for the regression that produced equation (4) can be found in Table 1 below. However, it was later found that the units for the result were in Kpsia, instead of psia, so the statistics were revisited for this paper and equation (4r) generated as a replacement.

(4r) $Ln(P_{max}) = 1.2273 * Ln(Den_0) + 6.807$

Table 1 Closed Bomb Data

Init T, C	Init P, psia	Pmax	Init Den, lb/ft3
150	22000	43000	30.77
150	22000	68300	30.77
250	4500	20100	11.82
250	4500	18700	11.82
250	2250	7500	6.36
250	2250	8570	6.36
150	4500	32900	16.2
150	4500	32300	16.2

The data for the regression that produced equation (5) can be found in Table 2 below; includes 13 plant incidents within DuPont, it's joint ventures, or it's Licensees over a 40 year period.

Table 2 Violence Coefficient Data

Initial P, psig	Initial P, barg	Initial T, C	Violence Coefficient, K'	Lab or Plant Data	Relich-Kwong Initial Density, lb/ft3	BWR-LK Initial Density, lb/ft3	Initial Density, lb/ft3 ref 7
9000	621	170	2.2	Lab	23.47	22.8	22.944
12700	876	150	8.3	Lab	27.4	26.4	26.603
14000	966	150	3.2	Lab	28.2	27.1	27.352
22000	1517	242	9.7	Lab	29.15	27.6	--
22000	1517	299	8	Lab	27.73	26.1	--
22000	1517	283	10	Lab	28.11	26.5	--
27000	1862	243	7	Plant	30.82	29.1	--
26000	1793	249	4.8	Plant	30.37	28.7	--
13000	897	215	10	Plant	25.22	24.2	24.738
24000	1655	253	7	Plant	29.61	27.9	--
11000	759	222	5	Plant	24.98	22.5	25.352
3770	260	142	0.54	Plant	15.92	15.9	16.022
3770	260	260	0.47	Plant	10.81	10.2	10.87
4002	276	288	0.63	Plant	10.62	10.3	10.458
3886	268	280	0.48	Plant	10.55	10.2	10.367
3712	256	260	0.57	Plant	10.67	10.4	10.703
3654	252	264	0.51	Plant	10.41	10.1	10.421
3770	260	254	0.73	Plant	10.99	10.7	11.053
2600	179	140	0.57	Plant	11.78	11.7	11.871
9000	621	360	1.18	Lab	16.99	16.1	--
9000	621	276	0.91	Lab	16.59	18.6	18.785
2250	155	250	0.176	Lab	6.78	6.7	6.698
2250	155	250	0.143	Lab	6.78	6.7	6.698
12700	876	350	10.2	Lab	20.94	19.8	--
14000	966	375	3.2	Lab	21.35	20.1	--
22000	1517	370	12	Lab	26.11	24.4	--

Additional regression work has been done on this data also to determine the effect of using different equations of state to determine the initial gas density of the system. In doing this update it was noticed that regression of the same data could not duplicate the original equation, the reason for which is unknown. So, four equations were generated by regressing the data in Table 1, using the different sets of density data from either actual data tables (from Ref 7), or two equations of state (Relich-Kwong and Benedict-Webb-Ruben--Lee-Kessler). The results of the regression are shown below:

Original	$\mathrm{Ln}(K') = 2.663 * \mathrm{Ln}(Den_0) - 6.896$
RK	$\mathrm{Ln}(K') = 2.692 * \mathrm{Ln}(Den_0) - 7.060$
BWR-LK	$\mathrm{Ln}(K') = 2.735 * \mathrm{Ln}(Den_0) - 7.084$
Actual $Density_7$	$\mathrm{Ln}(K') = 2.450 * \mathrm{Ln}(Den_0) - 6.514$

To determine if the difference in these equations is significant, 5 different densities were input to each equation in the "Peak Pressure Estimation" method discussed below. For each case the same rupture disk area of 31.7 in^2 and volume of 355 ft^3 were used. The standard deviation for these results was about 5% indicating that the four equations provide about the same result. So, it is recommended that the equation regressed on actual density data be used (equation (5r) below, since it is based on actual data. If the original equation is used, or an equation of state is used to estimate ethylene density, the results will not change enough to provide an unsafe rupture disk sizing result.

(5r) $\quad Ln(K') = 2.450 * Ln(Den_O) - 6.514$

To test this peak pressure algorithm, (equations (1) – (5)) a computer simulation was written to solve equation (3) for small time increments, changing the differentials to differences leading to equation (6) below:

(6) $\quad P_{t+1} = P_t + (2.3 * K' * P_t - 0.147 * P_t^{1.5} * S) * t_d$, where

t_d = calculation time increment, sec

The results are shown in Figure 1 (1987 incident – 140°C, 2600 psia, 355 ft^3 vessel, with 31.7 in^2, 3300 psia burst pressure). The graph in Figure 1 indicates a good fit to the actual event. The plot shows pressure constant until t = 0.0, then rising based on equation (6) without any venting until $t = t_r$, where the rupture disk bursts. At $t = t_{max}$, the pressure rise portion of equation (6) drops out of the equation since there is no longer any ethylene to decompose. The pressure at time t_{max} is the peak pressure. The times t_r and t_{max} are calculated by equations (7) and (8) below. Simulation of other events at different operating conditions, not shown, indicate a good match for peak pressure. Therefore. the equations above provide a good estimate of ethylene decomposition peak pressure given the initial conditions.

Figure 1: Ethylene Decomposition Incident - 1987

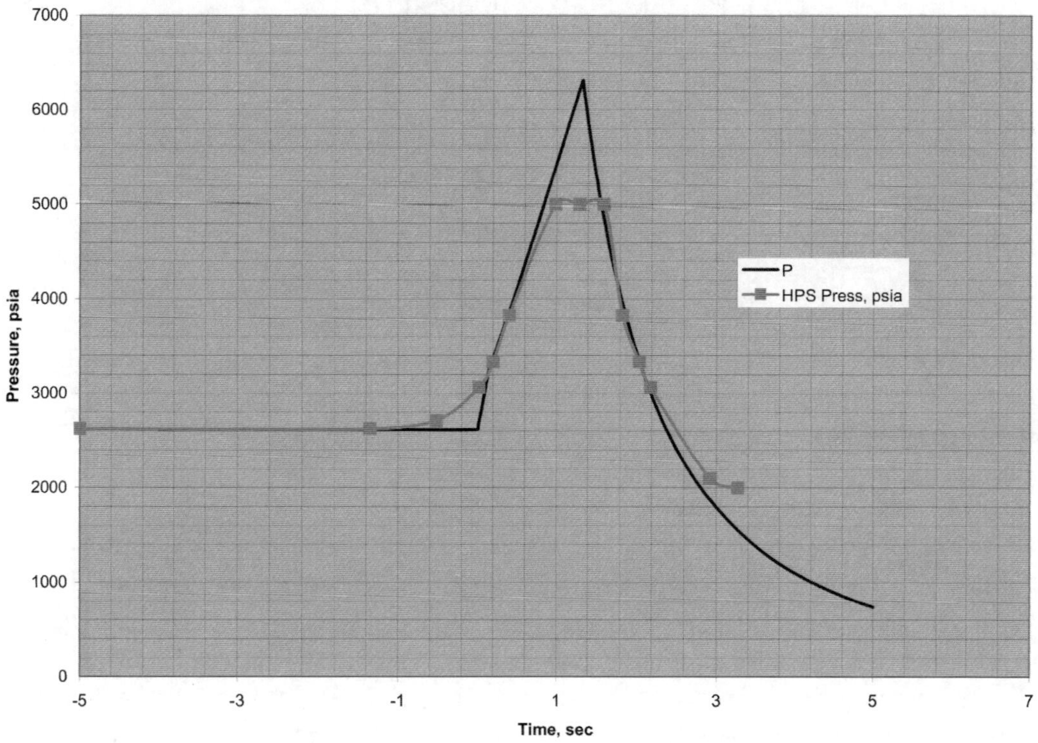

Peak Pressure Estimation

If the value of equation (3) is greater than zero, a more rigorous calculation has to be made. This can be done by unsteady state calculations as discussed above (equation (6)), or by integrating equation (3). In order to estimate the peak pressure during a decomposition event, the rupture disk burst time and the duration of the event need to be calculated. To do this, equation (1) is integrated and solved for t_{max} to get equation (7).

$$(7) \qquad t_{max} = \frac{\ln\left(\frac{P_{max}}{P_o}\right)}{2.3 * K'} \text{ , where}$$

P_{max} = Closed Bomb Pressure, psia,
or pressure at which reaction is completed without a relief device.
P_o = Initial pressure, psia
t_{max} = Decomposition duration, sec

By substituting the rupture disk burst pressure (P_r) for the closed Bomb Pressure (P_{max}), the time to rupture disk burst (t_r) is calculated by equation (8).

$$(8) \qquad t_r = \frac{Ln\left(\frac{P_r}{P_o}\right)}{2.3 * K'}$$

The amount of time the rupture disks are venting until peak pressure is reached is calculated by equation (9).

$$(9) \qquad t_v = t_{max} - t_r \text{ , where}$$

t_v = venting time to peak pressure, sec

The times calculated above are then used in an integrated (rearranged) version of equation (3): equation (10) which is used to calculate the peak vented pressure (P_v). The integration is done between the rupture disk burst pressure (P_r) and the vented pressure (P_v), from t_r to t_{max} (or zero to tv, tr is reference). This peak pressure is the maximum pressure expected in a vessel during an ethylene decomposition.

$$(10) \qquad P_v = \frac{X^2}{Y^2} \text{ , where}$$

$$X = 2.3 * Pr^{0.5} * K' * e^{0.5*K'*2.3*tv}$$
$$Y = 2.3 * K' - 0.147 * P_r^{0.5} * S + Z$$
$$Z = 0.147 * P_r^{0.5} * S * e^{0.5*K'*2.3*tv}$$

For the incident conditions above (1987) equation (10) yields a peak pressure of 6497 psia (K'=0.636), which agrees with the unsteady state method results shown in Figure 1.

EFFECT OF COMONOMERS

A small amount of data is available for Vinyl Acetate decompositions from a DuPont internal report. As shown in Table 4 the violence coefficient for the 80% VA system is more than 1 order of magnitude lower. A fit to this data combined with the ethylene K' data gives the following:

$$Ln(K') = 2.663\ Ln(Eth\text{-}Den_o) - 6.896 - 2.63066\ (Wt.Fr.VA)$$

This is the ethylene K' as a function of initial density with the addition of a term that corrects for VA concentration. The effect that this has on an ethylene decomposition is that K' for 18% VA, for example, is approximately 1/2 of the ethylene K'.

More data is needed to accurately do these calculations for systems with Vinyl Acetate, or other comonomers, in them. The major limit here is that the effect of VA on the closed bomb pressure is not known. So, with the significant reduction in K' with increasing vinyl acetate, and no closed bomb pressure data, sizing for ethylene only is recommended.

Table 4
80% Vinyl Acetate - Ethylene Decomposition
Violence Coefficient Data

Initial Press. Psig	Initial Temp. ^oC	Initial Density lb/ft^3	Initial Eth.Density lb/ft^3	Violence Coefficient K'	Ethylene Predicted K'
3758	272	19.00	10.48	.036	.51
2420	330	12.49	6.02	.0047	.127
2880	302	15.06	7.58	.0063	.225
2865	329	14.22	7.11	.0046	.19
3085	361	14.15	7.94	.0315	.253

ALTERNATE METHODS:

There are at least two other methods of calculating decomposition required relief area. One recent paper presented a comparison of the DuPont method, discussed above (based on ref. 8 equations from 1992) and what was designated as "conventional method, which assumes a steady state mass flow rate that is sufficient to move energy from the vessel at a rate equal to the maximum rate at which it is input into or released within the vessel", and a modified DuPont method. See reference 9 for details.

REFERENCES:

1. Neumann, R. and G. Luft (Aug. 1973), "Studies of the Thermal Decomposition of Ethylene under High Pressure", Chemical Engineering Science 28, pp. 1505-1514.
2. Albert, Jens and Gerhard Luft (1998), "Runaway phenomena in the ethylene/vinylacetate copolymerization under high pressure", Chemical Engineering and Processing 37 (1998) pp. 55-59.
3. Albert, Jens and Gerhard Luft (October 1999), "Thermal Decomposition of Ethylene-Comonomer Mixtures under High Pressure", AICHE Journal Vol. 45, No. 10, pp. 2214-2222.

4. Bonsel, Harald and Gerhard Luft (1995), "Safety Studies on the Explosive degradation of Compressed Ethene", Chemical Engineering Technology 67, No. 7, pp. 862-864.
5. Watanabe, Hiromasa and Masaaki Takehisa (1982), "Studies on the Explosive Decomposition of Ethylene—Mechanism of Ethylene Decomposition under High Pressure", Koatsu Gasu 19, No. 12, pp. 605-612 (Department of Research, Takasaki Institute, Japan Atomic Energy Institute).
6. Watanabe, Hiromasa, H. Kurihara and Masaaki Takehisa (1972), "The Explosive Decomposition of Ethylene", Loss Prevention and Safety, Pachec '72, pp. 225-232.
7. (Reference Book) DeReuck, K.M., International Union of Pure and Applied Chemistry: ETHYLENE – International Tables of the Fluid State.
8. Sullivan, J.F. and D.I. Shannon (1992), "DECOMPOSITION VENTING IN A POLYETHYLENE PRODUCT SEPARATOR". PVP, Vol.238, Codes and Standards and Applications for High Pressure Equipment, ASME, 1992, pp. 191-195.
9. Darby, Ron (2006), "Relief Vent Sizing for Deflagrations", Process Safety Progress (Vol. 25, No. 2), pp. 130-134.

Using Industrial Wireless Instrumentation to Monitor Safety Showers in Chemical Plants

Gene Yon
Adaptive Instruments/Accutech
Hudson, MA 01749
jrogers@accutechinstruments.com

Abstract:

Safety showers in industrial facilities provide first-line emergency treatment for employees that have come in contact hazardous materials. They are normally combined with eye-wash stations for emergency treatment for eyes. In virtually every instance where an employee has triggered the use of a safety-shower or eye-wash station, that employee will be in need of help and assistance for further treatment. Rapid assistance is often critical in minimizing or mitigating any potential injury and getting help to an injured employee can often make the difference between a serious injury and a minor incident.

Unfortunately, the cost to install wired monitoring switches to multiple safety-showers in a major production plant can be extremely expensive. Wiring monitoring switches entails the cost of running the wires to the switches in the field, often through areas classified as Class I, Div I or Class I, Div II. Wired switches also entail the expense of dedicating individual discrete inputs into plant DCS systems. These costs rapidly add up. One recent facility estimated the costs for wiring safety-showers to be over $4,000 per point. Per point monitoring costs are generally reduced by 75% using wireless sensors compared to hard wire.

A brief summary of this safety application will be reviewed along with other safety application using wireless instrumentation such as on/off valve positioning, overfill protection, control valve status and pressure and temperature monitoring. This applications survey is intended to show that wireless solutions do have a place in safety compliance conformity at plants and refineries. When properly applied, this technology will solve difficult measurement problems not presently served with wired instrumentation or captured at all.

Tips for Successful Modeling of Relief Valve and Vent Emissions

Randy Hawkins
RRS Engineering
League City, TX 77573
donna@rrseng.com

Abstract:

One of the most common uses of dispersion models is to investigate the potential consequences of relief valve and vent emissions to the atmosphere. The model programs are used to predict the concentrations of toxic and/or flammable materials delivered to grade, the levels of thermal radiation delivered by jet and pool fires, and the overpressures resulting from vapor cloud explosions. The application of most dispersion model program packages to those purposes appear to be very straightforward; however, there are several "traps" that an inexperienced user can fall into. Potential problems related to source descriptions, meteorological specifications, and interpretation of output will be discussed, and simple solutions to each problem will be revealed.

Obtaining Realistic Human Error Rates for Process Hazard Analyses

John C. Wincek, CSP*
Croda, Inc.
8 Croda Way Mill Hall, PA 17751
570-748-7796
j.wincek@crodausa.com

Joel M. Haight, Ph.D., PE
Department of Energy & Geo-Environmental Engineering
The Pennsylvania State University
110 Hosler Building
University Park, PA 16802-5000

*Author to whom correspondence should be addressed.

ABSTRACT

While human error has long been recognized as a contributor to process incidents, Process Hazard Analysis (PHA) teams often struggle to reach consensus on the frequency of human errors might. This disagreement can be due to the varied backgrounds and experiences of team members, as well as the degree to which they believe human error is inevitable as opposed to simply being a reason for discipline. A process that combines research-based, quantitative estimates of human error rates with plant experiences is proposed for use during qualitative PHAs. Use of these quantitative estimates within a qualitative Process Hazard Analysis is described, along with one facility's successful implementation of this process.

Keywords
Human Factors, Operator Error

INTRODUCTION AND BACKGROUND

Process Hazards Analysis and Human Errors

From before the earliest days of the process safety management movement there has been a growing recognition of the importance of human errors in the causative events leading to an incident. The Center for Chemical Process Safety (CCPS) devoted an entire chapter to Human Factors in one of their first frameworks for Process Safety. [1]. Lees' Loss Prevention in the Process Industries devotes considerable time to this topic as well. [2]. Consideration of the contributions of human factors to process hazards is currently required by both the Occupational Safety and Health Administration's Process Safety Management standard [3] and the Environmental Protection Agency's Chemical Accident Prevention Provisions (Risk Management Plan). [4]

Despite available guidance [2] and [5], Process Hazard Analysis (PHA) teams can still struggle to agree on even a qualitative estimate of how often an employee will make a mistake. As with any team, the members bring with them a wide variety of beliefs, expectations and experiences (both professional and personal) that influence their view of human error. Opinions can vary from "they should pay more attention, or face disciplinary action," to "it is more a question of when, not if a human being will make a mistake." Some team members may try to relate an error under consideration to their own personal experiences. Others may recall errors that have occurred in their professional experience, in order to estimate error frequency, or they may conclude that any error, even one they have made themselves, is the result of laziness or inattentiveness, and can be avoided simply by "being more careful." The PHA team leader can feel cornered into accepting the opinion of the highest-ranking member of the team, or perhaps that of the most obstinate participant.

One chemical process industry facility experienced several human error-related incidents over a relatively short period of time. In one of the incidents, an operator was required to purge a pressure vessel with nitrogen by raising the pressure to 25 psig, then venting to atmosphere down to 1 psig. The operator accomplished this using hand valves on the nitrogen and vent lines on the mezzanine level of the vessel. This operation was to be repeated two times, finishing the purge by closing the vent valve, at which time the operator could return to the ground floor to charge the next reactant. The PHA team believed leaving the vent valve open after completing the purge operation was not a credible error because the operator 1) had to go to the mezzanine level to purge the vessel, 2) went to the mezzanine level only for the explicit purpose of purging the vessel, and 3) had no reason to leave the mezzanine level until the purge was completed. As might be guessed, the operator was called away on another task while the vessel was venting down from 25 psig, and he failed to close the vent valve after the purge was completed. This resulted in a release of toxic chemical, above the Threshold Quantity.

During the facility's investigations of this and other human error-related incidents, the PHAs were consulted to determine how the error potential had been assessed. In all cases, the team had either underestimated the frequency of the error, or determined that the error was not a credible occurrence. It was obvious that the PHA teams' assessments of what is "credible" required improvement. It was also surmised that human errors were occurring, with minor consequences, but were not being reported. This could have contributed to the PHA team's assessment of a low risk of human error.

Human Error Rates

To better understand human reliability at this facility, a literature search was performed in an attempt to quantify the error likelihood of many of the tasks that take place in this type of operation. Any time a quantitative human reliability analysis is undertaken, the analyst has to consider the use of a general industry average task-based error rate for each task or to collect site-specific error rate data. Each path has its advantages and disadvantages. A general industry average error rate is easier to obtain, but may not reflect the level of training or experience, the quality of the procedures or the effectiveness of the management systems of the particular operation being analyzed. The site-specific error data will reflect the performance of the site being analyzed, but the data are difficult to collect and it takes a significant amount of time to obtain enough data for it to be representative of actual system performance to an acceptable level of statistical confidence. [6], [7] and [8]

Given this dilemma, there are two routes an analyst can take. One is to collect enough site-specific error rate data to allow error rate prediction with some relative uncertainty through a Poisson, binomial or similar analysis depending on the resulting frequency distribution generated by the data. The other is a quantitative weighting process in which a PHA team mathematically "adjusts" the general industry error rates based on several performance shaping variables. These variables, partially noted above, could be the level and quality of the training program, the level of experience of the work force, the quality of the operating procedures, the level of distractions an operator experiences throughout a shift, etc. The weighting factors used are dependent upon those that are deemed to impact performance at the plant or operation being analyzed. The weighting factors were developed to allow the team to systematically and consistently adjust the general industry data to a level on which the team can agree. Each weighting factor was developed to generate as much as a one order-of-magnitude change in the general industry error rate if it is warranted. [9 and [10] While it is conceivable that with several weighting variables in use, an error-rate adjustment could be several orders of magnitude, it is likely that in most cases, the variables will work more to offset each other than to enhance each other. "Over adjustment" of the base error rates can be monitored and controlled by the PHA team leader.

The actual procedure used at the subject facility is outlined and detailed in the methods section below. The base data comes from several databases. Blanchard (2000) [11], Swain and Guttman, (1983) [12], Peterson, (1996) [13], Stevenson, (1991) [14], Sverdrup (website accessed 4 August 2004) [15], and Clemmons (2002) [16] provided much of the general industry base data used for this method at this facility. The literature review produced several useful task categories from which appropriate error rates could be determined as starting points for the analysis. These task categories and error rates are shown in Table 1 below.

TABLE 1 - BASE ERROR RATES FOR VARIOUS TASK TYPES

([11], [12], [13], [14], [15] and [16])

KEY	Category	Chance/ Opportunity
A	Crew reaction during air disaster	1.00E-01
B	Simulated military emergency	1.00E-01
C	Passive inspection (general walkaround)	1.00E-01
D	Monitor/Inspector fails to recognize initial error by operator	1.00E-01
E	Simple arithmetic error (without redoing calculation on separate paper)	3.00E-02
F	Missetting large manual valves (no indication except open or closed	1.00E-02
G	Technician "seeing" an out-of-calibration instrument as "in tolerance"	1.00E-02
H	General error of omission (no control room display-fail to reset valve after maintenance)	1.00E-02
I	Selection of switch dissimilar in shape or location to desired switch	1.00E-03
J	General errors of omission for items imbedded in a procedure	3.00E-03
K	General errors of commission, e.g., misread label and selected wrong switch	3.00E-03
L	Missetting large manual valves (controlled by procedure, keys, chaining, etc.)	1.00E-04
M	Selection of key operated switch rather than non-key switch (not decision error)	1.00E-04
N	Upper limit to credibility	1.00E-05
O	Two-person team (one do; one checks and then reverse roles)	1.00E-05
P	Observe chart - inappropriate switch activation	1.13E-03
Q	Read gauge - read incorrectly	5.00E-03
R	Read instructions - make procedural error	6.45E-02
S	Connect hose - Improperly connect	4.70E-03
T	Torque fluid lines - Incorrectly torqued	1.04E-04
U	Tighten nuts and bolts - Not tightened	4.80E-03
V	Install nuts and bolts - Not installed	6.00E-04
W	Install O-Ring - Improperly install	6.67E-02
X	Solder Connectors - Improper solder joint	6.46E-03
Y	Assemble connector - Bent pins	1.50E-03
Z	Assemble connector - Omitted parts	1.00E-04
AA	Close valve – Not closed properly	1.80E-03
AB	Adjust mechanical linkage - Improper adjustment	1.67E-02
AC	Install Line orifice - Wrong size installed	5.00E-03
AD	Machine valve port - Wrong size drilled and tapped	2.08E-03
AE	General rate/ high stress/ dangerous activity	2.00E-01
AF	Checkoff provision improperly used	5.00E-01
AG	Error of omission – 10-item check off list	1.00E-03
AH	Carry out Plant Policy - no check on operator	5.00E-02
AI	Select wrong controls	3.00E-03
AJ	Operator fails to recognize undesired position of valves	5.00E-01

METHODOLOGY

Error Rates for Facility-Specific Conditions

Because the base human error rates were obtained from a broad range of working conditions (industry types, management systems, shift patterns, etc.) it was believed by the analysts at this facility that the rates should be adjusted to better fit the working conditions in the subject operations by using Error Rate Modifiers (ERMs). For each task where an error could occur, the PHA Team assigned an ERM between 0.1 and 10 for each of the following:

- Time Pressure to complete the task (i.e. the task must be completed in a small window of time)
- Experience level of the operator performing the task
- Distractions or other tasks being performed simultaneously
- Adequacy of procedures
- Fatigue due to 12-hour rotating shifts

Each Error Rate Modifier, selected to capture the conditions that make a specific facility and/or task unique or different from the "average," can have the effect of changing the base error rate by one order of magnitude in either direction. A value of 1 is assigned if a facility considers itself "average" with respect to a particular modifier, or if the particular modifier is thought to have no impact on the task at this site. The modifier values were selected as a means to adjust base error rates by up to one order of magnitude in either direction because it is thought that this is a conservative approach that could be supported by both experience and reasonableness in the absence of local data. The teams agreed and as can be seen below, most of the analysis teams remained conservative even within the conservative constraints of the two order of magnitude range of variation.

The adjusted rate is obtained by multiplying the Base Error Rate by each of the modifiers:

$$\mathbf{AER = BER \times ERM_1 \times ERM_2 \ldots ERM_n} \qquad \text{(Equation 1)}$$

Where: AER = Adjusted Error Rate with the units chance per occurrence
BER = Base Error Rate with the units chance per occurrence
ERM = Error Rate Modifier (unitless)

If the team believes that one of their ERMs would affect the error rate favorably (meaning it would lower the base error rate), they might assign a value of 0.3 to reduce a rate from, for example, 0.01 to 0.003 (base error rate (0.01) x error rate modifier value (0.3) = adjusted error rate 0.003. If the team believes that the experience at the site warrants an increase in the error rate, they might assign a modifier value of 5, increasing the error rate (Base Error Rate (0.01) x Error Rate Modifier value (5) = Adjusted Error Rate (0.05). Since there are several factors that could influence the error rates, and since some of them could positively affect the error rate and some could negatively affect the error rate, it is possible that the effect of one modifier will offset the effect of another modifier. See "Use During PHAs" below for further description and an example to illustrate how this process works.

In the author's experience, PHA teams at the subject site typically assign values between 0.5 and 3 for individual modifiers, and it is rare that more than 2 of the modifiers are assigned a value other than one for a particular task. It has also been recognized that the value of a modifier such as Operator Experience can change over time, even without a change to the process. This could be due to employees gaining experience over time, as well as promotions and employee turnover.

Use During PHAs

During a PHA, such as a Hazard and Operability Study (HazOp), the study team typically identifies several human error opportunities in a particular operation. The risk ranking is done using a traditional risk assessment matrix methodology. An example from an actual PHA is shown in Figure 1, where the analysis team assigns a Risk Ranking (columns S, F and RR) for both the existing system and the system as modified by the team's recommendations. The analysis team first assigns values for S (Severity, ranging from 1 to 5, with 1 representing the most severe consequences) and F (Frequency, ranging from A to E with A representing the most frequent occurrence) under the heading "Risk Matrix." A Risk Matrix is used to plot the intersection of the severity and frequency values, producing an RR (Risk Ranking) value between 1 and 25, with 25 representing the highest risk. If recommendations are made, the team recalculates the Risk Ranking as if the recommendations have been implemented, producing a new Risk Ranking value that reflects the risk reduction achieved by the recommendation. These new Risk Ranking values are entered in the last three columns of the PHA worksheet shown in Figure 1, and provide management a measure of the value received if the recommendation is implemented.

For each potential error, the team agrees on the most appropriate error type (error category) and its corresponding Base Error Rate from Table 1. The team then assigns values for each modifier, based on the circumstances and performance shaping factors [13] and [17] surrounding the specific task under consideration. For documentation purposes, the PHA scribe enters this information into a spreadsheet like the one shown in Figure 2, which automatically computes the Adjusted Error Rate. The adjusted error rate can then be multiplied by the number of times the task is performed in a year, to obtain the number of errors (or probability of error – since it is known how many times/year a task is performed, the probability can be computed) per year.

EPY = AER x OPY **(Equation 2)**

Where: EPY = Errors (or probability of error) per year
AER = Adjusted Error Rate (from Equation 1, with the units chance of error per opportunity)
OPY = Opportunities per year (number of times the task is performed/year)

Depending on the circumstances, a PHA team may want to identify the number of times per year a specific task is performed, or broaden the scope to include the number of times a type of operation is performed. For example, a HazOp was recently conducted on a batch process that manufactured 100 batches per year. Each batch required the operator to reposition 42 hand valves. Having any of the valves in the incorrect position could result in the release of a flammable liquid. The PHA team, when developing the expected human error rate for valve positioning errors, chose to group them together as 4,200 valves repositioned per year. This provided a clearer picture to management, showing that the number of opportunities for a single human error (miss-positioning a valve) to cause a flammable liquid release was quite high.

VALIDATION OF RESULTS ACROSS SEVERAL PHA STUDIES

A concern one might have about using any new methodology is validating that new methodology against known and accepted standards. This sometimes presents a problem, because a new method may not have established standards against which it can be compared. While the method presented here is new, there are data available from three studies performed at this site using this method, which help to support more extensive and continued use.

Since the base error rates are selected from a list of task descriptions and error rates from the research literature, and this list is confined to basic human actions, there is good consistency (approximately 90%) across the three studies in selecting base error rates. For tasks from these three studies where the operator may select the wrong valve, all three study teams consistently chose the same base error and error rate for operating an incorrect valve. Since the number of batches run per year for each product is factual and measurable, there is accuracy and consistency from this variable in the calculation.

There are differences in the weighting factors selected by the three separate analysis teams. Since each of the three studies evaluated were done on three different processes, each with different operators, different objectives, different materials and different procedures, differences in the weighting factors are expected. This is precisely the reason for using weighting factors in the first place. Even with these differences, the selected weighting factors only varied within a relatively small range (from 0.5 – 3). This resulted in approximately 75% of the final error rates remaining within one order of magnitude of the error rates for similar actions across the three studies. This conservative weighting approach appears to be a function of a conscientious and conservative work force.

As with any study, time will help to improve the consistency and overall validity of this method as more analysts become comfortable with and knowledgeable in its use, and as more data become available. Currently, the authors consider this method to show promise and they will continue to evaluate its efficacy and validity. It is expected that opinions given in a qualitative analysis will vary, but the consistent framework provided by the base error-rate list and determining the number of times an action is taken over time lend both structure and consistency to a process that allowed wide variation using prior error estimation techniques. Since use of this method began, concern about underestimating error rates seems to have diminished at this facility.

INCORPORATING QUANTITATIVE ERROR ESTIMATES INTO A QUALITATIVE PHA

Figure 3 - Human Errors from a PHA Revalidation using this technique

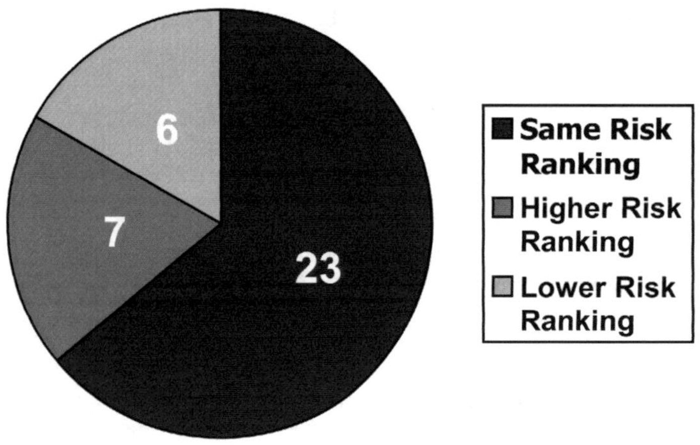

While the process described above produces a quantitative frequency estimate for the error being considered, the most frequently used PHA techniques are qualitative in nature. However, the PHA team now has a reasonable estimate of how often the human error will occur. Given this as a starting point, the team can now qualitatively assess the mitigation provided by the identified safeguards.

The PHA team must be careful not to double-count safeguards during the PHA. For example, consider a task where an operator forgets to close a valve on a 10-item checklist. If the base error rate AG (Error of omission – 10-item check off list) is used, the use of a checklist cannot be considered as.

Figure 1 – Sample HazOp Worksheet

Node: 8. Positioning valves in preparation for a new batch

Design Conditions/Parameters: All valves are hand valves.

Drawings: 201-6-001; 201-6-002; 201-6-008; 201-6-009

Deviation: 1. No

Causes	Consequences	Safeguards	Risk Matrix			Recommendations	Responsibility	S	F	RR
			S	F	RR					
1. Operator fails to close drain valves listed in steps 5.2.3.1.1	1. Release of hexane, hexane vapor and/or hexane/oil mix to floor. Potential hexane overexposure.	1. Operator completes Start-up checklist, which includes checking position of these valves.	2	C	17	7. Consider providing additional assurance that checklist valves are positioned correctly. (automation, self-closing valves, interlocks.....)	Engineering Manager	2	D	12
	2. Release of hexane, hexane vapor and/or hexane/oil mix to floor. Potential fire.	2. On feed and hexane tanks, operator is present while charging the tank - open valve would be readily apparent and quickly corrected.	2	C	17					
		3. Operators wear static protective clothing.								
		4. SRO is a Class I, Div II area.								
		5. Employees wear half-face respirators with OV/AG cartridges during hexane charging and drumming.								
2. Operator fails to close drain valves on the HSS skid.	1. Release of hexane, hexane vapor and/or hexane/oil mix to floor. Potential hexane overexposure.	1. Operator completes Start-up checklist, which includes checking position of these valves.	2	C	17	7. Consider providing additional assurance that checklist valves are positioned correctly. (automation, self-closing valves, interlocks.....)	Engineering Manager	2	D	12
	2. Release of hexane, hexane vapor and/or hexane/oil mix to floor. Potential fire.	2. Employees wear half-face respirators with OV/AG cartridges during hexane charging and drumming.	2	C	17					
		3. Operators wear static protective clothing.								
		4. SRO is a Class I, Div II area.								

another safeguard (as shown in the Safeguards column in Figure 1) that further reduces the risk – the checklist use is already accounted for in the base error rate

DISCUSSION AND RESULTS

The author's facility first used this technique to re-evaluate the risks associated with a process from which a toxic release occurred. The probability of this release occurring had been underestimated (qualitatively) during

Figure 2 – Sample Human Error Rate Calculation Spreadsheet

				Values Range from 1 to 10							
PHA Line Nmbr	Error Description	Cat	Base Error Rate	Time Pressure	Exper	Distract	Proced.	12 hr Shift - Fatigue	AER	OPY	Prob or Frequ per Year
8.1.1	Operator fails to close drain valve	J	3E-3	1	1	2	1	1	6E-3	100	0.6
7.1.1	Operator fails to close drain valve on HSS skid.	AG	1E-3	1	2	1	0.5	2	2E-3	400	0.8
16.5.8	Operator closes inlet valve instead of outlet valve.	AI	3E-3	1	2	2	1	1	1.2E-2	200	2.4

PHAs conducted prior to the incident. Of the 36 human errors evaluated, 19 percent resulted in an increased (higher risk) Risk Ranking, and 17 percent resulted in a decreased (less risk) Risk Ranking (See Figure 3). These changes were primarily due to a change to the Frequency component of the Risk Ranking. When these results were reviewed with employees and managers, both groups felt that the human error frequency estimates were realistic, and in line with their experiences and expectations. Based upon the new Risk Ranking values, the analysis team made several new recommendations to improve process safety, targeting the areas of increased risk. Had this technique been used during the original PHA, the revalidation team felt that several of the original recommendations, targeted to risks whose risk rankings were lowered during the revalidation, would not have been made. In effect, use of this technique during the original PHA would have assisted in allocating resources to those potential hazards that truly present the most risk. The author has since led one PHA revalidation and two new PHAs using this technique. The results from these PHAs have been similar to those described above.

Since the subject facility began using this technique, several human error-related incidents have occurred. However, when the PHAs (completed using this technique) were examined during the incident investigations, the investigation teams felt that the revalidation team had appropriately treated the error potentials (although in some cases, recommendations made by the PHA team to decrease the error potentials had not yet been implemented).

SUMMARY AND CONCLUSIONS

There is a perception among the facility's PHA team participants that the use of this technique increases the amount of time spent conducting a PHA. During one recent PHA, the analysis team (which was experienced in this technique) spent approximately one out of seventeen study hours calculating Adjusted Error Rates and determining the expected number of Errors per Year. In the author's opinion, however, prior to this technique being implemented, two to three hours per PHA were being lost to debate over appropriate error rates. Further experience has shown that a PHA Team accustomed to using the technique will spend less than two minutes determining an Adjusted Error Rate for a specific task, including time spent selecting a Base Error Rate and assigning Error Rate Modifiers.

It is key to the success of the technique that facility management endorses its use, as well as the base error rates. Using base error rates from accepted research will instill confidence in those that must approve their use. Showing examples of the technique's impact on existing PHAs can give a very clear understanding of the technique and its effects. Giving the PHA teams the ability to adjust the base rates, using these or other adjustment factors, gives managers the assurance that the error rates used in the study will reflect actual plant experience.

It appears to the author that most disagreement on human error arises not from the base error rate, but from team members trying to recall how many times in the past an error has occurred. Everyone can remember the errors that caused a catastrophe, or even a ruined batch of product. But the errors that "could have," or "would have but for...." can either go unreported or are quickly forgotten, and are therefore not considered by the team. By first establishing a base error rate (the operator will only forget to open the valve once in one thousand times), applying modifiers to reflect plant conditions and circumstances surrounding the specific task, then factoring in the number of times the task is performed in a year, PHA teams clearly understand and quickly reach consensus on how and how often the error might occur.

REFERENCES

1. Plant Guidelines for Technical Management of Chemical Process Safety (Revised Edition), (1992), Center for Chemical Process Safety/AIChE, Online version available at:http://www.knovel.com/knovel2/Toc.jsp?BookID=795&VerticalID=0

2. Lees' Loss Prevention in the Process Industries, Hazard Identification, Assessment and Control (3rd edition/edited by Sam Mannan), (2005) Elsevier, Inc

3. United States Department of Labor, Title 29 Code of Federal Regulations, 1910.119, Process Safety Management of Highly Hazardous Chemicals

4. United States Environmental Protection Agency, Title 40 Code of Federal Regulations, Part 68, Chemical Accident Prevention Provisions

5. Center for Chemical Process Safety, *Guidelines for Hazard Evaluation Procedures*, American Institute of Chemical Engineers, New York, NY, 2nd ed., 1994

6. Haight, J.M. and Kecojevic, V., Automation vs. Human Intervention – What is the Best Fit for the Best Performance? *Process Safety Progress Journal, American Institute of Chemical Engineers*, 24, 1, 45-51, 2005

7. Parasuraman, R. and Riley, V., "Humans and Automation: Use, Misuse, Disuse and Abuse," *Human Factors; The Journal of the Human Factors and Ergonomics Society*, 39, 2, 230-253, 1997

8. Molloy, R. and Parasuraman, R., "Monitoring an Automated System for a Single Failure: Vigilance and Task Complexity Effects," *Human Factors; The Journal of the Human Factors and Ergonomics Society*, 38, 2, 311-322, 1996

9. Mumaw, R. J., Roth, E. M., Vicente, K. J., Burns, C. M., "There is More to Monitoring a Nuclear Power Plant than Meets the Eye", *Human Factors; The Journal of the Human Factors and Ergonomics Society*, 42, 36-55, 2000

10. Kirlik, A., "Modeling Strategic Behavior in Human-Automation Interaction: Why an "Aid" Can (and Should) Go Unused," *Human Factors; The Journal of the Human Factors and Ergonomics Society*, 35, 2, 221-242, 1993

11. Blanchard, A., *Savannah River Site Generic Data Base Development – A Technical Report*, United States Department of Energy, Westinghouse Savannah River Company, Aiken, SC, 2000

12. Swain, A.D. and Guttmann, H.E. Handbook of Human Reliability – Analysis with Emphasis on Nuclear Power Plant Applications, NUREG/CR-1278, US Nuclear Regulatory Commission, Washington, DC, 1983

13. Petersen, D. *Human Error Reduction and Safety Management*, Van Nostrand Reinhold, New York, NY, 3rd ed. 1996

14. Stephenson, J. System Safety 2000 A Practical Guide for Planning, Managing and Conducting System Safety Programs, John Wiley and Sons, New York, NY, 1991

15. Sverdrup, (website accessed 10 August 2004)

16. Clemens, P.L. Human Factors and Operator Errors, Conference Presentation/Website posting, 2nd ed. February 2002

17. Center for Chemical Process Safety, *Guidelines for Preventing Human Error in Process Industries*, American Institute of Chemical Engineers, New York, NY, 1994

Connecting Process Safety Performance to Organizational Culture - a Root Cause Approach

Walter L. Frank
ABS Consulting
Wilmington, DE 19808
wfrank@absconsulting.com

and

Steve Arendt
ABS Consulting
Houston, TX 77060
sarendt@absconsulting.com

Abstract:

Many companies have implemented process safety management (PSM) systems, pursued continuous improvement strategies, adopted performance metrics, and audited themselves until they are exhausted - and their process safety performance is still not where they want it to be. Because accidents, near misses, management system failures, and unsafe acts all have a common denominator - they are the product of human action and inaction - efforts to fix these problems will continue to be imperfect until we understand and correct the underlying cultural causes that drive human performance and nonperformance.

This paper presents a way to correlate process safety technical performance results to cultural root causes. The basis for this groundbreaking approach is an extension of root cause analysis that directly correlates PSM performance dysfunctions to assessments of employee and management process safety culture using intermediate "cultural causal factors." Case history examples are used to illustrate the method. By connecting the dots between PSM (non)performance and culture issues, effective and long-lasting corrective actions can be implemented to treat the causes, rather than the symptoms.

Explosion Protection by Using the Database CHEMSAFE

Maria Molnarne, Federal Institute for Materials Research and Testing,
(Bundesanstalt für Materialforschung und –prüfung, BAM),
D-12200 Berlin, Germany, Tel.:+493081043452, Fax: +493081041227
Maria.Molnarne@bam.de

INTRODUCTION

According to the European Occupation Safety and Health guideline (especially the EC hazardous materials directive 98/24/EC on the protection of the health and safety of workers from the risks related to chemical agents at work and the directive 1999/92/EC on minimum requirements for improving the safety and health protection of workers potentially at risk from explosive atmospheres) the employer is obliged to identify and evaluate the risk on fire, explosion and exposure to dangerous substances at the working place. For this purpose they need reliable, easily available and expert-evaluated data to estimate the risk not only on toxicity but also on fire and explosion. The database CHEMSAFE [1] delivers these data, mostly for pure flammable substances, but also for a number of mixtures.

SAFETY CHARACTERISTICS

European, American and international standards describe the test methods for safety characteristics, like explosion limits, maximum explosion pressures and maximum rates of pressure rises. In the US the wording "Flammability limits" etc. is used instead of the European terminology "Explosion limits, explosion ranges or explosion regions". The comparison of hazardous properties of different substances becomes possible by using the same test methods. This has been the main goal for collecting, evaluating and storing these measured data in a database for twenty years. Now the database CHEMSAFE contains rated safety characteristics on explosion prevention and protection for approximately 3000 different flammable gases, liquids and dusts and over 500 mixtures, such as explosion limits, flash points, ignition temperatures etc. CHEMSAFE is a joint project with three project partners from Germany: BAM, Berlin, PTB (Physikalisch-Technische Bundesanstalt) in Braunschweig and DECHEMA (Gesellschaft für Chemische Technik und Biotechnologie e. V.) in Frankfurt am Main. The recommendations in the database are given by the experts of two Federal Institutes of Germany: BAM and PTB. This database is available all over the world as online version through Internet (see http://www.dechema.de) and through STN International (see www.cas.org or www.stn-international.de) and as in-house version for PCs and workstations in German and English, developed by the DECHEMA, Frankfurt/Main. The in-house version also uses graphical representation of the data in x-y or in triangular diagrams (see Figure 1).

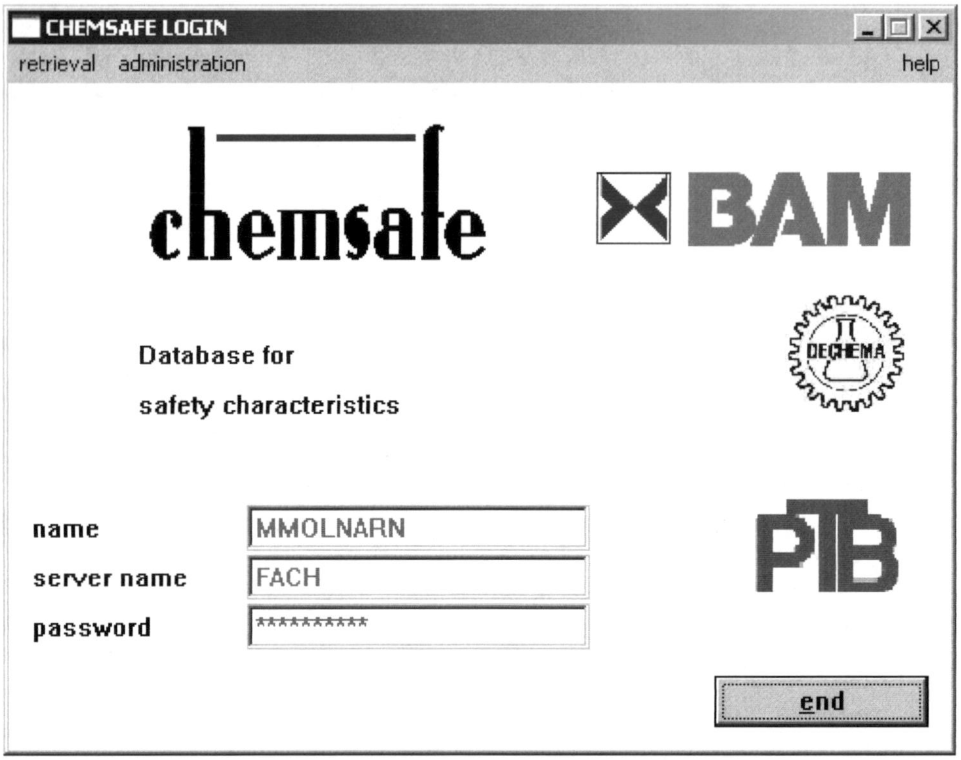

Figure 1: In-house version of the retrieval system CHEMSAFE

In order to avoid explosive atmospheres in accordance with flammable substances, the solution generally applied in industrial processes is purging with inert gases. Reliable measurement data are required concerning the explosion range of such ternary systems. Using measured explosion limits data by different inert gas contents, the program TRIANGLE [2] is able to interpolate the explosion limit curves, to calculate the liming values of a ternary system - e. g. limiting oxygen concentration (LOC), maximum permissible amount of flammable gas - in order to avoid an explosion and plot them in a triangular diagram (see Figure 2). Table 1 shows the acronyms and their meanings, applied in CHEMSAFE and in the handbook "Safety Characteristics" [3]. These diagrams can be used, for example, for the estimation of flammability of gas mixtures, which is very important information for customers from chemical and process industries or also for shipping companies.

With the help of limiting values, e. g. LOC, a process engineer can control the oxygen amount in the gas mixture in order to avoid the dangerous explosive atmosphere in an industrial process.

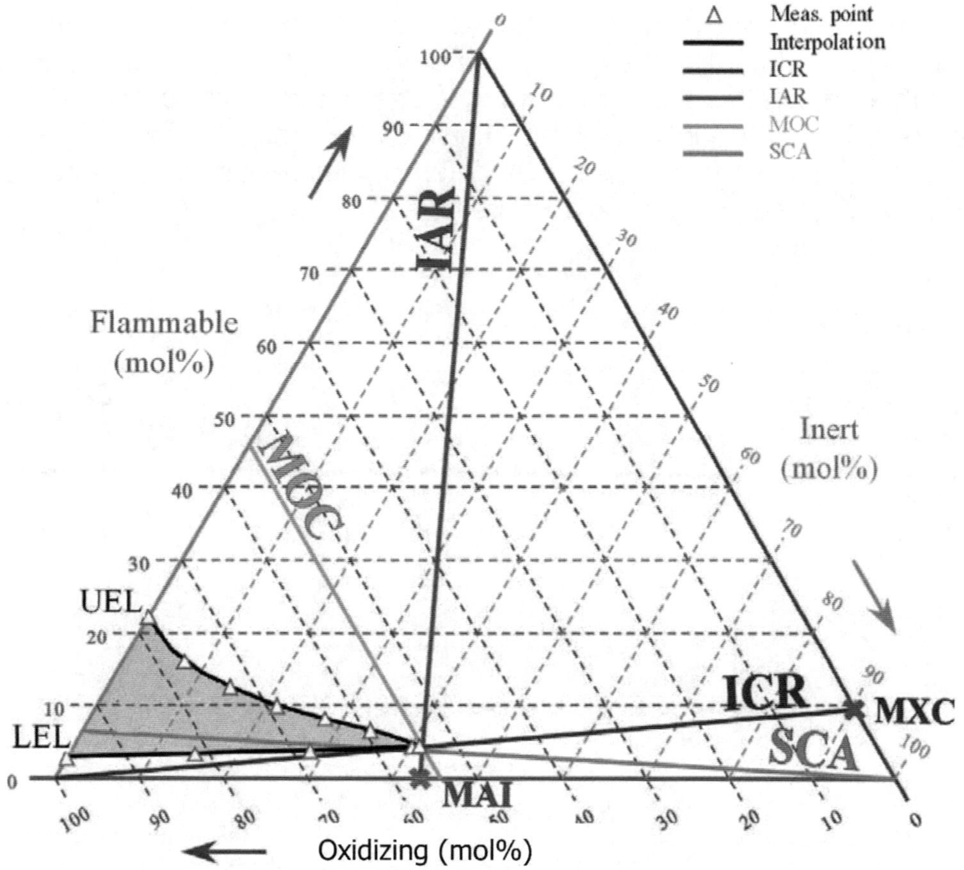

Figure 2: TRIANGLE-Plot with Limiting Values of the Explosion Range of a Ternary System Containing a Flammable, an Inert and an Oxidizing Gas

Table 1: Characteristic Values of the Explosion Range of a Ternary System Containing a Flammable, an Inert and an Oxidizing Gas

Acronym in CHEMSAFE	Description
LEL	Lower explosion (flammability) limit
UEL	Upper explosion (flammability) limit
ICR	Minimum inert gas / combustible ratio
IAR	Minimum inert gas / air (oxidizing gas) ratio
MXC	Maximum permissible amount of combustible
MAI	Minimum required amount of inert gas
MOC	Maximum oxygen (oxidizing gas) content
or LOC	or Limiting oxygen concentration
SCA	Stoichiometric concentration in air

Comment: we use combustible as synonym for flammable in these definitions.

Sometimes the users of the database like to know how the temperature and the pressure (or both) influence the explosion range of a gas mixture. In these cases it is important to plot the different explosion ranges of a gas mixture containing flammable, inert and oxidizing gases in the same triangular or Cartesian diagram, using the measured data from measurements under different initial conditions. Typical examples of such ternary diagrams are shown in Figures 3 - 4.

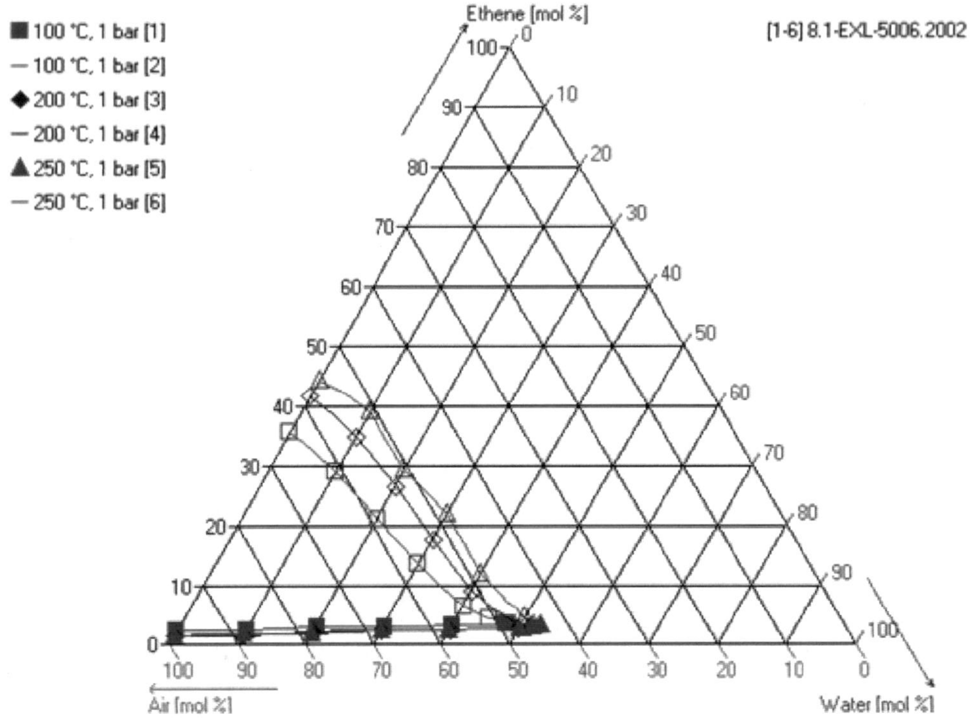

Figure 3: Influence of the Initial Temperature on the Explosion range of the Ternary System Ethene/Water Vapor/Air

The database CHEMSAFE – includes, among others, over 200 gas mixture systems measured at initial conditions up to 100 bar and 673.15 K. These data can be used to develop estimation methods for safety characteristics [4, 5, 6 and 7]. The explosion range and the characteristic values of many ternary systems containing flammable gas, inert gas and air, were stored in the database.

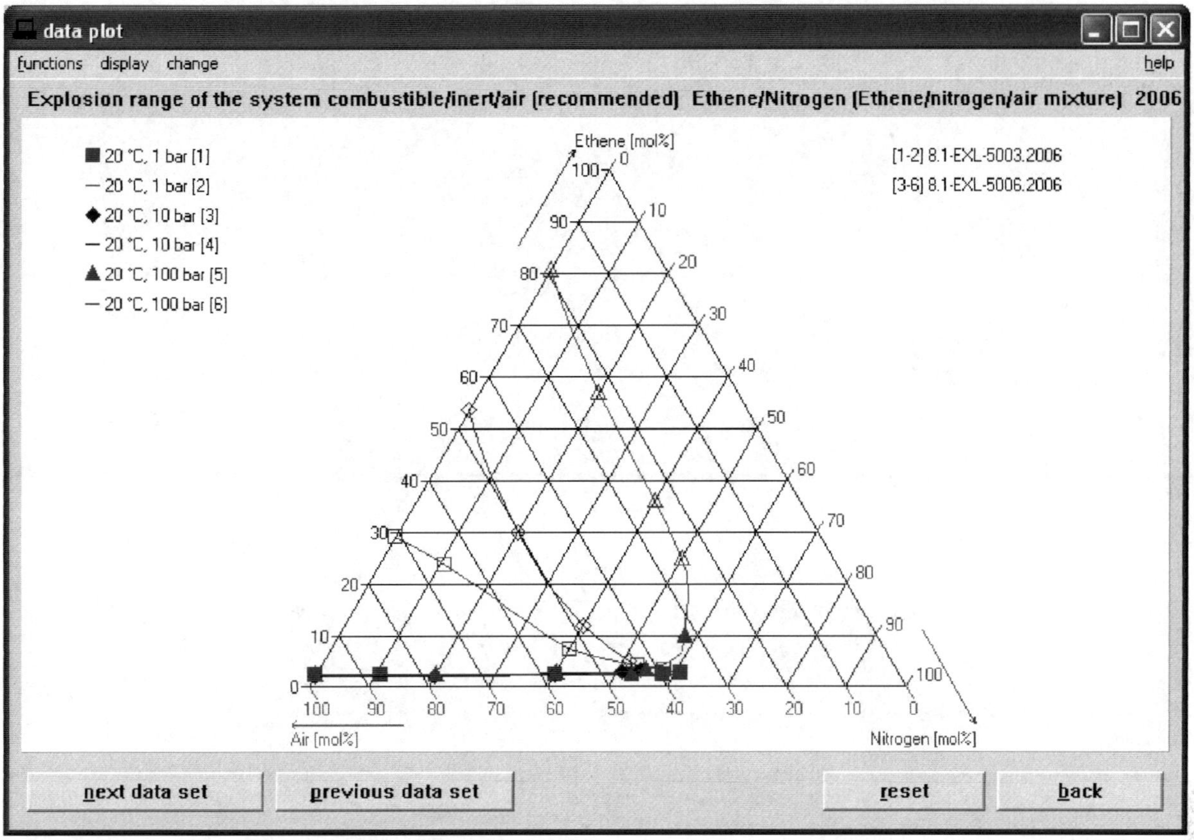

Figure 4: Influence of Pressure on Explosion Range of the Ternary System Ethene/Nitrogen/Air (Source: BAM, CHEMSAFE)

The explosion limits and the flash point are known for many flammable substances, but the limiting oxygen concentration and the other limiting values of these ternary systems are rather rare. It is also mentioned from Mashuga and Crowl in their publication (Process Safety Progress, 1998) that "literature sources of flammability diagrams are limited in quantity and vary of quality". Therefore it is important to point out, that in CHEMSAFE one can find this information.

EXPLOSION PROTECTION IN CHEMICAL INSTABLE GASES

Ethylene oxide (EO) has been investigated in detail in the BAM. The explosion prevention method "inerting" can be used only in a restricted manner with such gases. The upper explosion limit is 100 mol%, which means that even the pure substance can decompose in an explosive manner. For inerting such systems the knowledge of the maximum permissible amount of flammable gas (MXC) and the section point of the explosion limit curve with the inert gas axis is of particular interest. The MXC corresponds to the burning reaction of EO in a mixture with air but the axis point corresponds to the decomposition reaction of EO in a mixture with carbon dioxide, without any air. This can be seen in Figure 5.

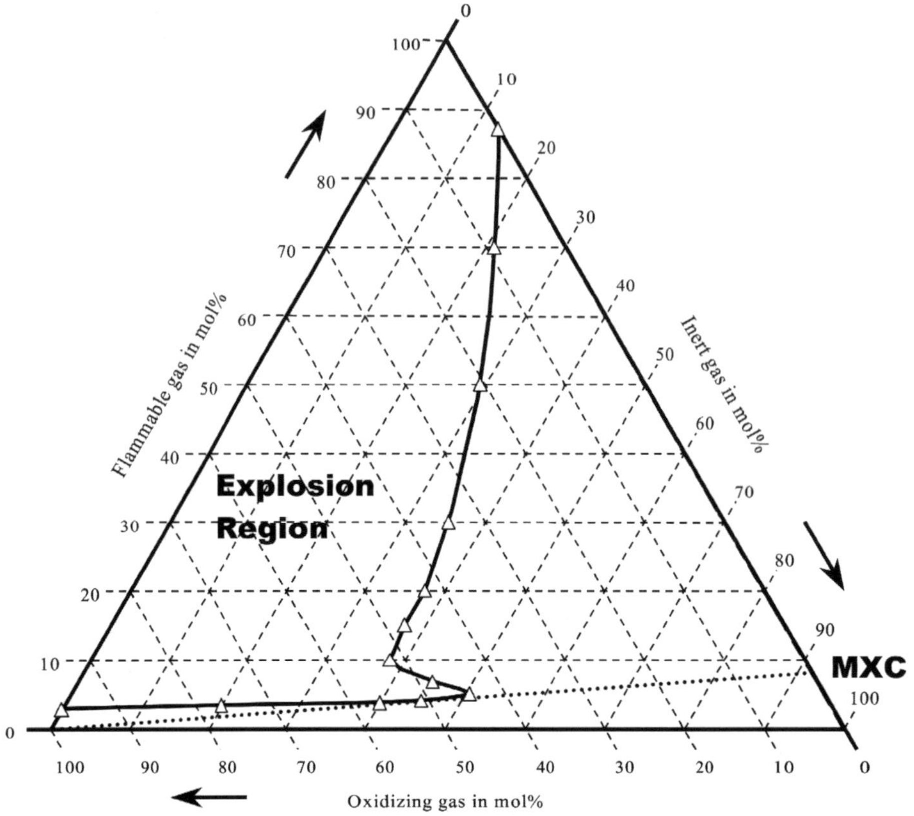

Figure 5: Triangular Diagram of the System Ethylene oxide/Carbon dioxide/Air (Source: BAM, CHEMSAFE)

SAFETY CHARACTERISTICS MEASURED IN SPECIAL CONDITIONS

Most of the material safety data sheets and safety data tables which are available contain data which are mostly measured at room temperature and at atmospheric pressure as initial conditions. The crucial advantage of CHEMSAFE is its data being at elevated initial conditions, and the supply of triangular diagrams of flammable mixtures including explosion regions of flammable – inert - oxidizer systems, as well as data of decomposable gases and gas mixtures. Furthermore the database CHEMSAFE supplies the following:

- Stability pressure limits for chemical instable gases
- Data for safety characteristics of the constructive explosion protection as P_{max}, $(dp/dt)_{max}$ and K_G values.

- Safety characteristics data at sub-atmospheric pressures
- Using oxidizing agents besides air, like pure oxygen, N_2O, NO, Cl_2

The proportion of these data in the database is growing permanently from year to year. The BAM and the PTB investigate such systems systematically in different research projects (e. g. the European Research project SAFEKINEX, see "www.safekinex.org") by mostly standardized testing methods. These data are very useful for the chemical industry while the process conditions have been changed from using air to applying other oxidizing agent and elevated pressures and temperatures.

CALCULATION METHODS

Calculation methods are also included in the databank for

- vapor pressures,
- experimental safe gaps for mixtures,
- flammability and oxidizing potential of gases and gas mixtures (according to the ISO 10156 (1996) and ISO 10156-2 (2005)) [8, 9, and 10].

Estimation of the Flammability of Gas Mixtures Using Explosion Diagrams

Estimation methods which enable the determination of gas mixture flammability are important technical aids for safety engineering. Using these methods, the number of tests can be reduced to a necessary minimum. The international standard ISO 10156 [8] published a calculation method for the estimation of the flammability of gas mixtures containing different flammable and inert gases in the air. The calculation method is based on the so-called Tci value - maximum flammable gas content for which a mixture of the flammable gas (i) in nitrogen is not flammable in air (see Table 2, ISO 10156 [8]). During the past years the BAM developed an experimental method to determine Tci (MXC) values using the apparatus according to the German Standard DIN 51649-1 [11]. The MXC values given in the database CHEMSAFE correspond to the Tci values of ISO 10156 if the inert gas in the ternary system is nitrogen. So it was possible to compare the new experimental data with the older ISO values. The calculation method was modified, due to replacing the ISO Tci values with the experimentally determined MXC-values according to the DIN 51649-1, and the "tube" test method of the EN 1839 [12] respectively and reported it [6, 7, 13, 14, and 15].
This modified calculation method is already included in the in-house version of CHEMSAFE (see Table 2).
Table 2 shows that some ISO values are significantly higher than CHEMSAFE data. This can result in wrong classification and labelling and even in accidental fires and explosions. The ISO 10156 is just under revision and Germany has proposed the use of evaluated Tci values from CHEMSAFE.

Table 2: Comparison of the Tci values from the ISO 10156 and from the database CHEMSAFE (in mol%)

Flammable gas	Tci ISO 10156 [8] based on [16] in mol%	Tci CHEMSAFE (MXC) [1] in mol%
Methane	14.3	8.7
Ethane	7.6	4.7 by DIN EN 1839-T
Propane	6.	4.1 by DIN EN 1839-T
Butanes	5.7	3.4 – 3.6
Ethylene	6.	4.1
Acetylene	4.	3.0
Propylene	6.5	4.4 by DIN EN 1839-T
1-Butylene	5.5	3.3
Hydrogen	5.7	5.5
Ammonia	-	40.1
Ethylene oxide	3.1	4.8

FUNCTIONAL CORRELATIONS BETWEEN SAFETY CHARACTERISTICS

Test results for safety characteristics of the same compound can sometimes be used to check the reliability of the measured data. In addition to the minimum ignition energy, the maximum experimental safe gap, which in turn is correlated with the maximum normal burning velocity, can also serve as a measure of ignitability at nearly point-ignition sources.

It has been proved for a great variety of substances that minimum ignition energy, minimum ignition current and maximum experimental safe gap are correlated with one another so that the classification of substances is preserved when classification is made according to each of these three quantities (see Table 3).

Table 3: Correlation between Minimum Ignition Energy (MIE) and Maximum Experimental Safe Gap (MESG)

Flammable gas	MIE in mJ	MESG in mm
Hydrogen	0.017	0.29
Acetylene	0.019	0.37
Diethyl ether	0.19	0.87
Propane	0.24	0.92
Methane	0.29	1.14
Ammonia	14.	3.18

It can be seen clearly in the Table 3, that the gases have the same order by MIE as by MESG.

CONCLUSION

For explosion prevention measures one has to know the safety characteristics of the hazardous substances under atmospheric and other process conditions like explosion limits, explosion points, and minimum ignition energy. Constructive explosion protection needs additional information about the maximum explosion pressure, maximum rate of explosion pressure rise and the maximum experimental safe gap of a hazardous system for designing constructive safety devices, to reduce the impact of an explosion on the surrounding. In the database CHEMSAFE one can find both, rated safety characteristics for atmospheric conditions needed for Material Safety Data Sheets and characteristics for process conditions needed for process design and operation. The most data in CHEMSAFE are measured according to the German and European Standards but some of them by ASTM. The user is supported by a help text about the test methods, definitions of safety characteristics and the possibility of the conversion the units. In addition, CHEMSAFE can handle mixtures and supplies semi-empirical calculation methods for the most important characteristics.

ACKNOWLEDGEMENT

The author would like to thank the working group "Safety Related Properties of Gases" headed by Dr. Volkmar Schröder, BAM, Berlin, for the fruitful co-operation in the field of gas explosions.

References:

1. CHEMSAFE® – Database for Recommended Safety Characteristics, BAM, PTB, DECHEMA, Germany, Version 1.4.7, Update 2006

2. Molnarne, M., Bulin, M., Viczián, G., Kollár-Hunek, K.: TRIANGLE - ein Computer-Programm zur Darstellung und Auswertung der Explosionsbereiche von Dreistoffsystemen (in German, translated into English: "A computer program for presentation and evaluation of ternary systems"), Technische Überwachung, Springer-VDI Publishing Co., 44, pp. 32-35, 2003

3. Molnárné, M., Schendler, Th., Schröder, V.: Sicherheitstechnische Kenngrößen, Volume 2 "Explosionsbereiche von Gasgemischen" (in German, translated into English: „Explosion Ranges of Gas Mixtures"), NW-Verlag (Publishing Co.), Bremerhaven, Germany, 2003

4. Razus, D., Molnarne, M., Fuß, O.: Limiting Oxygen Concentration Evaluation in Flammable Gaseous Mixtures by means of Calculated Adiabatic Flame Temperatures, Chem. Eng. Process., 43, pp. 775-784, 2004

5. Razus, D., Molnarne, M., Movileanu, C., Irimia, A.: Estimation of LOC (limiting oxygen concentration) of fuel–air–inert mixtures at elevated temperatures by means of adiabatic flame temperatures, Chem. Eng. Process., 45, pp. 193-197, 2006

6. Schröder, V., Molnarne, M.: Flammability of gas mixtures – Part 1: Fire potential, Journal of Hazardous Materials, 121, pp. 37-44, 2005

7. Molnarne, M., Mizsey, P., Schröder, V.: Flammability of gas mixtures – Part 2: Influence of inert gases, Journal of Hazardous Materials, 121, pp. 45-49, 2005

8. ISO 10156, Gases and gas mixtures – Determination of fire potential and oxidising ability for the selection of cylinder valve outlets, Second edition, (1996) (under revision 2007)

9. ISO-10156-2: Gas cylinders - Gases and gas mixtures-Part 2: Determination of oxidizing ability of toxic and corrosive gases and gas mixtures, First Edition (2005)

10. Schröder, V., Mackrodt, B., Dietlen, S.: Determination of oxidizing ability of gases and gas mixtures", ASTM STP 1395 (2000)

11. DIN 51649-1, Bestimmung der Explosionsgrenzen von Gasen und Gasgemischen in Luft (1986) (in German, translated into English: Determination of explosion limits of gases and gas mixtures in air) (withdrawn due to DIN EN 1839)

12. European standard EN 1839, Determination of explosion limits of gases, vapours and their mixtures (2003)

13. Molnarne, M.: Estimation of flammability of gas mixtures in air containing flammable and inert gases - Testing of the calculation method proposed by ISO 10156 using data from CHEMSAFE, CHISA'96 Congress "12th International Congress of Chemical and Process Engineering, Symposium on Environmental and Safety Engineering", August 25-30, 1996, Prag, Czech Republic

14. Molnarne, M.: Estimation of flammability and limiting oxygen concentration of gas mixtures using the database CHEMSAFE, Internationaler Ausstellungskongress für Chemische Technik, Umweltschutz und Biotechnologie, ACHEMA 2000, Sektion Sicherheitstechnik, May 2000, Frankfurt/M., Germany (Proceeding) (2000)

15. Schröder, V., Dietlen, S., Schulz, S.: Classification of gas mixtures for application in process safety, storage and transport – An experimental method to determine Tci values, Proceedings of the 9th Int. Loss Prevention Symposium, Barcelona, Spain (1997)

16. Zabetakis, M. G.: Flammability characteristics of combustible gases and vapours, Bureau of Mines Bulletin 627, U.S. Government Printing Office, Washington D.C., 1965

Cumulative Benefit Analysis for Ranking Risk Reduction Actions

By

Fred L. Leverenz, Jr., Battelle Northwest, Richland, WA, fred.leverenz@pnl.gov
Julio Aysa, Battelle Mexico, Mexico City, Mexico, aysaj-c@battelle.org

BACKGROUND

In a recent project for a pipeline company, Hazard and Operability (HAZOP) studies were completed for several pumping stations. The goal of these studies was to identify and prioritize risk reduction actions (actions), based on the findings of the HAZOP studies. The HAZOP studies utilized a "scoring" system to provide relative measures of accident frequency and consequences. The challenge was to develop a method of ranking the findings in terms of optimal risk reduction.

CONDUCTING THE HAZOP STUDIES

Process Description

The general concept of a pipeline pumping station is relatively simple; that is, pumping stations add energy to the movement of a product through the pipeline systems. Stations are located starting at the source location, then at various locations where the additional pumping energy is needed to maintain flow of the product to the destination. These stations can have different types of pumps, depending on energy sources, for example electric pumps, gas turbine pumps, diesel turbine pumps, or even a combination. Given that the pipelines are anywhere from 12 inches in diameter to 48 inches in diameter, and the product is pumped from sea level over mountain ranges, the station pumps can be very large. A typical station among those in this study had a flow rate through the station of 500 thousand barrels per day. Knowing that one product in the pipeline system is crude oil, a quick calculation at $50/barrel shows the value of the product, and hence the importance of each station in maintaining the transportation of the product. Typical station operations included temporary storage of product in large storage tanks, starting, stopping, and switching pumps depending on operational needs, maintenance needs, and/or failures.

Scenario Development

The HAZOP studies were conducted in a typical fashion as presented in the ASME/CCPS training course[1]. The hazards for a typical pumping station include fires and explosions, and environmental releases. The HAZOP study method was applied to each typical station operation, for example 4 of 6 pumps in operation and 2 in stand-by and/or maintenance. The HAZOP study also included activities such as temporary diversion to a site storage facility, retrieval from the temporary storage facility, and switching of an operating pump to a stand-by pump. The scenarios were developed to explicitly identify the cause of the event, the consequence of the event, and each safeguard which could act to reduce the likelihood of the scenario. Table 1 provides an example of the qualitative description of a scenario. This scenario results from postulating a deviation in the HAZOP study of "low flow" from the station discharge. Specifically, the relief valve, used to protect the discharge line from overpressure, inadvertently fails open, allowing some of the product flow to fill the storage tank piped to the discharge of the relief valve.

Table 1: Example Scenario

Cause:	Pump discharge relief valve fails open
Consequence:	Crude diverted to storage tank, tank overfills, spills, and a fire occurs. The fire is unmitigated and becomes a "boil-over event" causing the fire to spread to the entire station.
Safeguards:	Storage tank level detector and alarm
	Site ignition source control measures
	Site fire suppression system (water and foam)

Risk Scoring

Each scenario developed during the HAZOP study was "scored" by the HAZOP study team. The cause was scored relative to its frequency, the consequence relative to its impact, and the safeguards relative to their reliability. The scoring system used was based on an order-of-magnitude scale. For example the score for the Cause frequency would be given a score from Table 2, the safeguard scores from Table 3 and the Consequence score from Table 4.

---

[1] *HAZOP Studies and PHA Techniques for Process Safety and Risk Management,* Short Course Number 157. http://catalog.asme.org/Education/ShortCourse/HAZOP_Studies_PHA_Techniques.cfm.

Table 2: Cause Frequency Scoring

Score	Order Of Magnitude Frequency	Qualitative Descriptor
1	About once per month	Expected to occur frequently or regularly
0	Once per year	Likely to occur occasionally; several times in plant lifetime
-1	10% chance of one per year	Probably will not happen more than once per plant lifetime
-2	1% chance of one per year	Not expected to, but could occur during plant lifetime
-3	0.1% chance of one per year	Would be surprising if happened during plant lifetime
-4	0.01% chance of one per year	Extremely remote, or not expected to be possible.

Table 3: Safeguard Scoring

Score	Probability of Safeguard Failure when Needed	Probability of Safeguard Success when Needed	Example
0	1.0	0.0	No safeguard, or safeguard very unreliable in this situation
1	0.1	.90	Single operator responding to an alarm with adequate time
2	0.01	.99	Single set of hardware (sensor, controller, response device), functionally tested on a regular basis.
3	0.001	.999	Passive safeguard (relief valve, rupture disk, fusible link)
4	0.001	.9999	Safeguard made up of fully redundant elements to achieve high reliability.

Table 4: Consequence Scoring

Score	Qualitative Descriptor	Unlikely, but possible, effect on 1 or 2 persons (about 10% chance)	Effects reasonably Expected to occur to 1 or 2 on-site personnel	Effects reasonable expected to occur to more than 2 persons onsite and/or anyone offsite	Impact on facility, environment, business.
6	Extremely High	⇒	⇒	Fatality(ies)	
5	Very High	⇒	Fatality(ies)	Immediate impairment; permanent health effects	Results in major environmental damage, loss of entire facility with long term outage.
4	High	Fatality(ies)	Immediate impairment; permanent health effects	Severe injury requiring hospitalization; significant lost time	Results in serious environmental damage, significant plant damage, and/or extended outage time
3	Intermediate	Immediate impairment; permanent health effects	Severe injury requiring hospitalization; significant lost time	Injuries requiring medical treatment	Results in measurable environmental impact, minor plant damage and/or outage time
2	Low	Severe injury requiring hospitalization; significant lost time	Injuries requiring medical treatment	Minor injuries; first aid cases	Unlikely to impact environment or damage facility
1	Very Low	Injuries requiring medical treatment	Minor injuries; first aid cases	Probably none	Probably None
0	Negligible	Minor injuries; first aid cases	Probably none	Probably none	Probably None

Table 5 shows how this scoring system might be added to a scenario (from Table 1).

Table 5: Scenario with Risk Scoring

Cause:	Pump discharge relief valve fails open	Score = -2
Consequence:	Crude diverted to storage tank, tank overfills, spills, and a fire occurs. The fire is unmitigated and becomes a "boil-over event[2]" causing the fire to spread to the entire station.	Score = 5
Safeguards:	Storage tank level detector and alarm	Score = 1
	Site ignition source control measures	Score = 2
	Site fire suppression system (water and foam)	Score = 2

In order to achieve a scenario risk score an algorithm needs to be utilized which combines the scores for each part of the scenario into a risk score. For the example used here, the scoring system was selected considering the advantage of a simple algorithm to determine a score for scenarios risk.

First, a typical definition of risk is:

$$Risk = Frequency \times Consequence \qquad (1)$$

In the case where the sequence Cause and Safeguards are evaluated separately, then the sequence frequency is given by (assuming safeguard failure probabilities are independent):

$$Frequency = Cause\,Frequency \times \prod_{i=1}^{\substack{Number\ of\\ Safeguards}} P(Safeguard\ Failure_i) \qquad (2)$$

Finally, in terms of Cause, Safeguards, and Consequence, the risk can be given by:

$$Risk = \left[Cause\,Frequency \times \prod_{i=1}^{\substack{Number\ of\\ Safeguards}} P(Safeguard\ Failure_i)\ \right] \times Consequence \qquad (3)$$

Now, given that the risk scores represent measures of orders-of-magnitude, equation (3) can be modified recognizing that multiplication is accomplished by addition of base ten exponents. That is:

[2] From NFPA 30: "**Boil-Over:** An event in the burning of certain oils in an open-top tank when, after a long period of quiescent burning, there is a sudden increase in fire intensity associated with expulsion of burning oil from the tank".

$$T = 10^{-1} \times 10^{-4} \times 10^{6} = 10^{(-1)+(-4)+(6)} = 10^{1} \tag{4}$$

The resultant risk score equation for a scenario, given that an order-of-magnitude score is the desired result becomes:

$$Risk\ Score = Cause\ Score - \sum_{i=1}^{\substack{Number\ of \\ Safeguards}} Safeguard\ Score_i + Consequence\ Score \tag{5}$$

For any given HAZOP study of a pumping station, the number of scenarios was in the hundreds. Total risk then would be the sum of the risk from each scenario. Hence the facility risk score would be found by:

$$Facility\ Risk\ Score = \log_{10}\left[\ \sum_{i=1}^{\substack{Number\ of \\ Scenarios}} 10^{Scenario\ Risk\ Score\ _i}\ \right] \tag{6}$$

For the example scenario in Table 5, the Risk Score = -2. Assume there are three other scenarios with scores of -1, -4, and -3 respectively. Then the Facility Risk Score would be:

$$Facility\ Risk\ Score = \log_{10}\left[10^{-2} + 10^{-1} + 10^{-4} + 10^{-3}\right] = \log_{10}[.1111] = -0.95428 \tag{7}$$

By utilizing a spreadsheet to record the qualitative scenario descriptions and scores these evaluations are completed during the recording process.

Action Item Recording

As each scenario was developed, and documented, the need for risk reduction was discussed by the team in terms of possible risk reduction actions. Each action item was recorded qualitatively in the spreadsheet. Further, each action item was also included in each scenario as an additional Safeguard with a score of zero (because, at the time of the HAZOP study, the safeguard does not exist), using a code such as A1 for action item 1, A2 for action item 2, etc. Including the action items in this way provides a mechanism for their evaluation in the subsequent analyses for risk importance. Table 6 shows an example of the inclusion of an action item.

Table 6: Scenario with Inclusion of Action Item

Cause:	Pump discharge relief valve fails open	Score = -2
Consequence:	Crude diverted to storage tank, tank overfills, spills, and a fire occurs. The fire is unmitigated and becomes a "boil-over event" causing the fire to spread to the entire station.	Score = 5
Safeguards:	Storage tank level detector and alarm	Score = 1
	Site ignition source control measures	Score = 2
	Site fire suppression system (water and foam)	Score = 2
	A1-Automate fire detection and startup of fire suppression	**Score = 0[3]**

ANALYZING THE RESULTS

The development of the set of accident scenarios for each facility was completed at the facility with facility staff using the above approach. Depending on the complexity of the facility and its operations, the number of scenarios for each facility varied from two to four hundred, all recorded in a spreadsheet. The next step was to analyze these scenarios to identify:

- the scenarios that had the highest risk,
- the accident Causes, Safeguards and Action Items that are the most important to risk, and
- providing a list of the Action Items based on Cumulative Risk Benefit.

Risk Reduction Worth Analysis

Risk Reduction Worth (RRW) has been used in the analysis of nuclear power plant risk for many years. It was introduced as a tool to aid in the analysis of process plant HAZOP studies as reported in other papers.[4,5] RRW is a measure of the importance of

---

[3] The action item is included in the scenario with a score of "0" which indicates its failure probability is "1" because, in reality, this safeguard does not exist.

[4] G. Bradley Chadwell and Fred L. Leverenz, Jr., "Importance Measures for Prioritization of Mechanical Integrity and Risk Reduction Activities", 33rd Annual Loss Prevention Symposium, March 1999.

[5] Fred L. Leverenz, Jr., "Using Risk Analysis to Identify Cost Effective Improvements to a Semiconductor Fabrication Chemical Distribution System," ASME International Mechanical Engineering Congress and RD&D Expo, November 2003.

an accident Cause or Safeguard relative to reducing risk. That is, given the RRW for every Cause and Safeguard for a facility, and that the list of RRWs is sorted from largest RRW to smallest, the Causes and/or Safeguards at the top of the list are those that would reduce the risk by the greatest amount given a reduction in Cause frequency or increase in Safeguard reliability. Given risk is for a facility is expressed as an equation, such as Equation (3), the RRW of each Cause and Safeguard can be determined mathematically. For simplicity, assume that all Causes and Safeguards can be defined as "basic events", and designated as BE_i, that is the ith basic event. Then $P(BE_i)$ is the probability that the basic event event BE_i occurs. Then, RRW is defined as:

$$RRW_i = \frac{Facility\,Total\,Risk}{Facility\,Total\,Risk\ when\ P(BE_i) \to 0} \tag{8}$$

As a side note, which may help understand RRW, RRW is a normalized measure of the rate-of-change of facility risk with respect to the rate of change of a basic event probability. That is, it can be shown that:

$$RRW_i = \frac{1}{1 - IM_i} \qquad \text{where} \quad IM_i = \frac{\partial(Facility\,Total\,Risk)}{\partial P(BE_i)} \bullet \frac{P(BE_i)}{Facility\,Total\,Risk} \tag{9}$$

As defined, RRW results in a numerical value that is always > 1. For a situation where the RRW is very low, RRW ≈ 1.0000. Further, the RRW considers not only the importance of decreasing the probability of the Cause or Safeguard failure; it also considers the number of scenarios that include the Cause and/or Safeguard.

Before continuing, a simple example will be presented to solidify the concepts to this point. Table 7 provides an example from a HAZOP study with several evaluated scenarios. Tables 1 through 4 are used to score the scenarios. Risk is given as value rather than as a score. Table 8 provides the results of the RRW evaluation of these scenarios using equation (8). The most important basic events to consider for risk reduction (in terms of their impact on total facility risk) are the Back-up Pump, and failure of Pump 1. Next on the RRW list is the Action Item 1, to improve fire suppression (together with ignition source control). Note that Fire Suppression is also listed, but since the Action Item 1 is the improvement of the Fire Suppression, the listing of Fire Suppression is not as interesting. Next on the list is the Storage Tank Alarm.

As can be seen the RRW analysis provides valuable information on what to do to first to reduce risk. For example, improve the reliability of the Back-up Pump. However this list does NOT provide information on what to do after improvement of the Back-up

Pump, even though the next thing on the list is the failure of Pump 1. The importance of Pump 1 to risk reduction is a function of the improvement of the Back-up Pump reliability. That is, the RRW of every basic event can change, given a change in one of the basic events. This leads to the need for development of an additional analysis tool.

Cumulative Risk Benefit Analysis

In any risk assessment where risk reduction is part of the analysis, it would be very surprising if only one issue was identified for risk reduction. In fact often the question involves how many of the possible risk reduction actions should be completed, and with what priority. Given, as concluded above, that the risk reduction actions are interrelated in the risk model, a process is required to allow these interrelationships to be evaluated. Figure 1 shows a process for accomplishing this goal. In essence, the risk analysis is completed with an initial RRW for Causes, Safeguards, and Actions. Then the initial candidate for risk reduction is identified from the top of the list by the risk analyst and the design team. A specific solution is postulated to the extent that a revised Cause frequency, Safeguard reliability, or contribution by an Action Item can be identified. For example, in the RRW list from Table 8, suppose the design team determines that a second back-up pump should be considered. This second back-up pump would have the same reliability as the existing pump (a Score of 2).

Next, again in Figure 1, the RRW calculation is revised, using equation (8), with the additional back-up pump included in those scenarios that have a back-up pump. The resultant RRW results would be as shown if Table 9.

Figure 1: Evaluation of Multiple Action Risk Benefits

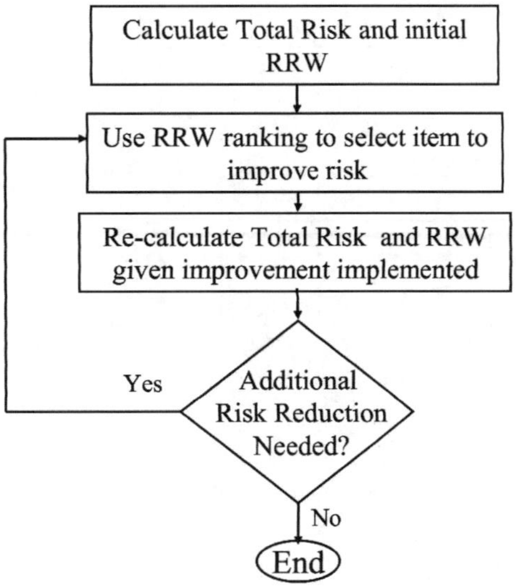

Table 7: Example HAZOP Study Scenarios

Date: 1-Jan-07

Node: Node 1

Design Conditions: Pump Crude to next station.

Scenario Code	Parameter	Guide Words	Deviation	Cause	Frequency	Consequence	Eval. Consec.	Safeguards	Safeguard 01 Reliability	Safeguard 02 Reliability	Safeguard 03 Reliability	Safeguard 04 Reliability	Risk	Comments
E-1	Flow	No	No Flow	Pump 1 fails	-1	Loss of station output	2	Back-up Pump	2				1.0E-01	
E-2	Flow	Less	Less Flow	Relief Valve 1 fails	-2	Overflow of storage tank with boil-over fire	5	StgTankLvlAlm, Ignition Source Control, Fire Suppression,A1 Improve Fire Suppression	1	2	2	0	1.0E-02	
E-3	Flow	Less	Less Flow	Leak in line	-3	Spill and small fire	4	Ignition Source Control, Fire Suppression,A1 Improve Fire Suppression	2	2	0		1.0E-03	
E-4	Flow	No	No Flow	Pump 1 discharge valve fails closed	-2	Relief valve opens and directs flow to storage. Overflow of storate with boil-over fire.	5	StgTankLvlAlm, Ignition Source Control, Fire Suppression,A1 Improve Fire Suppression	1	2	2	0	1.0E-02	
E-5	Flow	No	No Flow	Pump 1 discharge valve fails closed	-2	Relief valve fails and line ruptures with fire	4	Relief Valve 1 fails,Ignition Source Control, Fire Suppression,A1 Improve Fire Suppression	3	2	2	0	1.0E-05	

Table 8: Example HAZOP Study Risk Reduction Worth (RRW) Results

Cause/Safeguard	RRW	Total Risk	Cause/Safeguard Found in Scenarios:
Back-up_Pump	5.759638267	1.21E-01	E-1
Pump_1_fails	5.759638267		E-1
A1_Improve_Fire_Suppression	1.2101		E-2, E-3, E-4, E-5
Fire_Suppression	1.2101		E-2, E-3, E-4, E-5
Ignition_Source_Control	1.2101		E-2, E-3, E-4, E-5
StgTankLvlAlm	1.198000198		E-2, E-4
Pump_1_discharge_valve_fails_closed	1.09018018		E-4, E-5
Relief_Valve_1_fails	1.09018018		E-2, E-5
Leak_in_line	1.008332639		E-3

Table 9: Revision 1 of RRW Table

Cause/Safeguard	Results Given Second Back-up Pump	
	RRW	Total Risk
A1_Improve_Fire_Suppression	22.97	2.19E-02
Fire_Suppression	22.97	
Ignition_Source_Control	22.97	
StgTankLvlAlm	11.15	
Pump_1_discharge_valve_fails_closed	1.84	
Relief_Valve_1_fails	1.84	
Leak_in_line	1.05	
Back-up_Pump	1.05	
Pump_1_fails	1.05	

Note that given this initial risk reduction action is implemented and the RRWs are re-calculated, the order is changed. First, the team identified action to improve the Fire Suppression System is now at the top of the list. Further, the importance of the Back-up Pump has fallen to the bottom of the list, together with the Cause associated with the Back-up Pump safeguard scenario. In addition the total facility risk has been reduced from 1.2×10^{-1}/yr to 2.2×10^{-2}/yr. Now the risk analyst and facility management ask the question of whether or not the risk is at a tolerable value. It will be assumed here that that answer is "no" to continue the process; that is, repeat the steps in Figure 1 again. Obviously, the Action Item A1 identified during the HAZOP study must be considered next. Further assume that the design team determines that the best way to improve this system's response to the fire is to add a fire detection system at the tank that automatically initiated the fire suppression system, and that such a system would have a reliability Score of 2. Table 10 provides the result of this second risk reduction option.

First, as before, the risk was reduced; this time from 2.2×10^{-2}/yr to 1.2×10^{-3}/yr. Moreover, the total risk for implementation of the two risk reduction items is from 1.2×10^{-1}/yr to 1.2×10^{-3}/yr, that is, about two-orders of magnitude. Let's assume more risk reduction is desired. Examination of Table 10 now shows that the top of the list is the Back-Up Pump again, followed by the Fire Suppression System. At this point the design team can address these again, or investigate a new candidate. Suppose they want to consider the value of making the storage tank level system automatic instead of just an alarm. Assuming that this would improve this Safeguard by about an order of magnitude, the RRW table becomes as shown in Table 11.

Table 10: Revision 2 of RRW Table

Cause/Safeguard	Results Given Improved Fire Suppression System	
	RRW	Total Risk
Back-up_Pump	5.55	1.16E-03
Pump_1_fails	5.55	
A1_Improve_Fire_Suppression	1.22	
Fire_Suppression	1.22	
Ignition_Source_Control	1.22	
StgTankLvlAlm	1.21	
Pump_1_discharge_valve_fails_closed	1.09	
Relief_Valve_1_fails	1.09	
Leak_in_line	1.09	

Table 11: Revision 3 of RRW Table

Cause/Safeguard	Results Given Auto Response to Talk Level	
	RRW	Total Risk
Back-up_Pump	32.73	9.82E-04
Pump_1_fails	32.73	
A1_Improve_Fire_Suppression	1.03	
Fire_Suppression	1.03	
Ignition_Source_Control	1.03	
StgTankLvlAlm	1.02	
Pump_1_discharge_valve_fails_closed	1.01	
Relief_Valve_1_fails	1.01	
Leak_in_line	1.01	

The interesting result here is that there was no change in the order on the RRW table, however, the RRW values were changed significantly. In addition the risk was only slightly reduced, from $1.2x10^{-3}$/yr to $9.8x10^{-4}$/yr. Note now that the importance of the Back-up (or the Cause of the accident, Pump 1 failure) has dramatically increased. Indeed, if further risk reduction is necessary, these pumps must be further addressed.

The process outlined and demonstrated here has been termed by the authors the Cumulative Risk Benefit Analysis, because, as is evident, it takes into account the cumulative risk benefit from sequentially considering risk reduction actions, based on a priority that always considers actions at the top of the RRW values. In addition to providing a cumulative risk benefit result, it all ensures that the issues selected for risk reduction action are those that cause the risk to be reduced by the greatest amount. Note, for a more complex, realistic problem, the actions to be considered might be on the order of tens of issues, rather than one or two, in which case this process become very important and useful.

REALISTIC EXAMPLE

As discussed above the approach provided here is appreciated more in a realistic HAZOP study and analysis. In a recent project for a pipeline company, HAZOP studies were completed for several pumping stations. The goal of these studies was to identify and prioritize risk reduction actions, based on the findings of the HAZOP studies. The Cumulative Risk Benefit Analysis process (Figure 1) was applied. Figure 2 provides the results from the Cumulative Risk Benefit Analysis for one facility.

Figure 2: Cumulative Risk Reduction for Plant 1

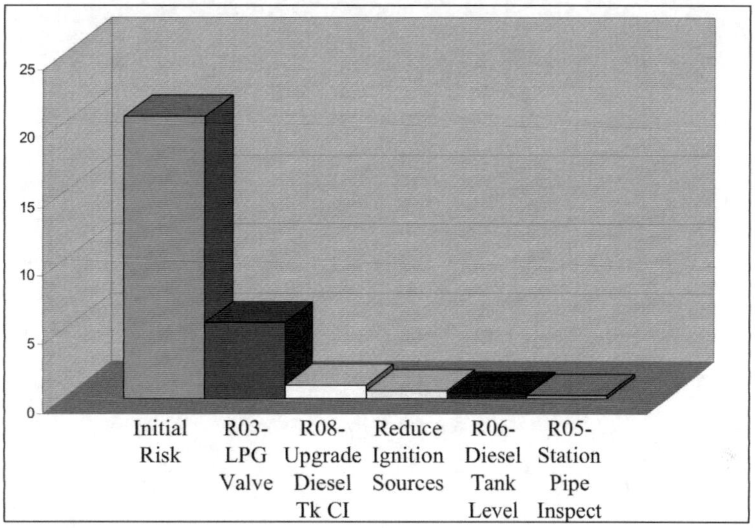

An immediate benefit was the selection of an optimal set of actions to reduce risk to a tolerable level. It also showed that even though numerous findings may have resulted from the study, only a few need to be implemented to achieve the plant's goals for risk reduction. Indeed, the station has already identified by "walk-around" and lessons-learned from other stations, numerous opportunities for risk reduction, but had no way to prioritize their actions. The study completed for their facility showed that by implementing only a few items, most of the risk reduction desired could be achieved.

CONCLUSION

The Hazard and Operability (HAZOP) study approach, and other similar methods, are very effective ways to qualitatively identify a comprehensive set of accident scenarios for a facility. If these analyses are modified to incorporate a simple system for evaluating relative risk, such as an order-of-magnitude scoring system, the resultant study can be a very powerful input to developing risk reduction strategies. By adding the concept of Risk Reduction Worth evaluations for all accident Causes, Safeguards, and proposed Action Items, an analyst can then formulate a strategy to select the minimal set of risk reduction actions that maximizes risk reduction. One strategy for doing this involves the iterative evaluation of RRW after postulation of risk reduction actions, until the residual risk reaches a tolerable value, termed Cumulative Risk Benefit Analysis. This concept was developed for the evaluation of a set of pipeline pumping stations, and provided valuable insight into how to reduce risk in a sensible, prioritized fashion.

Making HAZOP the method of choice for identification and assessment of process plant hazards

Philip W. Beall (E-mail: beall.philip@adlittle.com) and
Geoff Stevens (E-mail: stevens.geoff@adlittle.com)
Arthur D Little Limited, Science Park, Milton Road, Cambridge, CB4 0XL, UK

Abstract

Hazard and Operability Study (HAZOP) has long been established as one of several techniques for risk identification. Many firms incorporate HAZOP procedure into their safety management approach for both new build facilities as well as modifications to existing plant and some have come to regard the method as a 'gold standard' for comprehensive risk identification. However, the reality is that the technique has practical pitfalls and can be far from comprehensive when implemented without the necessary rigor or independence from interests of licensors or contractors.

This paper examines reasons for shortcomings in HAZOP studies and discusses training approaches designed to strengthen HAZOP recording and analysis in the future. A method of calibrating the HAZOP output from a completed study against industry wide loss expectation is also summarized. The extension of this simulation approach to the benefit-cost prioritization of alternative engineering solutions for new build or for existing plant upgrading at turnaround will be described.

Introduction

Hazard identification is one of the key steps in systematic Safety Risk Management, which can be illustrated in diagrams such as Figure 1.

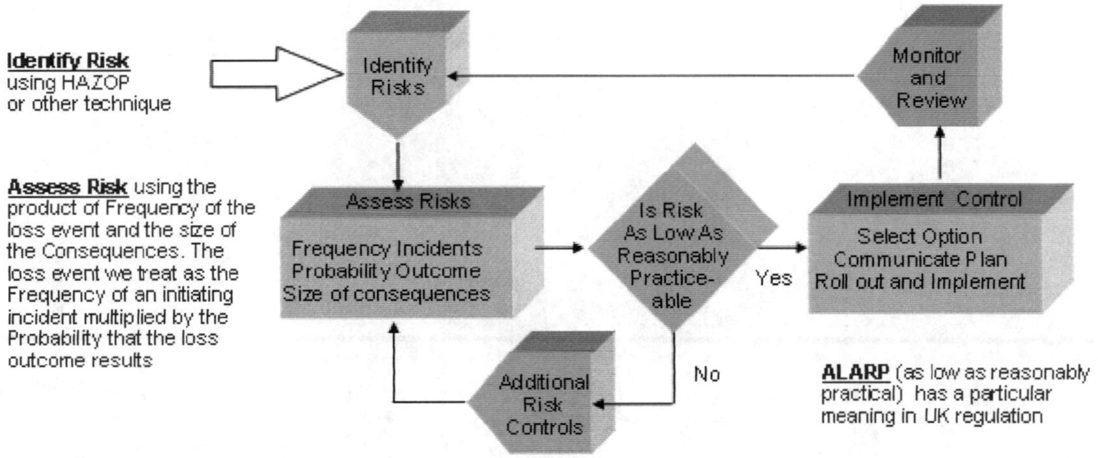

Figure 1: Block Flow Diagram showing Identification and Assessment of Risk as part of Safety Risk Management

Source: Arthur D Little

Process Hazard Analyses (PHAs) are frequently used within the oil and petrochemical industries to identify process risks inherent in new or existing processes. PHAs are a thorough, orderly, systematic approach for identifying, evaluating, and controlling the hazards of processes involving highly hazardous chemicals. This paper argues that the separate steps, (identification, assessment, implementation, monitoring) are interdependent and benefit from being conducted in a complementary manner, for example to a similar level of detail with a balanced allocation of resources.

A requirement for PHA is to perform a hazard evaluation using a methodology appropriate to the complexity of the process and must identify, evaluate, and control the hazards involved. Varieties of methods are available:

- What-if,
- Checklist,
- Hazard and operability study (HAZOP),
- Failure mode and effects analysis (FMEA),
- Fault tree analysis

of which HAZOP is generally recognized as very demanding in time and resources.

HAZOP studies in practice

One of the main characteristics of executing a successful HAZOP is teamwork. The U.S. Occupational Safety & Health Administration (OSHA) for example has said that the process hazard analysis is best performed by a team with expertise in engineering and process operation[1]. The team should include at least one member who has experience with and knowledge of the process being evaluated and one member knowledgeable in the specific analysis methods being used. In HAZOP this guidance is followed by forming a team representing specialist disciplines, facilitated by a Chairman experienced in the HAZOP approach and a Recorder who notes the team discussion and the conclusions reached (see Figure 2) [2,3].

Figure 2: HAZOP Team during a pause in proceedings

Source: Arthur D Little

The HAZOP approach is systematic and stepwise investigation into possible hazardous deviations from the design or operating intent. Figure 3 illustrates the steps in the method.

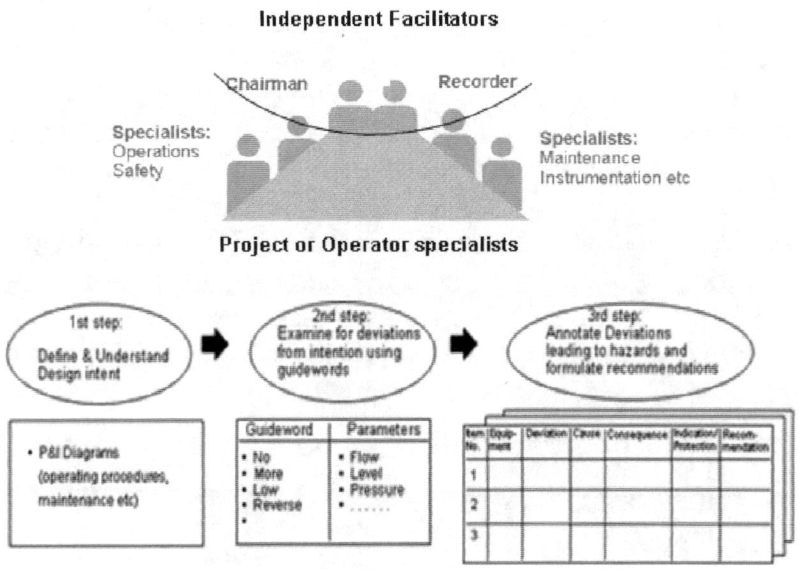

Figure 3: Schematic Diagram of HAZOP Team structure and Steps in working method

Source: Arthur D Little

It is important to realize that a HAZOP study is not a 'one-time' event. The HAZOP method is often applied at several times during the development of a project and can be used for operating plants for example when modifications are planned or when a major turnaround is anticipated including safety related upgrading (see Figure 4). For different project phases, the HAZOP methodology can be applied to a wide range of documentation including operating manuals once these are put in diagrammatic form.

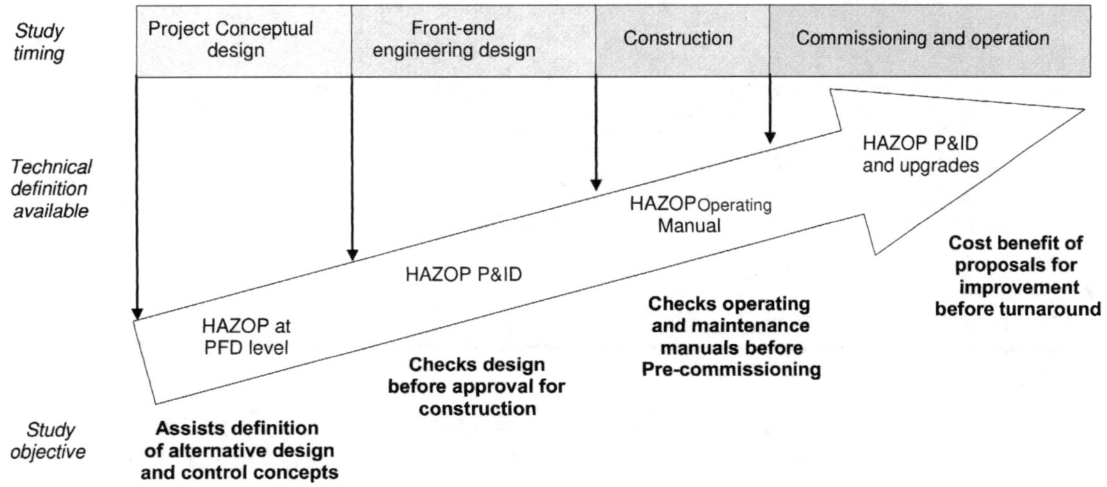

Figure 4: Project phases and application of the HAZOP approach

Source: Arthur D Little

Outcome from HAZOP studies: HAZOP Record

Although much attention is paid to the conduct of HAZOP studies, for example the use of guidewords and the sequence of analysis of nodes, we have concentrated on the output from the study. This is because the HAZOP study is not an end in itself, but part of systematic Safety Risk Management. How the output of the HAZOP informs Risk Assessment and plays its part in the Risk Management is our principal concern.

As well as their role in safety risk management, the HAZOP team meetings provide a great opportunity for knowledge transfer between Licensors and Operators if the HAZOP is facilitated to allow adequate time for questions and answers. Figure 5 illustrates such a session where the Licensor is explaining to the HAZOP team some of the protective features of the design.

Figure 5: Explanation of design intent as part of HAZOP meeting

Source: Arthur D Little

Is it useful for a summary of these discussions to form part of the HAZOP record? From the point of view of an engineering contractor, seeking to manage the design of a unit using limited resources may not seem essential. For the plant owner however, it is important that the management of operations is based on a thorough understanding of the technology (prompt, appropriate operator intervention can play as important a part as plant engineering in achieving safe reliable operations).

A comparison of HAZOP records illustrates these different approaches. Figure 6 illustrates the style of recording facilitated by the engineering contractor. This example discusses hazards associated with furnace tube leaks in a fired heater in a refinery plant. The record lists the safeguards in terms of the engineering features provided by the contractor and implies that these are considered satisfactory with no recommendations for improvement. To someone who did not attend the HAZOP it is difficult to gauge the extent to which the proposed arrangements were challenged by the HAZOP facilitation.

Parameter: Flow

GW	DEVIATION	CAUSES	CONSEQUENCES	SAFEGUARDS	REF#	RECOMMENDATIONS	BY
		88. Tube burst in H-	88.1. Uncontrolled fire in H	Proper design of heater piping and material selection of piping. Presence of pipe wall temperature measurement. Regular inspection of heater piping and flame patterns. Trip of H- at low flow.			

Figure 6: Example from contractor facilitated HAZOP

Source: Confidential Note, original record did not include an entry in deviation column

The second example (see Figure 7) is a record of a comparable discussion during a HAZOP facilitated independently from the owner or contractor. The more detailed record shows the HAZOP was facilitated to include Licensor's transfer of experience with similar plants elsewhere. The record includes a detailed discussion of operator emergency response and suggestions to revise Emergency Shut Down (ESD) logic. The discussion also includes references to options for operator training to improve the reliability of response.

Causes	Consequences	Safeguards	Recommendations
Full bore rupture of furnace tube	Reverse flow from the reactor; recycle gas compressor and feed pump are still running. Potential to lift the catalyst and some bubble caps in the reactor because of the reverse pressure gradient	Emergency guidelines provided P101 to be stopped thus blocking the feed to the reactor IZ003 trips the main burner IZ005 pilot trip IZ018 opens either UZV15 or UZV16 (Emergency depressuring valve is activated) Close FV007 and UZV001 stop the oil flow MUG compressor runs in a kick back to stop makeup of H2 Use of make up compressor to supply nitrogen	R 159.1 Provide revised ESD logic providing a total plant shutdown implementing the function described in the emergency guidelines. Develop the shut down using a hierarchy of trips (e.g. IZ001 for plant ESD)
		A dedicated simulator has not yet been purchased. Elsewhere such simulators are considered useful to improve the quality of training and reduce the chances of mistankes in emergency situation.	R 159.2 Evaluate if it is useful to introduce snuffing steam as indicated in emergency guidelines. May be preferable to maintain the flame to ensure release is ignited and not use of snuffing steam
			R 159.3 Provide total heater trip (shut main burners as well as pilots). In this furnace the pilots contribute considerable heat input and should be closed in an emergency situation

Figure 7: Example from independently facilitated HAZOP

Source: Confidential, the record was made under the deviation "Reverse Flow"

A similar discussion record in Figure 8 (for a different plant, different licensor and different HAZOP team) shows that comparable issues were raised but a different response was elicited from the team. In this example changes to the Emergency Shut Down functionality were not felt necessary by the team but the record contains a summary of the proposed operations response.

Causes	Consequences	Safeguards	Recommendations
Small leak at furnace coil	Maybe TAH from TI 135 Readings at AT001 or AT 003 may differ from usual	Regular inspection is recommended. Gradually reduce main firing maintaining pilots and reduce pressure gradually until nitrogen can be connected to inlet of recycle gas supply. Switch from hydrogen gas recycle to nitrogen recycle and purge catalyst circuit Maintain combustion air flow. No recommendation for an exclusion zone around the heater	R 138 Evaluate if it is desirable to cut the firing when the feed pump is stopped (IZ12)
Full bore tube rupture	Feed is fed to the fie box and the reactor circuit reverse flow depressures to the fire box	Manually depressure the unit and stop fuel supply to the furnace (main and pilot). Maintain the combustion air fans to purge fire box. Shut in recycle gas compressor and feed pump, MUG compressor. Activate snuffing steam	

Figure 8: Example from another independently facilitated HAZOP

Source: Confidential, the record was made under the deviation" Reverse Flow"

The record illustrated in Figure 6 is limited in its comments and suggests the HAZOP team placed a reliance on design. The record does not indicate if the team considered the deviation "reverse flow" from the reaction section into the firebox (one of the practical hazards to be managed, which are not always recognized by operators). The record mentions flame inspection but without recognition of the hazards of high-pressure furnace inspection (at least one fatal incident is known during such a procedure).

By comparison, the independently facilitated HAZOP records discussion on the protection against reverse flow and summarizes licensors advice on emergency response actions expected of the plant operators. Such a record can be written only if the HAZOP is facilitated to encourage this information transfer from the Licensor and experience with other plants of the type under review. Such facilitation is easier if the HAZOP is lead by someone independent of design contractor, free of any stake to minimize deficiencies, which may be expensive or time-consuming to address.

When comparing these two styles of HAZOP facilitation, it must be remembered that the record is the tangible output from many person-hours of HAZOP team discussion and the main guide to the quality of the work for those who were not present. For example, the HAZOP record for the study of a new plant may be reviewed by regulatory authorities as part of the facility permitting process. In representations to the authorities, the owner will draw from the use of recognized technology, implementation by experienced contractors and in addition (where the HAZOP has been facilitated by a third party) independent review (see Figure 9).

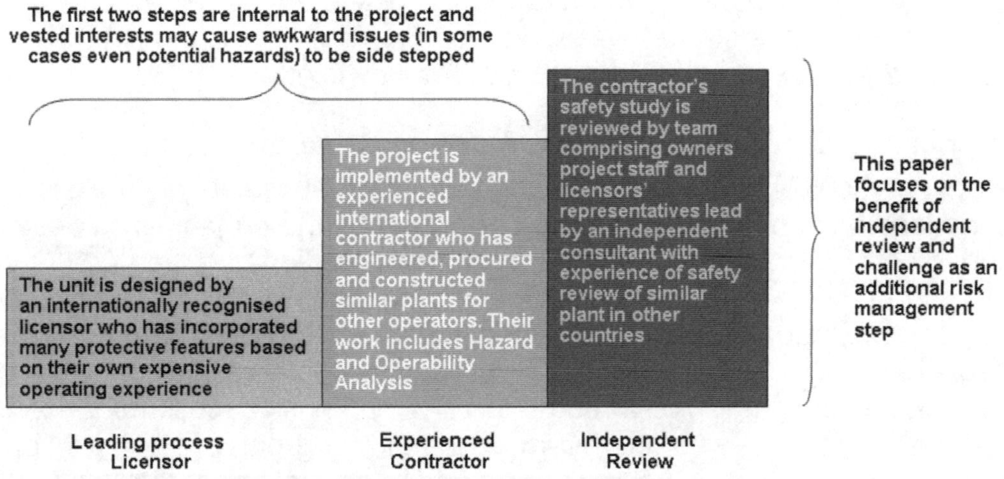

Figure 9: The role of independent HAZOP facilitation in permitting process

Source: Arthur D Little

If the HAZOP record is to be used as evidence of independent review, the documentation must be complete and convince those not at the HAZOP that a thorough examination took place.

Like other aspects of the conduct of a HAZOP, training specifically with attention to the role of the recorder can be useful in raising the standard of study documentation. In training courses used for our own staff as well as for clients, we include sessions where trainees are given a section of P&ID to HAZOP and their record is contrasted in a plenary session with an idealized version developed from previous case-work. We have extended this by developing HAZOP templates, which discuss the typical hazards and protections, employed.

Item	Deviation	Cause	Consequence/ Implication	Indication/ Protection	Answers/Comments
10.1.1	High Flow	Full open of upstream flow control valve	High flow leads to low furnace exit temperatures and an increase in firing to compensate	Furnace firing controlled by TIC from common exit from furnace	
10.1.2	Low Flow	Closure upstream control valve	Overheating furnace tube with risk of tube rupture	FAL on furnace feed or TAH on exit. If no operator response, FSLL on inlet to each pass or TSHH on exit to trip furnace main burners	
10.1.3	No Flow	Trip of feed pump	Risk of tube rupture for low flow	Confirm if other flow eg recycle gas sufficient to sustain furnace tubes	
10.1.4	Reverse flow	Tube rupture	Backflow from upstream reactor or column causing major fire	If furnace acts as reboiler for a column and service is clean consider check valve to prevent reverse flow especially on lines of 6" or less. If reactor consider blowdown from downstream to release flammable mixtu	Depending on circumstances may be desirable to maintain recycle gas compressor in operation and substitute hydrogen for nitrogen to purge reaction circuit including furnace. Snuff furnace with steam as soon as insufficient flame to sustain combustion
10.1.5	Loss of containment	Leak in tube	Assess size of leak to determine correct action.	Avoid emergency shutdown and attendant thermal shock which may make a small leak worse. Reduce firing but maintain some flame close to the source of small leak. Manually reduce pressure and substitute nitrogen for hydrogen. Snuff furnace when ignition no longer supported	

Figure 10: Training Template for Flow deviations for fired heaters

Source: Arthur D Little

Training templates have been developed for various types of equipment such as pumps, compressors, exchangers, and so forth. These templates are developed by taking the best from several previous case studies. They are used strictly for training HAZOP recorders and must not be used as a short cut during a HAZOP study itself.

Outcome from HAZOP studies: Priority Recommendations

As well as the HAZOP record, it is common to produce lists of HAZOP recommendations each of which has to be closed out by the project team engineers. The timeliness and completeness with which the recommendations are carried out varies from study to study depending on the number of recommendations (which can exceed 1,000 in a large study) and the resources available (which can vary from a large project team to spare time available to process engineers). One approach is to involve the members of the HAZOP team, after completion of the HAZOP study to carry out a risk assessment exercise using a Risk Assessment Matrix. Once each recommendation (or group of similar recommendations) has been assessed to estimate the risk before and after implementation, the recommendations can be prioritized in order of risk reduction. A further elaboration of the ranking approach evaluates the ratio of the estimated risk reduction to the cost of implementation of the recommendation. Provided the risks are below the intolerable threshold, this approach is consistent with reducing risks as low as reasonably practicable. Recommendations with comparatively high cost or little risk reduction tend to come lower in the ranking. The approach relies firstly on grouping recommendations into main issues and then preparing a Risk Register for each of the issues. The register contains an extract from the HAZOP record, provides the details of responsibility for implementing the recommendation and assesses the risk as follows:

- The frequency of the initiating event (for example release of a flammable vapor)
- The probability that the initiating event will lead to a loss
- The magnitude of the loss

The assessment of Frequency, Probability and Consequence is carried out by members of the HAZOP team using risk assessment matrixes as illustrated in Figure 11. Each matrix covers five ranges and enables a rapid assessment of risk.

Frequency		
Range	Low	High
1	1.0E-04	1.0E-03
2	1.0E-03	1.0E-02
3	1.0E-02	1.0E-01
4	1.0E-01	1
5	1	10

Implementation Cost		
Range	Low	High
1	1,000	5,000
2	5,000	25,000
3	25,000	100,000
4	100,000	250,000
5	250,000	1,000,000

Probability		
Range	Low	High
1	1.0E-03	1.0E-02
2	1.0E-02	5.0E-02
3	5.0E-02	2.5E-01
4	2.5E-01	9.9E-01
5	9.9E-01	1.0E+00

Consequences		
Range	Low	High
1	1	100,000
2	100,000	500,000
3	500,000	2,500,000
4	2,500,000	15,000,000
5	15,000,000	50,000,000

Figure 11: Example of a Risk Assessment Matrix for Frequency, Probability, Consequence and Implementation Cost

Source: Arthur D Little

It is also important to recognize that loss needs to consider a variety of types of consequence including loss of corporate reputation that can result from a major incident. The categories are:

- Safety loss (extent of possible injury to third parties or staff)
- Property loss (extent of damage to equipment and cost for reinstatement)
- Business Interruption (the commercial losses from reduced output following the incident)
- Environmental Liability (resulting from environmental impairment following the incident)
- Reputation Loss (the potential effects of a serious accident on Corporate Reputation for example on market capitalization)

The assessment is recorded in the risk register as illustrated in the following figure:

Client	Client Gas oil project								Revision	1.1
Project No:	20238	Status	Basic proforma adjusted as discussed in Location on 11/11/04						Date	14-Nov-04
Risk Description	Failure of hydrogen quench line								Author	GCS
Initiating event	Line rupture or leak followed by ignition may result in intense flame capable of severe damage									
Development										
Consequences	No specific shutdown sequence has been proposed to deal with this situation.									
Safeguards	1. Shut C8101 to reduce pressure and inventory in the hydrogen line. 2. Keep P8101 running to maintain a cooling hydrocarbon flow through the reactors. 3. Cut H8101 furnace to reduce the temperature of the hydrocarbon reactor inlet stream. 4. Depressure system. 5. Cut C8102 to reduce pressure and inventory in the hydrogen line.									
Recommendations	Include scenario in emergency procedures and develop instructions which minimise exotherm risk without allowing fire to persist longer than necessary and without exposing staff to hazard of entering HP area									
Assessment	Before implement			After implement			Calculation Before implement	Calculation After implement	Plant Benefit	Assessment Notes
	Frequency	Probability	Consequence	Frequency	Probability	Consequence				
Safety										
Property Loss	3	2	4	3	1	4	3,953	559	3,394	
Business Interruption	3	2	5	3	1	5	19,764	2,795	16,969	
Environmental Impairment										
Reputation Benefit										
Cost of Implementation	2		Rank			Benefit Cost Ratio	2.5	Total benefit	20,363	
Proposed Action										
Assigned to:									Date Due	
Actual Resolution:										
Completed by									Date Completed	
Approved by									Date Approved:	
Implementation Comments										

Figure 12: Example of a Risk Register

Source: Confidential

The completion of the Risk Register for all groups of similar recommendations allows the recommendations to be listed:

- In descending order of risk reduction (difference in annualized loss before and after implementation of recommendation)
- Grouped into recommendations with Risk reduction / Cost ratio greater than and those less than 1.0

The following graph (Figure 12) illustrates the result of such a prioritization showing the risk reduction (annualized $ loss averted) by implementation of 24 recommendations which were assessed with benefit/cost ratio greater than 1.0.

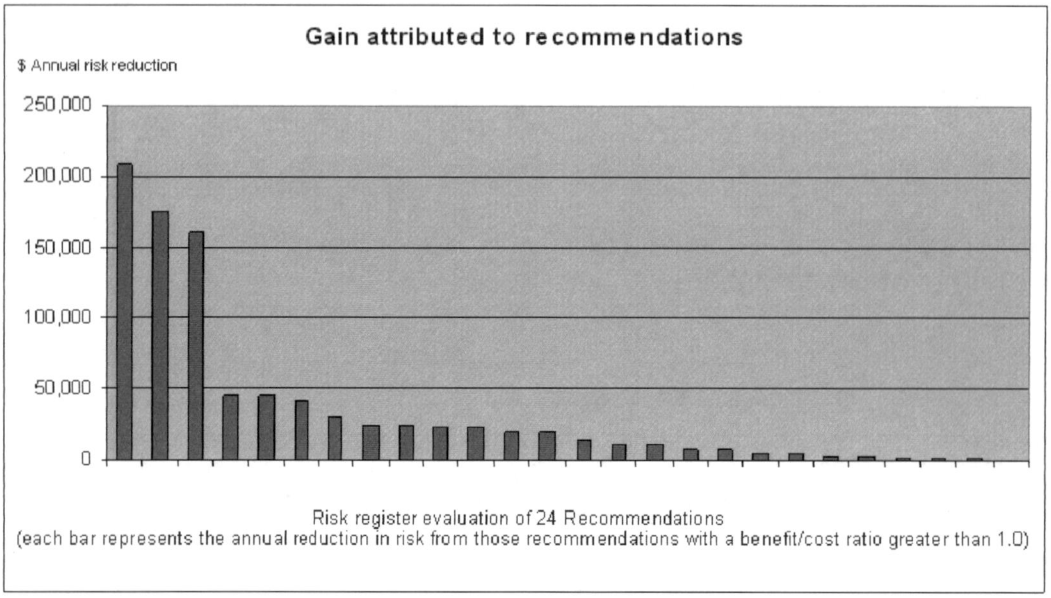

Figure 13: Loss aversion ranking for recommendations

Source: Confidential

The assessments recorded in the risk register can be used as input to a Monte Carlo simulation of the loss expectation for the plant in question using techniques detailed elsewhere[4]. If these simulations are carried out for Loss categories limited to the insured aspects of plant operations, the Loss Profile from the HAZOP study compared to losses recorded for accidents elsewhere in the industry (or if there are available, claims history for such plants).

We have conducted this exercise for a number of plants and find it a useful way of cross checking that hazards identified and risks assessed are reasonable given Loss History elsewhere in the industry. The Loss profiles also serve to clarify to Management the relative positioning of their plant in relation to others in the industry and the potential value (in terms of loss aversion) were a revamp to be conducted (for example at the next scheduled turnaround). This technique, which includes Business Interruption, Environmental Impairment and Loss of Corporate Reputation, can help managements reach an objective understanding of the potential benefits of investing in upgrading of existing facilities, especially those that have been in operation for many years.

References

1) 29 CFR Part 1910 (February 1992) *Process Safety Management of Highly Hazardous Chemicals* Federal Register Vol 57 No 36 pages 6356 to 6417
2) Center for Chemical Process Safety (1985) :*Guidelines for Hazard Evaluation Procedures* American Institute of Chemical Engineers 2-14
3) Chemical Industries Association (1990):*A Guide to Hazard and Operability Studies*
4) G.C. Stevens, A. Verna and M. Marchi, (October 2001) "*Getting full value from HAZOP. A practical approach to the assessment of risks identified during studies on process plants*" IChemE Hazards XVI Manchester England

Selected Examples of Design Optimization Using An Interactive Work Process And Software Tools

M. D. Muhlheim, J. W. Cletcher, II; Oak Ridge National Laboratory; Oak Ridge, TN 37831

S. Flanagan, J. L. Hynek, Isograph Direct; Newport Beach, CA 92660

R. J. Stack; Dow Chemical; Hillsdale, MI 49242

Abstract

Risk assessment involves evaluating both the likelihood and consequences of system failures. An interactive risk assessment of a design requires that both the likelihood and consequences of system failures be quickly available to the design team. The consequence assessment focuses on the key failures and failure scenarios which are the input for the likelihood assessment process. While a screening evaluation of protection layers can provide general design guidance, often a more detailed evaluation of the safety and reliability of a design option can result in a design optimization. Probabilistic Risk Assessment (PRA) is an analytical method used to estimate the likelihood of failure of a system and to determine what the most likely contributors are to that failure once the consequence assessment has identified the key events for analysis. Space, nuclear, medical, chemical and defense industries are among those that have used PRA methods for assessing risks and/or reducing the costs in designing, upgrading, manufacturing, assembling, and operating components, systems, or facilities. When applied at an early stage of a project, PRA can be a valuable design tool. Current PRAs, however, are generally performed to demonstrate safety and are often unsuited for applications aimed at making design or operating decisions. Conventional risk assessment (PRA) tools allow integrated design teams and safety risk (PRA) analysts to make comparisons at design "freeze" points to address and mitigate the risk drivers. Unfortunately, the comparisons are made in isolation—that is, one at a time, and often the opportunities for some late stage design modifications are often limited. Using a work process that encourages an interactive design approach and using software that permits rapid evaluation of different design configurations allows for a real time quantitative evaluation of a design and rapid feedback to the design team. This paper focuses on the interactive work process and examples of actual applications with software tools that allow risk analysts and designers to easily determine the probabilistic implications of different design configurations and operating conditions in various combinations to reduce, control, or eliminate risk by quantitatively identifying risk drivers as the design develops and allow users to do an interactive evaluation of design changes. Because of the ease in evaluating alternate component or system arrangements, dramatic increases in reliability were observed with atypical, unusual, or simply different design configurations compared to the designs using "proven" reliability design practices. The lessons learned can be counterintuitive and significant. With ever shorter design timelines, the work process together with the software tools enable an interactive design optimization from a safety and reliability perspective and also consider the costs of different options. In the interest of being better, faster and less expensive, this is clearly a 21st century analysis method.

Introduction

Historically, the risk assessment process has usually been used after the design of a system is finalized to see if the design meets acceptable risk criteria. In many cases, design changes driven by a risk assessment can require costly redesign and project time delays. This has in some cases led to the use of design rules or guidelines that add to the cost and complexity of a project. For every rule there can be an exception where blindly following the rules will lead to a sub-optimal design. Space, nuclear, medical, and defense industries are among those that have used risk assessment methods for assessing risk; increasing safety margins and optimizing costs in the design, fabrication, construction, testing, and operation of components, systems, and facilities.

When applied at an early stage of a project, Risk Assessment can be a valuable design input. The interactive modeling of potential scenarios and system failure consequences can alert the design team to the potential risk associated with a general design and suggest direction for reducing the consequence part of the risk. However, once a general design is established, the optimization of the risk reduction is normally achieved by minimizing the likelihood of process or system failure scenarios. Especially with complex systems, a simple layer of protection analysis may provide a path for risk reduction but this result may not be a risk based design optimization; a more detailed modeling can provide optimization.

PRA is an analytical method used to estimate the likelihood of failure of a system and to determine what the most likely contributors are to that failure. Current PRAs however, are generally performed to demonstrate safety and are often unsuited for applications aimed at making design or operating decisions.[1] In making design decisions, if comparisons are made using layer of protection screening or even simplified fault trees, the models may not accurately represent the system's functioning or malfunctioning.[2] Valid design or operation comparisons require detailed assessments. Developing a detailed PRA during the design phase of a project is an iterative process where the models are developed at design "freeze" points. However, as the design matures, the PRA results continue to reflect a previous design and not the current design. An example of the process is:

- development of a PRA model,
- use of the model to identify weaknesses,
- quantification of PRA benefits of alternate design and operational strategies, and
- adoption of selected design and operational improvements.

This process was followed by Westinghouse during the design of its AP600 nuclear reactor. As expected, the scope and detail of the AP600 PRA model increased as the plant design matured. Although the iterative process resulted in a number of design and operational improvements, the process spanned five stages and six revisions.[3, 4]

This iterative process can continue throughout the design cycle and into the certification process. For example, when considering the final design approval for the CE System 80+ nuclear reactor containment design, the Nuclear Regulatory Commission (NRC) requested that the designers evaluate design alternatives that would help mitigate the consequences of severe accidents.[5] Sixty-three design alternatives were considered, and twenty-seven were quantified. This back-end approach is expensive and time-consuming. A much better work process approach would be to use a dynamic risk assessment and PRA coupled to the design to reduce, control, or eliminate risk by quantitatively identifying risk drivers as the design develops.

Westinghouse, in its design of the International Reactor Innovative and Secure (IRIS) plant, brought the use of a dynamic PRA to reality by using a safety-by-design approach to cover all aspects of safety at the very beginning of the design process.[6] Rather than evaluate design alternatives on a system-by-system basis, one change at a time, and discarding previous evaluations, the different system and component configurations were evaluated in various combinations for the entire range of initiating events. Thus, unlike other design "optimization" programs, Westinghouse used its PRA program to evaluate different types of plant systems and components, in different operational states, that were interchangeably evaluated as the design progressed. Because each "new" plant configuration may or may not improve plant safety or may marginally improve safety at great cost or operational flexibility, all previous design alternatives were retained to maximize the benefits of a risk-based design. More importantly, because of the ease in evaluating alternate component or system arrangements, dramatic increases in reliability were observed with atypical, unusual, or simply different design configurations compared to the designs using "proven" reliability design practices. This contributed greatly in a two order-of-magnitude reduction in the base-case core damage frequency for IRIS. The design lessons learned from the dynamic PRA for IRIS were then applied by Westinghouse to the AP1000 even though it was undergoing design certification review by the NRC.[7]

Innovative Approaches to Probabilistically Evaluating Alternative Designs

As part of its safety-by-design approach and commitment to reduce the core damage frequency (CDF) to as low as realistically possible for its IRIS plant, Westinghouse had many team members focused on safety and reliability. One of the team members was the Oak Ridge National Laboratory (ORNL); its task was to develop an interactive software tool that would enable an interactive design work process. By allowing components, modules, and data to be interchanged at will with the probabilistic effect immediately apparent, the optimization of design from a safety and reliability perspective could be done. ORNL's program, called risk-based design optimization tool (RBOT), is coupled to *FaultTree+* and allows users to evaluate design changes; new modeling approaches, methods, or theories; modeling uncertainties and completeness; physical assumptions; and data changes on component, cabinet, train, system, and facility bases.

With RBOT, changing the design is as simple as picking a design alternative from a scroll-down menu (Fig. 1). The default (base case design) and currently selected design options are highlighted; graphical displays provide additional indicators of the option(s) chosen. With the scroll-down menu, many design alternatives, including unconventional alternatives, can be easily evaluated. The basic RBOT computer program automatically inserts the correct fault tree module(s) for the design configuration chosen into the PRA model in *FaultTree+*,[8] relinks the correct support systems, adjusts the common-cause failure probabilities on the train and system levels, and accounts for changes in recovery actions and human error probabilities. *FaultTree+* then recalculates the parameters of interest and transmits the results back to RBOT. The resulting fault tree is the same as that which would result if those choices were the original base case. Any or all of the design options can be made in any order using RBOT. Because all choices remain available, no information is lost and the design can be returned to any previous state. As this process optimizes the likelihood of system failures, the optimized design must be re-evaluated with interactive consequence analysis tools to check that the risk is minimized.

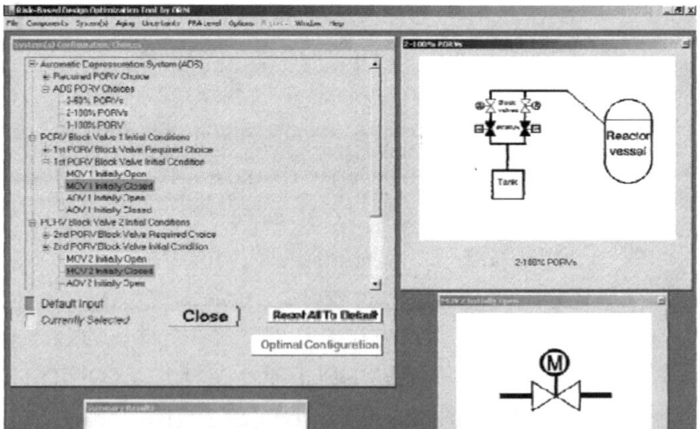

Fig. 1. ORNL's RBOT allows users to choose design alternatives from a scroll-down menu.

Consequence analysis can prompt looking at additional design alternatives. If additional design alternatives are desired and are not available in the scroll-down list of options modeled in RBOT, they can be easily created and added. Because all of the current and previous design, modeling, and data sets are still available, the safety ramifications of design options are evaluated very early, feedback on design alternatives is immediate, and true optimization of the plant design can be achieved. Thus, even with complicated systems and alternatives, designers can not only observe if the system is more reliable, but can also understand *why* the system is better or worse.

Examples of Interactive Design Optimization

IRIS: The IRIS design is aimed at achieving four major objectives: proliferation resistance, enhanced safety, economic competitiveness, and reduced waste. The safety approach of Generation II reactors [current nuclear power plants (NPPs)] can be defined as coping with accidents/consequences by active means, while the improved approach of Generation III reactors (e.g., AP600) is coping by inherently safer passive means. The design of IRIS, a next-generation NPP, promotes safety to the next level through safety-by-design that eliminates most accident initiators while the consequences of the other remaining accident scenarios are rendered acceptable.[9]

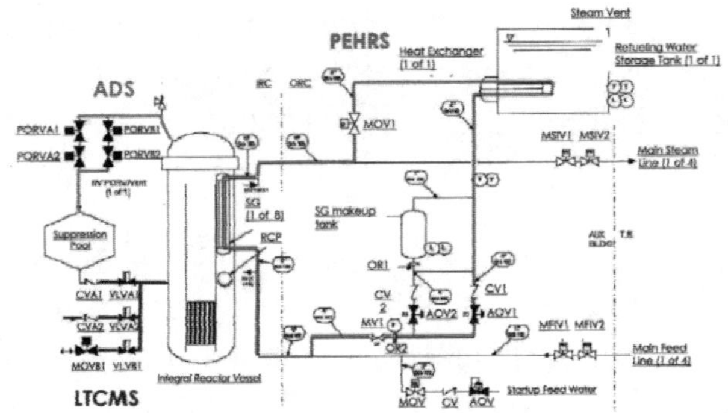

Fig. 2. Safety systems for the IRIS: (PEHRS is the passive heat removal, ADS is auto-depressurization, and LTCMS is the long term core makeup).

Figure 2 shows the safety systems for the IRIS reactor. The two systems that were evaluated using RBOT were the automatic depressurization system (ADS) and the emergency heat removal system (EHRS). The ADS is a pressure relief system with two parallel trains for blowing down into the suppression pool. There are four trains of emergency heat removal. The number of trains required for operation are dependent upon the initiating event and varies from 2–4. The EHRS is a natural circulation system that removes heat via the steam generator. The flow path is from the steam generator, to the refueling water storage tank, through a main and/or bypass leg, back to the steam generator.

A collection of design alternatives for various systems for IRIS were identified through the review of current and Generation III NPP designs. For example, a review of operating NPPs showed that typical relief systems (similar to IRIS's ADS) consist of one or two trains. Generation III NPPs have more complex systems such as two trains with four stages of pressure relief. Further review shows that the block valves for the power-operated relief valves (PORVs) may be set in the initially open or closed position.

A variety of system arrangements, component types, and component initial positions can be used in selecting a design alternative. In the past, design rules or guidelines would have been the basis for selected a design from the many possibilities. With interactive software, 40 unique design alternatives could be quickly evaluated. The following choices were made available in RBOT for optimizing the ADS:

- the choice of block valves and PORVs were air-operated valves (AOVs) and motor-operated valves (MOVs) (other type valves could have been added at any time),
- the initial position of the block valves (AOVs and MOVs) could be open or closed,
- the number of trains for pressure relief could be 1 or 2, and
- the capacity of each valve, whether AOV or MOV, could be 50% or 100%.

In this example, designers to evaluated 40 unique design alternatives for the ADS. Evaluating the failure modes then provided clues on how the system could be improved. Using this example, by clicking a single "button" to change the design of the system, designers learned that for those events requiring pressure relief that

- with only one block valve, the initial position of the block valve does not matter—failures of the PORV(s) dominate the reliability of the system;
- 2–50% capacity PORVs do not provide sufficient redundancy and are in fact worse than having just one PORV because the failure of either PORV/block valve arrangement causes the system to fail;
- true redundancy (2- vs. 1-100% capacity valve) improves reliability by an order of magnitude and not the square of the PORV's failure probability because of shared systems and components;
- for those events requiring pressure relief, it is 4–6 times better to have the block valve initially open, where it has to spuriously transfer to the closed position than to have it initially closed and having to open. For many system arrangements, the dominant contributor to system failure is the common cause failure (CCF) probability of the block valves to open. The simplest way to reduce the CCF probability of the block valves' failing to open is to have the block valves already open, and
- for the ADS, it is observed that having the block valves originally open improves system reliability 20% and reduces the CDF 3.5%. Once the CCF probability of the block valves failing to open is eliminated, the type of block valve (e.g., MOV or AOV) is inconsequential.

- removing MOVs from consideration because their failure probability is two to three times that of AOVs[13–16] would be wrong. A more detailed evaluation shows that the CCF probability for two AOVs is typically greater than that for two MOVs.[13–15, 17]

The other safety system evaluated using an interactive process was the EHRS. Identifying design alternatives for the EHRS was a challenge because it was already a relatively elegant system in its simplicity. The original design used natural circulation cooling (i.e., no pumps), and only 1 of the 2 AOVs needed to open to start the system. In fact, the designers purposefully used the most reliable valves they could—the AOVs. As such, identifying design alternatives for the EHRS was not originally considered because of the simplicity of the existing system design.

With an interactive tool, the number of trains in a system, or even component type or position—alternatives that would not normally be evaluated—were evaluated because of the ease of changing out components, adding trains, etc. This is contrary to past practices where typical alternatives were on big changes such as number of trains and cross-ties between trains.

Because AOVs have the best reliability characteristics for active valves, their use was expected to result in the most reliable system. Interactive tools provide easy evaluation of different components and their arrangement, so that designs with one or both of the AOVs to MOVs for each train of the EHRS could be evaluated. Changes automatically made to model for the valve options included

- changing the selected component(s) from AOV to MOV (or vice-versa if changing from MOV to AOV),
- changing the support system for those components from instrument air to electric power (or vice-versa if changing back to AOVs),
- changing the CCF probabilities on the train level, and
- changing the CCF probability on the system level (4 trains).

Surprisingly, using the higher reliability AOV valves actually made the system (and thus the plant) worse. Using the more unreliable MOVs actually improved system performance. This is because the CCF probability for the AOVs failing to open is larger than that for the MOVs, and the CCF probability of the valves was the largest contributor to system failure. Although the MOVs significantly improved the system, using a mix of AOVs and MOVs provided the most reliable system. In fact, a mixed-valve system improved system reliability by 42%; the CDF was reduced by 23%.

Surprisingly, the relatively few component choices for the ADS and EHRS resulted in a very large number of design alternatives. The 11 component choices—component type, position, and capacity—resulted in 160 design alternatives. The use of an interactive software tool allowed all of these alternatives to be evaluated probabilistically with ease and provided designers with insights into system behavior.

With the wealth of information available through the interactive software, design alternatives can now begin to consider production, operation, and maintenance costs. For example, if the use of a mixed-valve system significantly increases operation and maintenance costs, a decision could be made to maintain all MOVs in the EHRS because system unreliability would not increase appreciably. Thus, for the first time, plant safety, reliability and other attributes, such as cost, can be optimized at the same time.

Reflector Control Segments for a Space Reactor: Because of renewed interest in space reactor nuclear power systems, ORNL used its interactive software tool (RBOT) to evaluate the reliability of design alternatives for the reflector controls of an SP-100 type reactor (Fig. 3). The SP-100 was being developed using a lithium-cooled fast spectrum fission reactor with uranium nitride fuel that can produce 100 kW of electricity. Three mission profiles were under consideration:[10]

1. a high-altitude (>300-year orbital lifetime) mission, launched by a Titan IV launch vehicle, boosted by chemical upper stages into its operational orbit;
2. an interplanetary nuclear electric propulsion mission, started directly from a Shuttle parking orbit; and
3. a low-altitude (50-year orbital lifetime) mission, launched by the Shuttle and boosted by a chemical stage to its operational orbit, with subsequent disposal boost after operation.

Fig.3 Reactor power assembly SP-100.

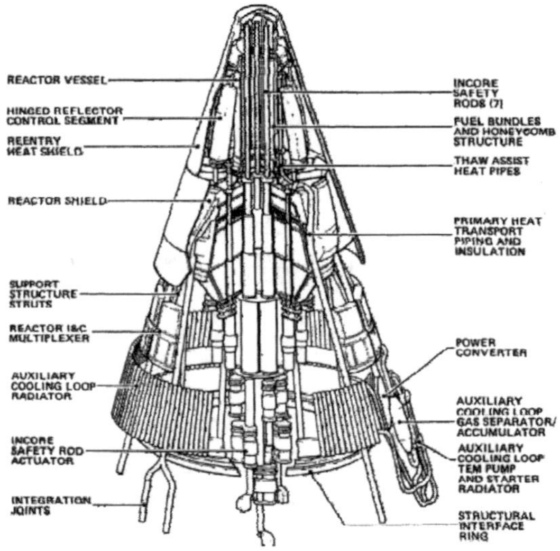

Reactivity of the reactor was to be controlled by 12 reflector elements, with additional shutdown margin provided by 7 internal safety rods. To operate successfully, four of the six movable reflector elements in the reflector control system had to move to ensure startup and shutdown. Three types of reflectors are typically considered for space reactors—petal, drum, and slide reflectors. Petal reflectors are hinged reflectors, drum reflectors rotate, and slide reflectors move up and down. When the reflectors are in their "in" position, they surround the reactor core and reflect the neutrons back into the core, thereby sustaining the chain reaction. Nine motor/reflector arrangements were evaluated using RBOT that ranged from one motor moving all six reflectors to six motors, one for each reflector. The motor-reflector arrangements evaluated were:

- 1–100% motor: one motor operates all six reflectors,
- 2–50% motors: two motors where each motor operates a bank of three reflectors, one motor is a swing motor that can operate either bank of three reflectors,
- 2–100% motors: each motor can operate all six reflectors (i.e., 100% backup),
- 3–33% motors: each motor operates a bank of two reflectors,
- 3–50% motors: two motors where each motor operates a bank of three reflectors, one motor is a swing motor that can operate either bank of three reflectors,
- 4–50% motors: two motors where each motor operates a bank of three reflectors, two backup motors that can operate one bank of reflectors,
- 4–50% motors: two motors where each motor operates a bank of three reflectors, two swing motors that can operate either bank of reflectors,
- 5–33% motors: three motors that each operate a bank of two reflectors, two swing motors that each can operate one of two banks of reflectors, and
- 6–17% motors: six motors that each operate one reflector.

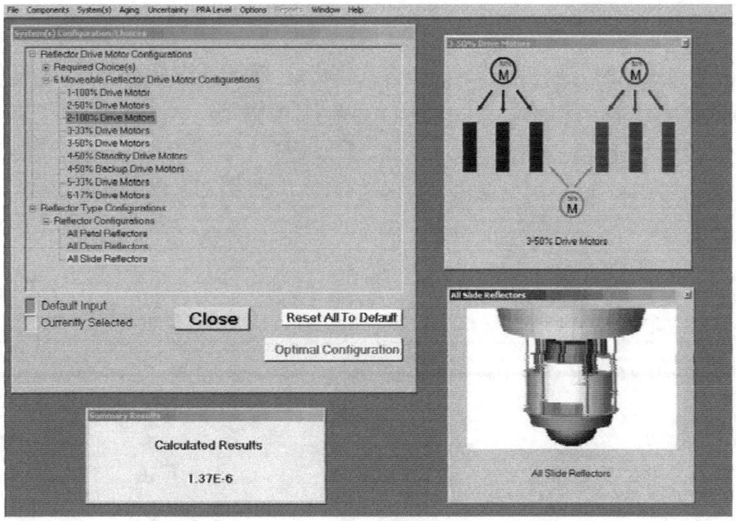

Fig. 4. Design alternatives for a space reactor power system's reflector control system.

Here the interactive software was used to evaluate, all of the motor-reflector arrangements given above (Fig. 4). Surprisingly, the five-motor configuration was the optimal system configuration. A review of why five motors is the best arrangement for six movable reflectors shows that in the five-motor configuration, four motors must fail for the system to fail (Fig. 5). A similar review of the other reflector motor options clearly shows that any of the other design options require fewer reflector motors to fail for the system to fail. For example, a system with six reflector drive motors—one for each reflector—requires three motors to fail for the system to fail (Fig. 6).

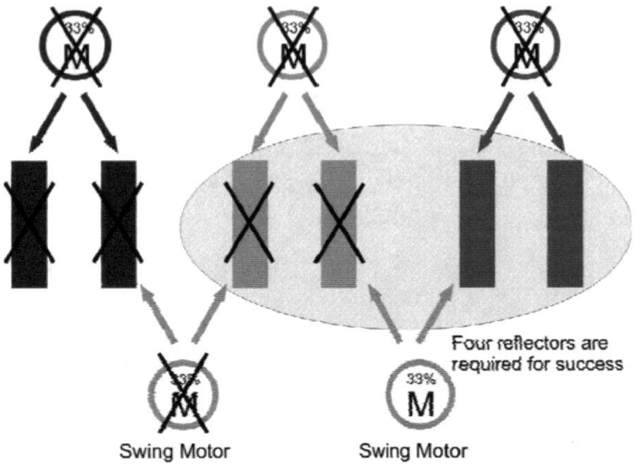

Fig. 5. The five-motor option for the reflector drive system requires the most motors to fail (four) for the system to fail.

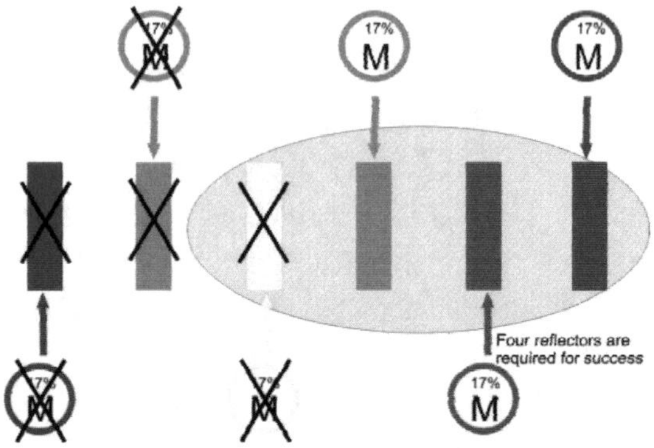

Fig. 6. Any motor option other than the five-motor option for the reflector drive system requires three or fewer motors to fail for the system to fail.

Chemical/Petrochemical Industry Example: In the chemical and petrochemical industry, risk analysis is used to look at the consequences and frequency of system failure scenarios. The consequence screening gives the critical failures that merit an analysis for design options to minimize the likelihood of failures. Fault tree analysis is used as a tool for improving new system design, for evaluating design upgrades for existing systems, and determining if systems meet established safety and reliability targets. Work processes such as Six Sigma can use fault tree tools to model the process to assess its safety and reliability capabilities and achieve breakthroughs in new designs. Using Fault Tree+ with a selection of fault tree modules that represent current best designs for these systems, models of a new design including these standard subsystems can be quickly evaluated. Putting these modules into an interactive tool gives the same benefits as shown in the previous examples.

A simple generic example—a high integrity isolation system involving the use of three valves in series—is used to show how different design options and assumptions provide different results. The design options are to use a control valve (CV), an MOV, or an AOV that fail closed in any combination except that only zero or one CV may be used for any option. This results in seven possible options (ten if there was no restriction on the CV usage) from this simple example (Table 1).

Valve Options using one CV			Valve options without using a CV			
Valve 1	Valve 2	Valve 3	Option	Valve 1	Valve 2	Valve 3
CV	MOV	AOV	4	AOV	AOV	AOV
CV	MOV	MOV	5	AOV	MOV	MOV
CV	AOV	AOV	6	AOV	AOV	AOV
			7	MOV	MOV	MOV

Table 1 Seven design options using zero or one CV

Using a interactive tool with a graphical interface to model and change out the valves, each of the design options can be quickly checked (Fig. 7). Because library models and data are used, a high level of expertise with the fault tree details is not required. For the example valve options given in Table 1, the data in Table 2 were used to evaluate which valve arrangement given in Table 1 would be the best performing isolation system (the mean time to repair for all seven options was 0.01 years with a test interval of 1 year).

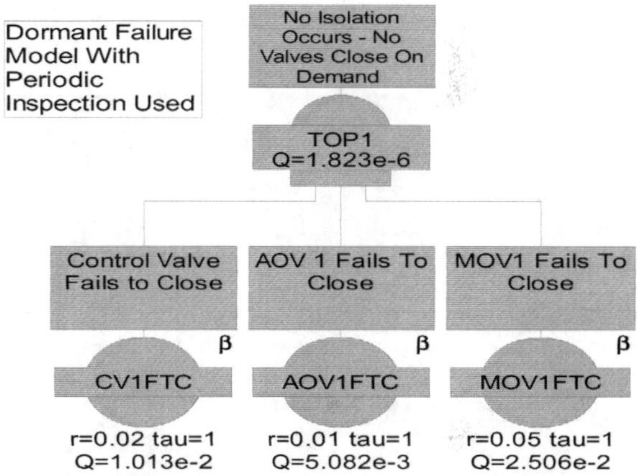

Fig. 7. One of 7 FaultTree+ library models available to the graphical interface in the sample problem.

The values for the failure rates and beta factors can be changed to check the sensitivity of the results. This can be done by having alternate values in the library that can be used in the fault tree models. Another variable that can be easily checked is extending the test interval while still meeting the safety targets. If a safety target of 1.0×10^{-5}/yr or less was specified, only three of the configurations would meet the target (Table 3). The common-cause failure contribution (%CCF) is a major factor in all the cases evaluated. These results agree with the IRIS example provided above.

The real power of an interactive design evaluation tool is evident when the many options for sensor and logic solver arrangements are also considered for closing these valves. The benefits of sharing sensor signals and logic solvers to close the valves as well as the use of higher or lower quality components and test intervals can be quickly evaluated. Again, some of the findings were surprising and led to a rethinking of some rule based design guidelines.

The emphasis of this paper is on the ease of use and time savings that can be achieved using an interactive software program for risk-based design optimization. With the short design timelines in the chemical industry, this tool makes what was previously almost impossible to be possible: design optimization from a safety perspective. In the interest of doing more fault tree studies with less resources, and being better, faster, and less expensive, this is clearly a 21st century analysis method.

Table 2 Data for valve design options for graphical interface library [a]

Valve	Failure rate (per year)*	Beta factor (failure to close)*
MOV	0.04	0.02
AOV	0.01	0.10
CV	0.02	0.05
CV to AOV or MOV	CCF	0.01
AOV TO MOV	CCF	0.01

[a]Note: The values used are not actual data

Table 3 Results for the 7 available design options in the graphical interface library

Design option	Unavailability	Contributor to CCF (%)
1	1.82×10^{-6}	70
2	1.16×10^{-5}	44
3	5.46×10^{-6}	94
4	1.34×10^{-5}	95
5	5.80×10^{-6}	44
6	5.08×10^{-4}	99+
7	5.17×10^{-4}	97

Other Uses of Interactive Fault Tree Modeling

Focusing on reducing the likelihood of internally initiated events could have minimal impact on the overall reliability characteristics of a facility. For example, externally initiated events are often significant contributors to an NPP core damage frequency (CDF).[11] Individual plant examinations (IPEs) for current-generation NPPs have shown that the CDF from seismic events are of similar magnitude to or even larger than the CDF from internally-initiated events and that 70% of the plants that performed IPEs proposed plant improvements based on their seismic analysis. In addition, 25% of the IPE external events submittals reported that the CDF from fire exceeds the CDF from internal events. Although none of the 70 IPEs for external events (IPEEE) submittals identified any high wind, flood, or other external-event-related vulnerabilities, one plant lowered its flood CDF by three orders of magnitude by improving door-penetration seals. Without considering external events from a scoping perspective to identify unique vulnerabilities, significantly reducing the CDF from internally initiated events may result in only marginally reducing the overall CDF.

The interactive software (RBOT) can be expanded to simultaneously evaluate external events, such as seismic, during the design phase by choosing a database whereby all seismically qualified components fail at their nominal failure rate while those not qualified are assumed to be failed. A fire or flood analysis inserts a database whereby all components affected by the fire or flood are assumed to be failed and the other components fail at their nominal failure rate. The same process would be used to evaluate rocket launches and their payloads. Military planes, ships, and armored vehicles can be evaluated for maneuverability or fire power by having an impact or explosion fail equipment and support systems in the vicinity of the anomaly. Similar to a seismic analysis, all other equipment fails at its nominal failure rate. Recovery actions to restore vital functions could be evaluated for impacts in different areas. The click on one of the menu options returns the analysis to its original state.

Through the interactive software, designers can now observe the effects of aging of systems and components by choosing the option where the failure probabilities are adjusted based on aging-related studies. The information available to RBOT from FaultTree+ provides users with a priority ranking of aged structures/passive components to determine their risk significance. Aging concerns are not limited to active components. New concerns with respect to aging arise because of the reliance on passive safety systems. Influences such as corrosion within piping could prevent natural circulation cooling in passive systems. In general, a system's operational mode is a factor influencing aging-related component failures because components in some standby systems display high aging fractions.[12] This is particularly important for spacecraft where maintenance is not feasible. The goal is not just to design a reliable system for today, but to design a system that will be reliable for the duration of the plant fuel cycle or mission length (i.e., 4 years). Those components with high risk significance are targets for exploring other design alternatives. Thus, by incorporating this information into the PRA models, designers can evaluate how aging components affect system and plant reliability (e.g., at 1, 2, and 4 years).

Human errors and recovery actions also contribute significantly in accident sequences—both in accidents occurring and in minimizing their consequences. The impact of operators can be assessed by selecting from the menu a database that sets all human error probabilities to zero or one for a comparison with the base case.

Early PRAs excluded failures of passive equipment such as pipes, wiring, and multiple check valves from the quantitative analyses because they generally represented a very small contribution to system failure.[12] Because the CDFs for next-generation NPPs (such as IRIS) should be one to two orders of magnitude lower than those for current-generation NPPs, passive failures may now be measurable contributors to the CDF. Conversely, if a passive system requires some active component for initiation, the exclusion of passive components may be realistic because the failure probability of the active component will dominate the system failure probability.

Conclusions

Including Risk Assessment input during design option evaluation can result in better design decisions. Evaluating design alternatives for complex systems using design rules for optimization can result in wrong decisions. Using an interactive design work process and software that enables risk evaluation of design options can provide the information for better design choices. The risk-based interactive tool (RBOT) has shown that a few component choices (11) yields many design alternatives (160) for consideration. Through the use of RBOT it was also learned that the "logical" design decision for optimization may in fact be the wrong decision. However, because no one actually evaluates these decisions, no one knows that these decisions were wrong. For example, removing MOVs from consideration because their failure probability is two to three times that of AOVs[13–16] would be wrong. A more detailed evaluation shows that the CCF probability for two AOVs is typically greater than that for two MOVs.[13–15, 17] Because CCFs dominate the failure modes of reliable systems, using the more "unreliable" valves results in a more "reliable" system. It was also learned that four out of six makes five. That is, the best arrangement for moving four out of six reflectors is using five motors. By maintaining all of the current and previous design, modeling, and data sets via the one-button architecture, the safety ramifications of design options are evaluated, feedback on design alternatives is immediate, and true optimization and understanding can be achieved. These insights and lessons learned helped the IRIS design team and PRA team to achieve a one hundredfold reduction in the base case core damage frequency.

Easily replaceable modules means that reviewers can assess new modeling approaches (e.g., working, failed, degraded); development and testing of new methods such as incorporating digital system failure or degradation of passive systems into a PRA; modeling uncertainties resulting from different modeling approaches, different modeling assumptions, different physical assumptions, and/or different levels of detail; and the sensitivity of assumptions, models, and data on the risk metrics of interest. More than one risk metric can be evaluated at a time such as internal and external events, safety and cost, reliability and cost, reliability and mass, reliability and ease of maintenance or operations. The power of the interactive software (RBOT coupled to FaultTree+) is that all of the design options are simultaneously available allowing internally and externally generated events to be evaluated at the same time.

Chemical plant applications also show that fault tree development time can be reduced by coupling a common database with the basic events and fault tree logic library already available with FaultTree+. This allows chemical systems to be evaluated quickly without being labor intensive by leveraging prior studies and data. As exemplified by the reduction in risk and increase in safety, any new complex design or design upgrade of a hazardous system or process would benefit greatly from using an interactive design work process and these cutting-edge risk evaluation tools.

Acknowledgment

Much of this work performed by ORNL was performed under funding provided by ORNL's Laboratory Director's Research and Development Program. ORNL is managed and operated by UT-Battelle, LLC, for the U.S. Department of Energy under contract DE-AC05-00OR22725.

References

1. A. D. Chambardel and L. Mange, "Completeness and Complexity of PSAs: Do They Need It?" Proc. Probabilistic Safety Assessment and Management II, San Diego, California, March 20–25, 1994.
2. A. Carpignano, "Use of System Simulation to Support Automated Fault Tree Construction," Proc. Probabilistic Safety Assessment and Management II, San Diego, California, March 20–25, 1994.

3. C. Haag et al., "The Use of PRA in Designing the Westinghouse AP600 Plant," Proc. Int. Topical Meeting on Probabilistic Safety Assessment PSA 96: Moving Toward Risk-Based Regulation, Park City, Utah, September 29–October 3, 1996, American Nuclear Society.
4. AP600 PRA, Revision 6, Section 59.2, "Use of PRA in the Design Process," Westinghouse, November 15, 1995.
5. R. E. Schneider, M. C. Jacob, and D. J. Finnicum, "Applications of Level 2 PSA Results and Insights in the System 80+ Reactor Containment Design," Proc. Int. Topical Meeting on Probabilistic Safety Assessment PSA 96: Moving Toward Risk-Based Regulation, Park City, Utah, September 29–October 3, 1996, American Nuclear Society.
6. Y. O. Mizuno and L. E. Conway, "Preliminary Probabilistic Safety Assessment of the IRIS Plant," Proc. ICONE10, 10th Int. Conf. on Nuclear Engineering, April 14–17, 2002.
7. T. Hayes, "APP1000 Plant Overview," IEEE Subcommittee on Qualification (SC-2), Clearwater, FL, November 10–11, 2003.
8. Isograph Reliability Software, FaultTree+ V11, 1986-2007.
9. M. D. Carelli and B. Petrovic, "Next Generation Advanced Reactor," Nucl. Plant J., May–June 2001.
10. D. R. Damon et al., *SP-100* Mission Risk Analysis, Volume I — Main Report, GESR- 00849, GE Aerospace, SP-100 Programs, San Jose Operations, San Jose, CA, August 1989.
11. Perspectives Gained from the Individual Plant Examination of External Events (IPEEE) Program, NUREG-1742, U.S. Nuclear Regulatory Commission, April 2002.
12. B. M. Meale and D. G. Satterwhite, An Aging Failure Survey of Light Water Reactor Safety Systems and Components, NUREG/CR-4747, U. S. Nuclear Regulatory Commission, July 1987.
13. Simplified Passive Advanced Light Water Reactor Plant Program, AP600 Probabilistic Risk Assessment, Rev. 7, Westinghouse Electric Corporation, June 28, 1996.
14. R. E. Ginna Probabilistic Risk Assessment Project, Report to the Nuclear Regulatory Commission in Response to Generic Letter 88-20, Rochester Gas and Electric Corporation, February 28, 1994.
15. R. E. Jaquith et al., Probabilistic Risk Assessment for the System80+ Standard Design, Rev. 1, DCTR-RS-02, ABB Combustion Engineering Nuclear Power, Windsor, Connecticut, March 1993.
16. R. J. Belles, J. W. Cletcher, D. A. Copinger, B. W. Dolan, J. W. Minarick, and M. D. Muhlheim, Precursors to Potential Severe Core Damage Accidents, 1997: A Status Report, NUREG/CR-4674, U.S. Nuclear Regulatory Commission, December 1998.
17. F. M. Marshall et al., Common-Cause Failure Parameter Estimations, NUREG/CR- 5497, U.S. Nuclear Regulatory Commission, October 1998.

QRA and ERS: an Integrated Approach.

Neil Prophet
ioMosaic Corporation
2650 Fountain View Drive, #410
Houston, TX 77057

Henry Ozog
ioMosaic Corporation
93 Stiles Road
Salem, NH 03079

ABSTRACT

Quantitative Risk Analysis (QRA) and Emergency Relief System (ERS) Analysis are both components of a successful Process Safety Management Program. While both studies often share the same information, they tend to remain separate, independent, studies. However, at facilities where relief valves can vent toxic and flammable materials directly to the atmosphere, these discharges can be a significant contributor to overall risk.

ioMosaic Corporation has recently performed studies combining QRA and ERS techniques, for major chemical companies. The integrated approach realized many benefits. Efficiency was gained in terms of data collection, data sharing and information management. The existing Emergency Relief System was analyzed for its adequacy; while the Quantitative Risk Analysis considered various hazard scenarios throughout the plant, including relief device discharge.

Using a case study, this paper outlines the methodologies used, and the risk mitigation options proposed. The original projects contain proprietary business information, which has been appropriately edited for the purposes of this paper.

INTRODUCTION

The historical reasons for the separation of QRA and ERS analysis are numerous, and vary from company to company; though there are a few common themes:

1. Budgetary and Departmental constraints. While it is common for ERS projects to be funded by individual operating units, QRA projects tend to be funded through central engineering, or risk management departments. Often the opportunity to combine such studies and share relevant resources is lost due to lack of communication, or bureaucratic budgeting systems.

2. Lack of specific expertise. Differing areas of expertise may lead an ERS design engineer to shy away from suggesting QRA; or vice versa. Even when this work is outsourced to contracting companies, the contracting firm may not possess expertise in both areas.

3. QRA is not regulated in the US and many other countries, while many aspects of ERS are regulated. There exist very well established and accepted design standards, practices and guidelines for both ERS analysis, and QRA.

   With ERS, the primary regulatory drivers in the US are:

   i. OSHA Title 29, Chapter XVII, Part 1910
   ii. ASME Section VIII
   iii. ASME Section I
   iv. ASME B31.1 and B31.3

   The established codes and practices for ERS include:

   i. API RP 520, Parts I & II
   ii. API RP 521
   iii. API STD 526
   iv. API STD 2000
   v. NFPA-30
   vi. CCPS, *'Guidelines for Pressure Relief and Effluent Handling'*

   QRA is not enforceable as a regulatory requirement in the USA. Furthermore, the use of phrases such as "probability of fatality" and "number of fatalities" tends to dissuade companies, from a liability standpoint. QRA does have well established guidelines, however, including:

   i. CCPS, *'Guidelines for Chemical Process Quantitative Risk Analysis'*
   ii. Committee for the Prevention of Disasters, *'Guidelines for Quantitative Risk Assessment' – (Purple Book)*

4. The tools and software for analysis are different and not readily transferable. ERS calculations tend to use different software tools compared with QRA calculations. If so, this information is not readily transferable, and would require significant duplication of effort. In the case study described in this paper however, SuperChems Expert™ enabled calculations to be performed on one integrated software platform, and significant efficiency gains were able to be realized.

REASONS FOR INTEGRATION

Although QRA and ERS studies are not commonly integrated, it is also worth considering the potential efficiency gains which can be achieved if they are integrated.

There is significant overlap of information, with both analyses requiring:

- Piping & Instrumentation Diagrams
- Process Flow Diagrams
- Heat & Material Balances

In addition, both analyses also require: mass flowrates, mixture compositions, operating temperatures and pressures, storage inventories, release elevations, nominal pipe sizes, and release orientation. Therefore, properly planned data gathering *and subsequent management of this data*, can significantly reduce duplication of effort, as the project progresses.

Finally, the contingency analysis that is performed during an ERS study can be useful in defining some of the release scenarios for QRA. Additionally, both QRA and ERS involve comparison of safeguards and mitigation techniques which often have overlapping (or interacting) effects on each other. The best alternative for ERS, taken as a standalone evaluation, may not be the most cost effective when overall QRA is considered.

BACKGROUND

This case study considers a specialty chemical manufacturing plant where it was possible to combine the QRA and ERS methodologies. The plant manufactures a variety of specialty chemicals which are utilized by a broad spectrum of industries. The plant is operated 24 hours a day, seven days per week, with over 500 full time employees at the site. The facility is rurally situated.

Onsite, the plant stores a wide variety of toxic chemicals in excess of the threshold quantities, some of which include: Anhydrous Ammonia, Acrylonitrile, Propylene oxide, and Methyl chloride. The plant also stores several flammable chemicals above threshold quantities as defined by OSHA regulations in the US.

THE QUANTITATIVE RISK ANALYSIS STUDY

The scope of the project originally set out to provide just a Quantitative Risk Analysis. However, as the study progressed, the initial findings led to a decision to combine approaches, and include ERS analysis.

The initial QRA objective was to determine both on-site and off-site risk, for the plant. Results would be presented in terms of both individual risk, and societal risk. Additionally, risk reduction opportunities were identified. As will be demonstrated in this paper, the risk reduction section led to the subsequent emergency relief systems design analysis.

The QRA was conducted using a typical approach:

- Hazard Scenarios Development
- Frequency Analysis
- Consequence Analysis
- Risk Determination

Hazard Scenarios Development

The scope of the QRA was limited to seven chemicals which were determined to pose the greatest risk. Based on previous process hazard analysis (PHA), and historical incidents at the facility, a list of scenarios was developed. The primary scenarios to be considered were:

- Vessel failure
- Line failure
- Relief device failure open
- Overfilling
- Hose failure
- External fire
- Contamination
- Mischarge
- Failure of heating or cooling medium
- Loss of inert gas

The final list of chemicals, and the number of relevant scenarios developed, is shown in Table 1 below:

Table 1: Chemicals included within QRA scope.

Chemical	Flammable	Toxic	Number of Scenarios
Chemical A	Y	N	7
Chemical B	Y	Y	14
Chemical C	Y	Y	11
Chemical D	N	Y	7
Chemical E	Y	Y	13
Chemical F	Y	Y	17
Chemical G	Y	Y	14

Frequency Analysis:

The frequency analysis of each scenario was conducted using the layer of protection analysis (LOPA) technique that is described in the CCPS publication '*Layer of Protection Analysis, Simplified Process Risk Assessment*'. There are advantages to using LOPA verses more traditional risk assessment methodologies such as fault tree analysis:

- LOPA facilitates the determination of more precise cause-consequence pairs, and therefore improves scenario identification.

- The frequencies and probabilities provided in the CCPS LOPA text reflect the most recent consensus of data from companies in the chemical process industries since the book was published in 2001. This data does not require extrapolation of old data to current equipment or data developed from other industries (i.e., nuclear) to the chemical industry.

For each scenario developed, the frequency of the event that produces the described consequences was determined. Each overall scenario frequency may consist of a number of events:

- The **initiating event** which causes an undesirable deviation from normal operation. This could be a human error, equipment or instrument failure or external event, such as fire. The initiating event may be expressed as an annual frequency or as a probability per opportunity. For example scenarios that have an initiating event based on human error are typically expressed as a failure rate per opportunity. The enabling event would then be expressed as the number of opportunities per year (frequency).

- An **enabling event** which is typically related to time the system is exposed to the undesirable deviation while it is in a hazardous state. An example would be failure of an unloading hose on a trailer due to the cab pulling away while still connected. If the trailer is empty when this happens, then no spill of hazardous chemical will result. The enabling event is typically expressed as a probability, but may also be expressed as a frequency when the scenario involves human actions that are done at repetitive intervals.

- The number of similar **units** that could experience the same scenario. These could be the number of vessels or equipment or length of piping.

- A **conditional event** that must occur for the undesirable consequences to result. An example is a failure to purge a vessel containing flammable material with nitrogen prior opening for maintenance. If the undesirable consequences are a fire, then the conditional event is the ignition of the flammable vapors in the vessel. A conditional event is included in the scenario frequency when the event is taken to a pre-defined consequence. Most loss of containment scenarios involving flammable materials do not include the conditional event in the frequency, but rather is developed as part of an event tree that evaluates all possible consequences including: no ignition, immediate ignition, and delayed ignition. This event tree analysis was also considered when setting up the QRA. A conditional event is expressed as a probability.

- An **independent protection layer (IPL) event** that either prevents or reduces the likelihood of the scenario. An IPL is expressed as a probability of failure on demand.

An IPL is a device, system, or action that is capable of preventing a scenario from proceeding to its undesired consequence independent of the initiating event or the action of any other layer of protection associated with the scenario. In order to be considered an IPL, the device, system or action must be:

- Effective in preventing the consequence
- Independent of the initiating event and the components of any other IPL
- Auditable through validation of design and effectiveness

A typical scenario frequency analysis, outlining the initiating causes and likelihood of failure is shown in Table 2 below.

Table 2: Sample frequency analysis.

Frequency Summary	Description	Value
QRA Reference	Chemical G External Fire	
Initiating Event (frequency per unit)	External Fire on a vessel containing Chemical G	1.00 E-02
Units	One Vessel	1
Enabling Event (probability)	Continuous Service	1
IPL 1	Emergency response to extinguish fire before contents of tank reaches runaway temperature of 200°C	1.00 E-01
IPL 2	Water sprays limit temperature of tank to 100°C	1.00 E-01
IPL 3	None	1
Scenario Frequency		**1.00 E-04**
Comments	1. No credit is taken for insulation on the tank, as it is not specified to be fire-rated.	
	2. It is assumed that the PSV was sized for fire, and is not adequate to relieve a runaway reaction.	

As described in the CCPS LOPA book, some items such as procedures, training and mechanical integrity would be factored into the initiating event frequency, but are not considered IPLs. Similarly, fire protection is typically considered in determining the duration of the event, except for external vessel fires which allow for emergency response before the vessel relief device opens. The vessel relief devices were not considered as IPLs, as the relief device discharging to atmosphere contributed to the risk calculation. Additionally, as stated in Table 2, it was assumed that the relief devices were sized adequately for fire; but not for runaway reaction, which would lead to loss of containment of the entire vessel.

This study used the approach as defined in the LOPA text which allows more than one IPL to be in the same Basic Process Control System (BPCS) or a BPCS IPL with a BPCS initiating event. This approach is based on the assumption that if a BPCS function fails, it is much more probable that the component that induced the failure is the detection device or the final control element and not the logic solver.

Where historical plant failure rates were available, this information was used in place of the generic values available in the CCPS LOPA text. Where neither source provided appropriate data, appropriate values were developed based on other proprietary sources or good engineering judgment.

Consequence Analysis

In conducting the Quantitative Risk Analysis, SuperChems Expert™ was used to model consequence results, risk results, and the subsequent Emergency Relief Systems modeling.

Local wind rose statistics were gathered, based over a three year period with sixteen wind directions; and dispersion results were generated for three different atmospheric stability classes: B, D, and F.

The discharge models providing source term calculations, calculated actual release duration based on storage inventory, and release rate. This release duration was then carried forward to the dispersion models. Existing mitigation measures, such as dikes, emergency response, and isolation, were accounted for when modeling release spread, and release duration.
Day and night on-site and off-site population and ignition sources were defined. On-site population information was provided by the client; off-site population was provided by the 2000 Census demographic data contained in LandView® 6.

The study considered consequences arising from thermal radiation, overpressure, and toxicity. Risk effects were modeled based on dose-based probit calculations. For the purposes of the consequence analysis, commonly accepted criteria were reported; including thermal radiation values of 4, 12.5, and 37.5 kW/m^2; and explosion overpressure values of 2.07, 6.9, and 20.7 kPa. For toxic effects, three Emergency Response Planning Guideline (ERPG) values were reported, signifying differing levels of harm based on a one hour exposure time.

Approximately 1100 possible hazard outcomes were generated, which considered release scenario, wind speed and direction probability, atmospheric stability category, and immediate or delayed ignition. These 1100 outcomes were then used in the risk determination.

Risk Determination

Both Individual and Societal Risk results were modeled, as shown in Figures 1 and 2. Onsite and Offsite effects were calculated, with results presented for daytime, night-time and combined.

Additional onsite risk statistics were also calculated showing the highest individual risk, average individual risk, fatal accident rate, and rate of death.

Where:

Individual Risk: "The risk to a person in the vicinity of a hazard. This includes the nature of the injury to the individual, the likelihood of the injury occurring, and the time period over which the injury might occur".

Average Individual Risk: "The average of all individual risk estimates over a defined population".

Fatal Accident Rate (FAR): is calculated from the average individual risk, and is normally used as a measure of employee risk in an exposed population. This is the number of fatalities occurring during 1000 working lifetimes (10^8 hours).

Average Rate of Death (ROD): is the estimated average number of fatalities in the population from all potential incidents.

Each of the 1100 scenario outcomes could also be sorted, and ranked in terms of their contribution to the overall risk. Ranking these outcomes would play a crucial role in determining the most appropriate risk mitigation options.

Figure 1. Overall Individual Risk Results. Base Case.

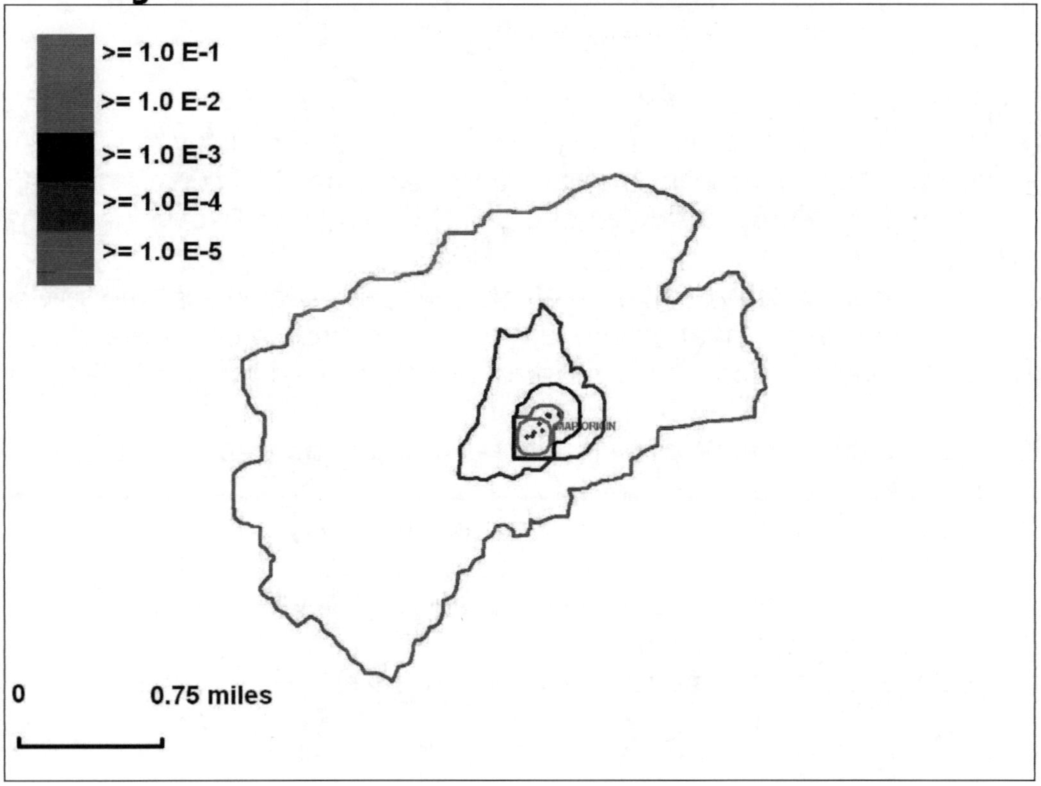

Figure 2. Overall Societal Risk Results.

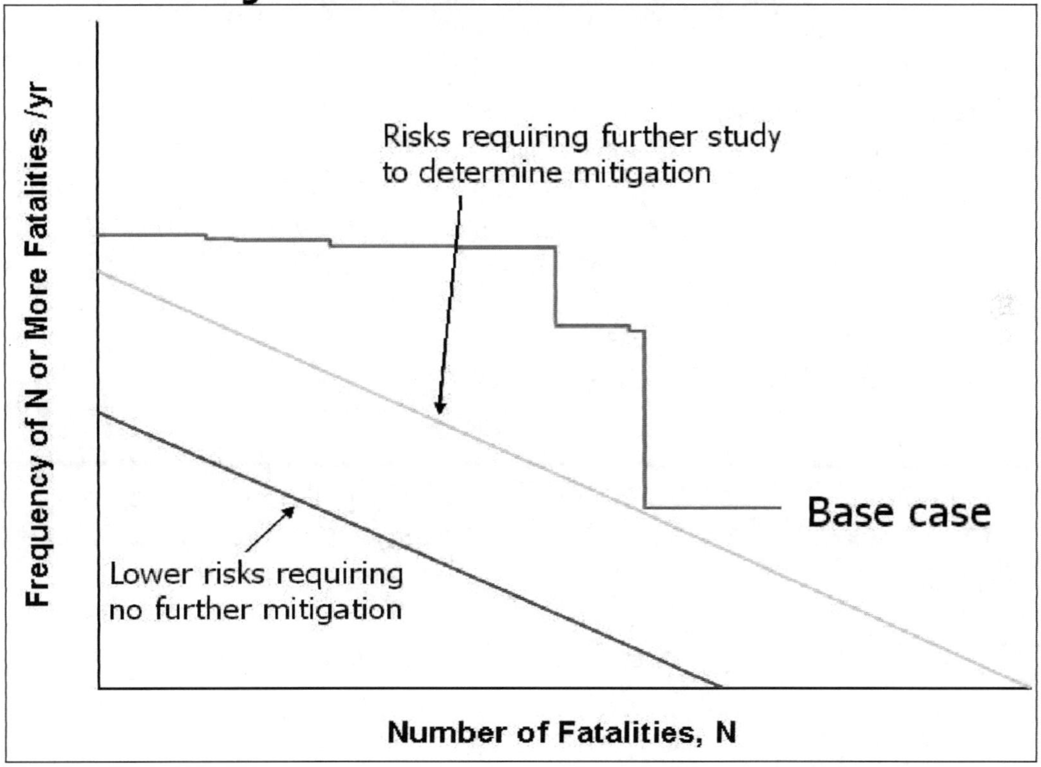

Risk Discussion

For both the Individual Risk, and the Societal Risk; the risk results exceeded the client's tolerable levels. In Figure 2, the Dutch Government risk criteria lines were overlaid, to give a basis for comparison. As can be seen, the calculated base case F-N curve significantly exceeds the risk criteria line representing "Risks Requiring Further Study to Determine Mitigation".

Scenario outcomes were ranked based on their risk, which revealed that the risk profiles were dominated by toxic releases to atmosphere, from pressure safety valves (PSVs), due to process induced runaway reactions, fire exposure, or fire induced runaway reactions.

The main PSV-related risk contributors were identified and prioritized:

- External fire leading to relief device actuation
- Contamination leading to relief device actuation
- Loss of heating and/or cooling leading to relief device actuation
- Relief Valve Failure Open
- Uncontrolled polymerization leading to relief device actuation

It is interesting to note that, for the seven chemicals considered in the analysis, the US EPA Envirofacts database of significant incidents lists six 'Stack or Point Emissions' occurring in the year 2004, with 900 kilograms of a toxic substance representing the largest release amount. It can therefore be deduced that relief device discharges can, and do, occur on a fairly regular basis

If the main ERS-related scenarios were mitigated/eliminated by installation of an effluent handling system consisting of a catch tank and an associated flare stack, the offsite risks could be reduced significantly. In order to successfully implement this mitigation measure, a full Emergency Relief System Design analysis was undertaken, to accurately determine all the relief loads which would affect the proposed effluent handling system.

THE EMERGENCY RELIEF SYSTEM DESIGN STUDY

The scope of the emergency relief system design study was limited to the items which would be routed to the proposed flare stack. Those items had been identified as significant risk contributors during the QRA phase.

The emergency relief system design study was performed, according to the following workflow:

- System-specific Emergency Relief Systems Evaluation
  - Overpressure contingency analysis
  - Determination of required relief rates
  - Capacity rating of existing relief system
- Effluent Handling System Evaluation
  - Common failure mode scenarios
  - Effluent Handling System Hydraulic Analysis
  - Catch tank design

System-specific Emergency Relief Systems Evaluation

API RP 521 provides very thorough guidance on overpressure contingency analysis, determination of relieving rates, and disposal systems. Table 2, of API RP 521 lists the following upset conditions, which may result in overpressure:

- Closed outlets on vessels
- Cooling water failure to condenser
- Top-tower reflux failure
- Sidestream reflux failure
- Lean oil failure to absorber
- Accumulation of non-condensables
- Entrance of highly volatile material
- Overfilling storage or surge vessel
- Failure of automatic controls
- Abnormal heat or vapor input
- Split exchanger tube
- Internal explosions
- Chemical reaction
- Hydraulic expansion
- Exterior fire
- Power failure

Using this list as a basis, it was possible to develop a matrix of overpressure scenario, cross-referenced to specific vessel types, as shown in Table 3:

Each individual emergency relief system was verified for adequacy. Where applicable, emergency relief system discharges were considered for either vapor, liquid, or two-phase flow. Inlet pressure drop was limited to 3% of the set pressure; the API recommended maximum value for both compressible and incompressible fluid flow. Outlet pressure drop for conventional type pressure relief valves was limited to the allowable overpressure for the contingency under consideration (usually 10% for process-related overpressures, and 21% for fire exposure).

Relief device outlet piping was redesigned in such a way that the discharge lines would connect together to form sub-headers, which in turn, connected to the proposed header. This header led to a catch tank, which would remove any harmful liquid toxic components, while safely burning the vapors via a flare stack.

For each relief device, the required flowrate, chemical composition, phase, temperature, and pressures for each piping segment; was taken forward to the Effluent Handling System Evaluation.

Table 3: Common contingencies cross-referenced to type of equipment

	Closed Outlets	Cooling Failure	Reflux Failure	Accumulation of Noncondensables	Mixing of Volatile and Hot Materials	Failure of Automatic Controls	Abnormal Heat or Vapor Input	Internal Explosions	Chemical Reaction	Hydraulic Thermal Expansion	Exposure to External Fire	Ruptured heat exchanger tube	Power / Utility Failure	Inadvertent Valve Opening	Transient Pressure Surges	Inbreathing/Outbreathing
Blower	☑												☑			
Boiler	☑										☑		☑			
Column - Packed	☑	☑	☑			☑	☑				☑		☑	☑		
Column - Trayed	☑	☑	☑			☑	☑				☑		☑	☑		
Compressor - Centrifugal	☑												☑			
Compressor - Positive Displacement	☑												☑			
Dryer	☑				☑	☑				☑	☑		☑	☑		
Filter	☑					☑				☑	☑		☑	☑		
Fired Heater	☑												☑			
Heat Exchanger - Aerial	☑					☑					☑		☑			
Heat Exchanger - Concentric Pipe	☑					☑				☑	☑		☑			
Heat Exchanger - Electric	☑					☑				☑	☑		☑			
Heat Exchanger - Other	☑					☑				☑	☑		☑			
Heat Exchanger - Plate and Frame	☑					☑				☑	☑		☑			
Heat Exchanger - Shell and Tube	☑					☑				☑	☑	☑	☑			
Pipe Segment	☑									☑			☑			
Pump - Centrifugal	☑												☑			
Pump - Positive Displacement	☑												☑			
Reactor	☑	☑				☑			☑		☑		☑	☑		
Storage Tank	☑					☑					☑		☑	☑		☑
Storage Vessel	☑					☑					☑		☑	☑		
Turbine	☑												☑			
Vessel	☑					☑					☑		☑	☑		

Effluent Handling System Evaluation

The overall goal of the Effluent Handling System Evaluation was to determine the adequacy of the system in the event of various common failure mode scenarios. Adequacy of the flare header, flare tip design, liquid separation equipment, and radiation intensity was considered for each identified common failure mode scenario.

The Effluent Handling System Evaluation also considered the various chemicals being released, to ensure there would be no undesired reactions occurring in the proposed catch tank.

Common Failure Mode Scenarios

A common failure mode scenario is considered a single event that may result in the overpressure of more than one emergency relief system impacting the effluent handling system. Typical common failure mode scenarios include loss of a single utility (cooling water, instrument air, steam, electricity, etc.) that adversely affects multiple systems. Note that for all common failure mode scenarios it was assumed that all control systems potentially venting to the flare are closed with no additional vapor load.

Common Failure Mode Scenarios were assessed for:

- External Fire
- Power Failure (total & partial)
- Cooling Failure (Glycol, Water)
- Utility Failure (Instrument Air, Steam, Nitrogen)

Details of one of the common fire scenarios are given below. A similar analysis was performed for each applicable common failure mode scenario.

External Fire

External pool fires were assumed to be circular and have a maximum diameter of 56 ft, equivalent to a maximum fire zone area of 2,500 ft^2, and encompassing major equipment items.

Table 4: Fire Scenario "1" Hydraulic Results

Relief Device	Device Type[1]	Protected Equipment	Set Pressure (psig)	Relief Requirement (lb/h)	Relief device capacity (lb/h)	Actual node flow (lb/h)	Percent of Max. Flow	Remains Adequate? (Y/N)
PSV-1	**C**	**TK-1**	**70**	**11000**	**12500**	**7000**	**56**	**No**
PSV-2	B	TK-2	15	11000	15000	14000	93	Yes
PSV-3	B	TK-3	15	6000	11500	7250	63	Yes
PSV-4	B	TK-4	15	11500	15500	13000	84	Yes

Note 1: C = Conventional, B = Bellows, P = Pilot

Based on the hydraulic analysis shown in Table 4, the relief capacity for relief device PSV-1 on TK-1 may be reduced beyond the relief requirement. To mitigate the potentially excessive backpressure concern:

- Consideration was given to reducing the relief requirement by taking credit for the maximum actual operating wetted area of TK-1; rather than the relief requirement being based on an assumption that the tank may operate 80% liquid full.
- Consideration was also given to PSV-1 being modified to a balanced-bellows type relief device.
- Additionally, the outlet piping could have been modified by increasing the line diameter, to alleviate the excessive outlet pressure drop.

It was also noted that the flow capacity through PSV-3 was derated significantly; though the relief device remained adequate.

Catch Tank Design

The main principle of the catch tank is to first separate and contain the liquid released, and then to pass the vapor for further treatment. Typical separation equipment that is used includes:

- Horizontal separators (gravity)
- Vertical separators (gravity)
- Cyclone Separators
- Quench tanks (open)
- Impingement separators

In the case study presented, a vertical separator was selected. The header catch tank was evaluated for all identified common failure mode scenarios. API RP 521 recommends a liquid retention time of 20-30 minutes with a maximum particle diameter at the outlet of the drum of 300-600 microns. The proposed catch tank was evaluated based on these criteria.

In order to meet the desired design criteria a vertical separator would require the following design conditions:

Table 5: Vertical Separator Sizing for Fire Scenario "1"

Catch Tank (Fire Scenario "1")	Required (ft)
HME: Mist eliminator adjustment. ft	0
HD: Minimum height required for disengagement. ft	3.35
HLN: Height from HLL to nozzle centerline. ft	1.35
TOTAL	4.70
HS: Height from NLL to HLL. Surge height. ft	0.50
HH: Height from LLL to NLL. ft	0.95
HLLL: Low liquid level height. ft	1.25
TOTAL	2.70
Separator length. ft	**7.40**
Separator diameter. ft	**3.70**

Flare Tip

Finally, for each Common Failure Mode Scenario, the conditions at the flare tip were assessed. The exit velocity is evaluated at the flare tip. It is generally recognized practice that the upper limit is Mach number 0.6. Based on the proposed flare elevation; the thermal radiation as shown in Figures 3 & 4, was found to be within the limits prescribed by API-521 presented in Table 6:

Table 6: Recommended Design Total Radiation

Permissible Design Level (kW / m^2)	Conditions
15.77	Heat intensity on structures and in areas where operators are not likely to be performing duties and where shelter from radiant heat is available
9.46	Heat intensity at any location to which people have access, with limited exposure time (few seconds) and where escape is possible
6.31	Heat intensity where emergency actions lasting up to 1 minute may be required by personnel with appropriate clothing, but without heat shielding
4.73	Heat intensity where emergency actions lasting up several minutes may be required by personnel with appropriate clothing, but without heat shielding
1.58	Heat intensity where personnel with appropriate clothing may be continuously exposed

Source: API Recommended Practice 521, 4th edition; Table 8

Figure 3: Vertical Radiation Profiles for Fire Scenario "1".

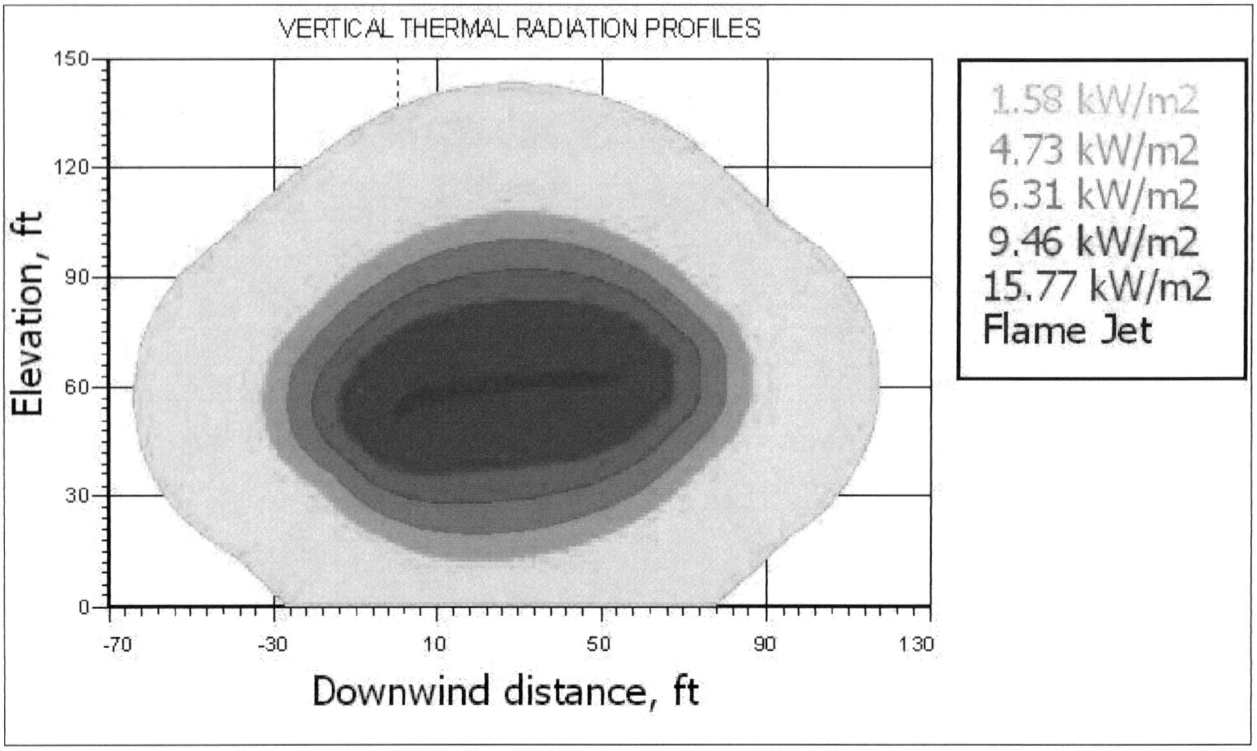

Figure 4: Projected Radiation Footprint for Fire Scenario "1".

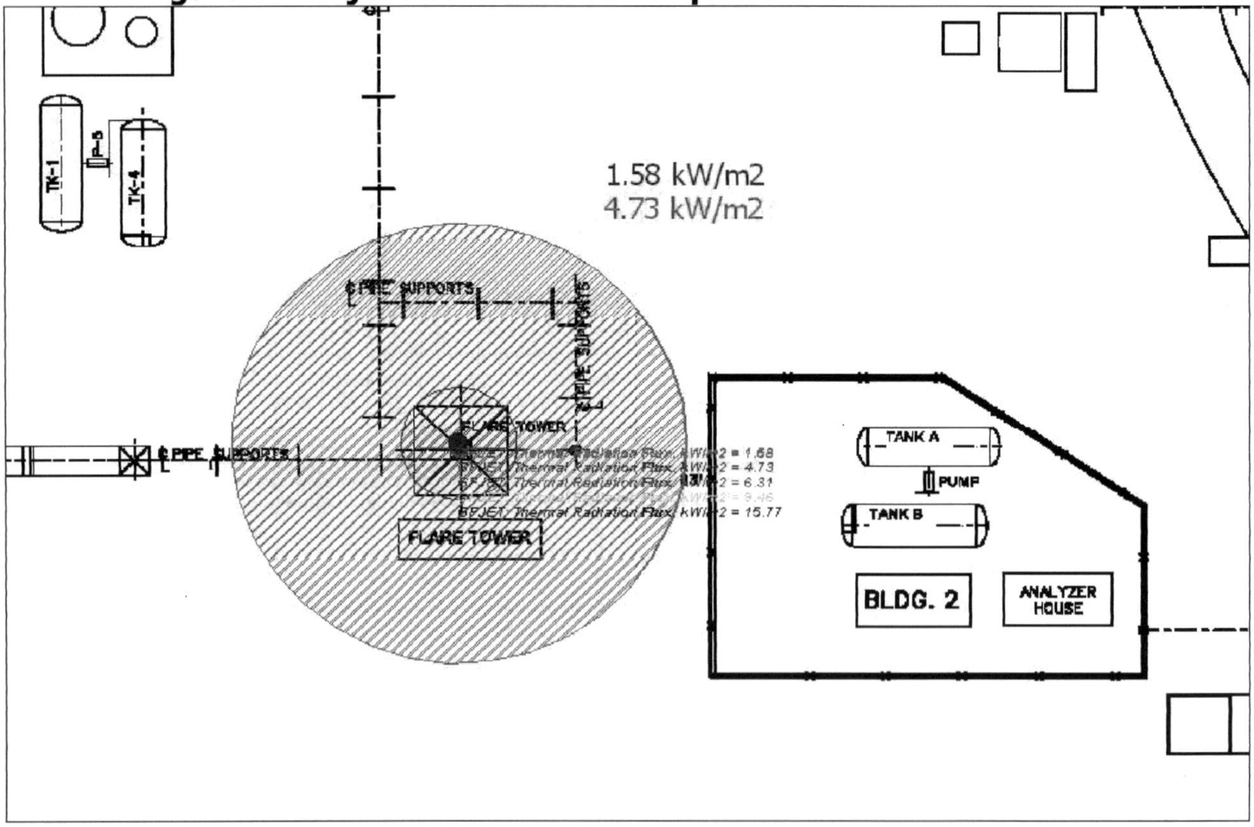

The Effluent Handling System Evaluation was repeated for each applicable common failure mode scenario, to ensure adequacy of the piping network and optimum design for the catch tank.

Upon completion of this stage of the project, it was possible to revise the Quantitative Risk Analysis accordingly.

Revised Risk Calculations

The existing QRA study was revised, with relief device discharge outcomes removed from consideration. Additionally, the catch tank was included within the study scope.

Revised risk results were now found to be as follows:

Figure 5. Overall Individual Risk Results. Mitigated Design.

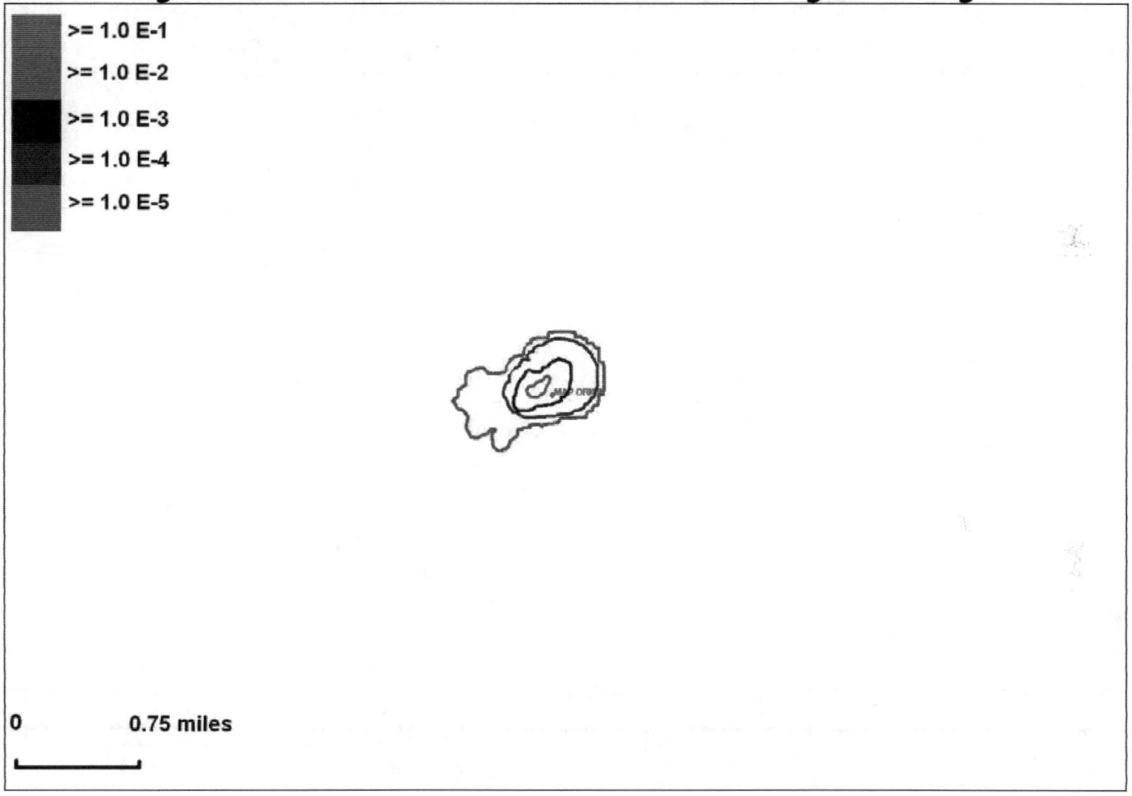

Figure 6. Overall Societal Risk Results. Mitigated Design and Base Case.

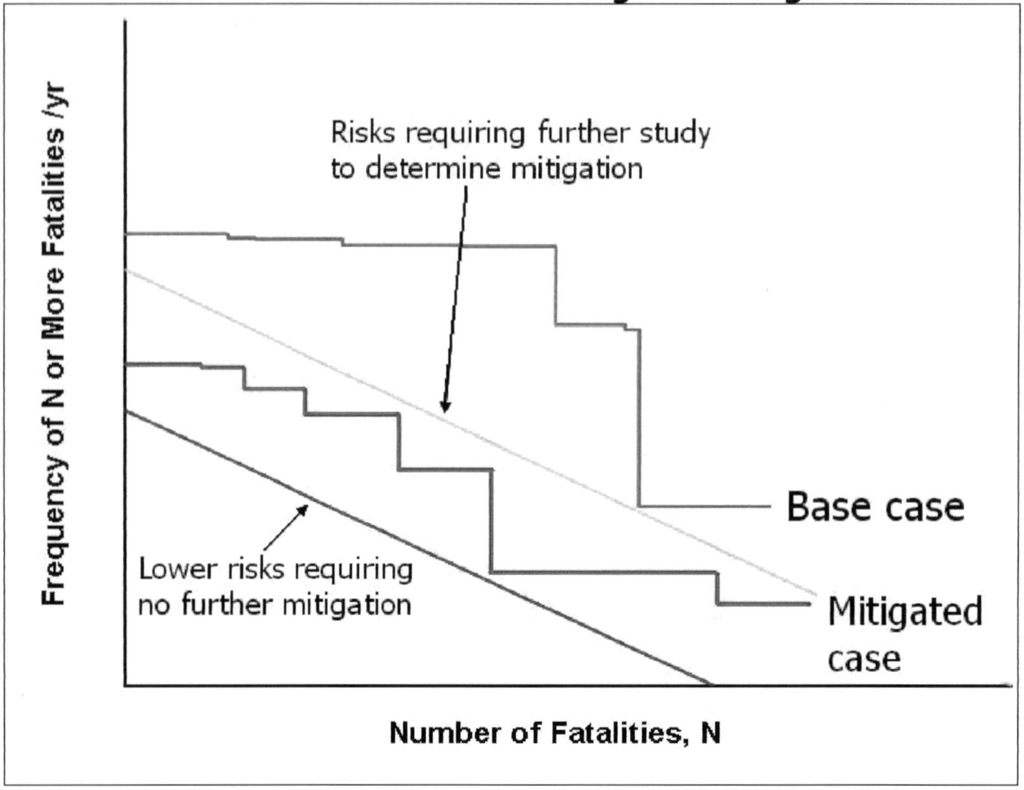

CONCLUSIONS

As can be seen in Figures 5 and 6, compared with the base case Figures 1 and 2; the implementation of a properly designed effluent handling system has significantly reduced the risk to within tolerable levels.

While the chemical industry is currently placing considerable focus on replacing atmospheric discharge relief devices with appropriate effluent handling systems, it is not often that their benefits are quite as obvious as demonstrated by this case study.

SuperChems Expert™ enabled these two studies to be combined efficiently on one integrated software platform. The client benefited by implementing significant risk reduction measures, while receiving up-to-date QRA and ERS studies, which can easily be updated should the existing arrangement ever change.

A combined QRA and ERS approach offers an innovative and accurate option to any company interested in reducing their risks.

References:

1. ASME Boiler and Pressure Vessel Code Section VIII – Pressure Vessels (2001).

2. API Recommended Practice 520 – Sizing, Selection, and Installation of Pressure Relief Devices in Refineries – Part I (2000).

3. API Recommended Practice 520 – Sizing, Selection, and Installation of Pressure Relief Devices in Refineries - Part II (2003).

4. API Recommended Practice 521 – Guide for Pressure Relieving and Depressuring Systems (1997).

5. CCPS (1998) "Guidelines for Pressure Relief and Effluent Handling" AIChE/CCPS, New York.

6. CCPS (1989) "Guidelines for Chemical Process Quantitative Risk Analysis", AIChE/CCPS, New York.

7. CCPS (2001) "Layer of Protection Analysis – Simplified Process Risk Assessment", AIChE/CCPS, New York.

8. Houston, Casey R. (2005) "Emergency Relief System Project Guidelines", ioMosaic Corporation.

9. Bellomo, P. J., and Stickles, R. P., (1995) "Select Design Bases for Emergency Relief and Other Process Safety Systems Based on Risk", proceedings from International Symposium on Runaway Reactions and Pressure Relief Design, pp. 631 – 647.

A Case Study of Safety Integrity Level (SIL) Assessment and Verification at Air Products and Chemicals: Electronics Division Product Line Evaluation and Analysis

John Day, Air Products and Chemicals, Allentown, PA
Hal Thomas, Air Products and Chemicals, Allentown, PA
James VanOmmeren, Air Products and Chemicals, Allentown, PA

ABSTRACT

With the adoption of IEC 61511 [1] and ANSI/ISA-84.00.01-2004 [2] standards, Air Products and Chemicals has made a concerted effort to provide a standardized approach to the design and implementation of Safety Instrumented Systems following the safety lifecycle model. This presentation provides a Case Study describing the methodology used during the Safety Integrity Level (SIL) Assessment and Verification of existing Electronics Division product lines. SIL Assessment was accomplished through the use of Hazard Identification; Likelihood, Consequence and Risk Analysis. Each of the Electronics Division Product's Safety Instrumented Functions (SIF) were identified during SIL Assessment, making use of both layer of protection analysis (LOPA) and consequence analysis. A target SIL was determined for each SIF through quantitative and qualitative analysis. SIL Verification was accomplished using Fault Tree Analysis in order to determine the average probability of failure on demand (PFDavg) of each SIF. This analysis was a joint collaboration between the Process Safety and Process Controls Engineering teams.

COMPANY BACKGROUND

The Electronics Division of Air Products and Chemicals is a $2 billion (FY2006) in sales global business focused on providing Specialty Gases and Chemicals, High Purity Equipment and On Site Services to the Electronics market. Our customers include manufacturers of Integrated Circuit, Memory Chips, Liquid Crystal Displays, and Light Emitting Diodes.

BACKGROUND

In 2003, Air Products began a proactive approach to develop internal work processes, tools, and training in anticipation of the pending adoption of ANSI/ISA-84.00.01-2004. This included the development of the following aids for use in SIL Assessment and SIL Verification:

1. Software tools
2. Global work processes
3. Global Engineering Standards and Procedures
4. Training materials
5. SIL application library
6. Standard templates for SIL documentation

Internally developed software tools have been deployed for use by Air Products Process Safety and Engineering Personnel. Fault Tree Analysis software (APTree) and Layer of Protection Analysis software (APLoPA) have been developed for quantifying risks, performing frequency analysis, and for verification of safety instrumented system performance. Failure rate data and frequency data for use by these programs was obtained through industry generic failure rate data books, 3rd party consultants, licensed databases, supplier safety manuals, supplier failure mode effects and diagnostic analysis reports, and internal process plant operating experience.

Basic engineering policies (BEP) for use in the design and implementation of safety instrumented systems were developed in order to establish and communicate the company's high level policy and strategy for the transition to the new industry standard. The BEP provided guidance for more specific and detail oriented engineering procedures and standards which were then developed.

Training materials were prepared using both internal and external Process Safety, Process Control, and Instrumentation Design experts. These materials covered various aspects of process safety analysis and were developed with specific Air Products terminology and work processes in mind. This training was rolled out to the company in a systematic approach.

An applications library was developed that contains typical instrumented protection systems designs that have been pre-analyzed to support production engineering as well as fundamental information such as FMEA reports, manufacturer safety manuals, etc.

After completion, SIL assessments were added to the library for future use to support management of change or to serve as a starting point for similar applications. This library has the benefit of helping to reduce engineering time and to facilitate consistent implementations across multiple business units. Standard templates for SIL assessment and verification reports were developed. These reports provide a consistent presentation format that aids the ease of use and understanding of the material.

CASE STUDY SPECIFICS

This case study describes the work process followed in the SIL Assessment and Verification of previously designed safety instrumented systems within our Electronics Division equipment product lines. The analysis examples presented in this case study are for illustrative purposes only.

Data Gathering

A summary of products for evaluation was prepared and grouped per application and known hazards. A number of equipment offerings were evaluated using the work process described in this case study. This includes equipment products listed in the following product groups:

1. GasGuard® QMAC : Analytical Systems / Product Quality Monitoring Systems
2. GasGuard® Bulk Specialty Gas Systems (BSGS)
3. GasGuard® High Flow Systems, Gas Cabinets, Valve Manifold boxes
4. Isomodule/Y Cylinder Heater Control Systems
5. GGT® Subatmospheric Gas Generators
6. Nitrogen Purifiers

These product groups have been designed by Air Products over the past 10 years for both internal use and for sale to outside customers. Hazard identification and consequence analysis had been performed during the development of the individual product lines. Instrumented protections were previously designed based on the hazard and consequence analysis. This information was collected and used in the SIL Assessment and Verification.

The steps outlined in the Safety Lifecycle[3] (Figure 1) were used as the basis for the evaluation. Some steps were bypassed as the products being evaluated were existing designs. For this case study, the existing Process Hazard Analysis (PHA) and risk assessment were examined and updated using layer of protection analysis[4,5]. A SIL target was defined for each SIF by back-calculating the required risk reduction needed to be achieved for that SIF in order to satisfy the Air Products' Corporate risk target. Each existing SIF was then evaluated to determine what performance could be achieved given its design, architecture, configuration and mechanical integrity program. Calculated performance was then compared to the SIL target (using maximum allowable PFDavg) to confirm that the safety instrumented function was adequate, coupled with the other layers of protection to meet our corporate risk target.

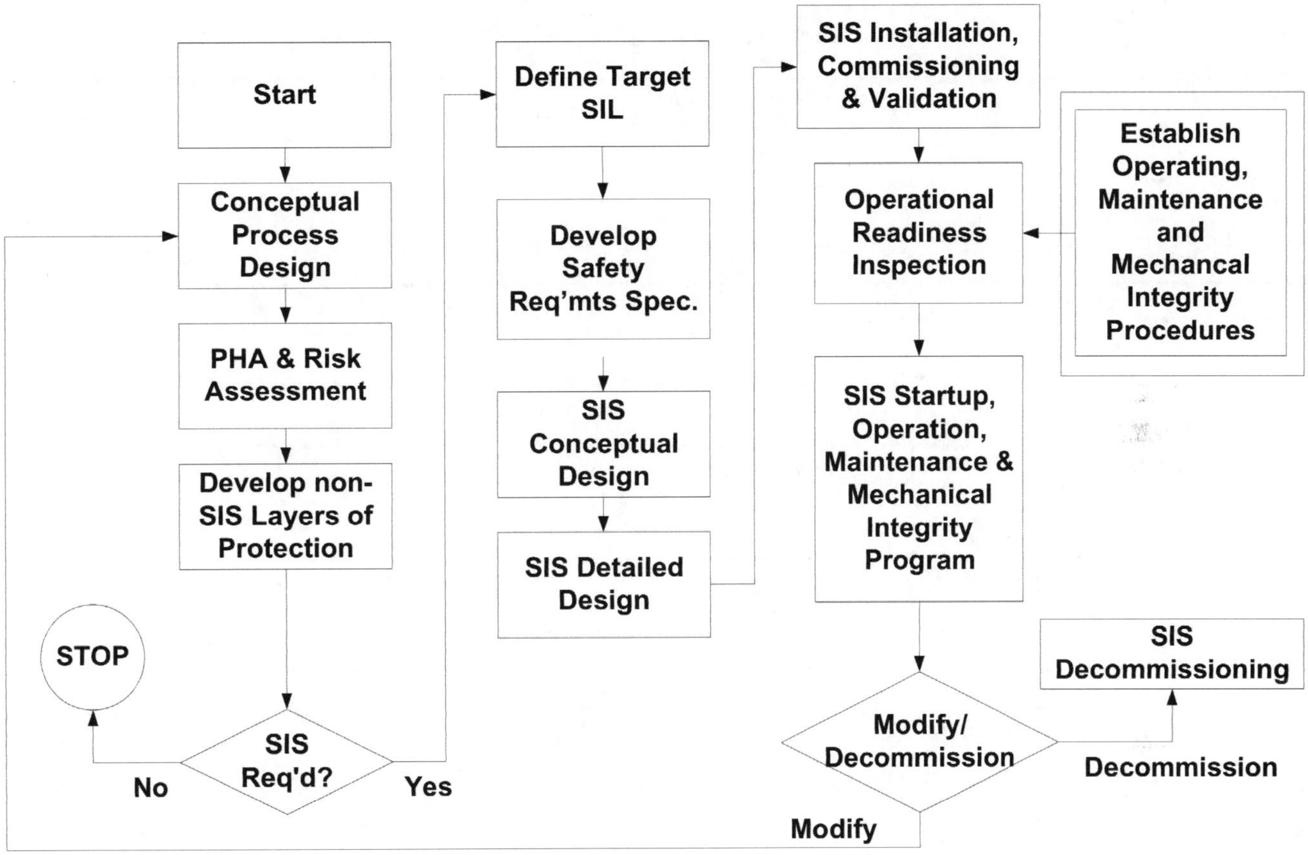

Figure 1: ISA 84.01 Safety Lifecycle

SIL Assessment

The goal of the SIL Assessment was to identify all Safety Instrumented Functions (SIF) and to determine a required performance level for each SIF. Air Products classifies risks into importance levels (L1 through L3). Only L3S safety instrumented functions were evaluated during this case study. It is Air Products' interpretation that the L3S classification corresponds to a SIF as defined in IEC 61511. Table 1 describes the importance level classifications used within Air Products.

Table 1: Importance Level Classification

Importance Level Class	Subcategory	Application Examples	Basis/Comments
L1	L1	• Regulatory Control • Equipment Interlocks where loss of equipment is not considered significant • Environmental Monitoring • Equipment Monitoring • Information/Status • Shutdown Pre-alarms • Shutdown associated with acceptable risk involving equipment damage, business interruption	• Maximum flexibility, minimum risk. • Financial risk acceptable to the responsible business area.
L2	L2S	• Emergency Response–Fire Detection Backing Up Automatic Sprinklers • Protection Required to Comply with Federal, State, or Local Regulations • Safety, Health, and/or environmental protection deemed important, but not meeting the definition of critical safety protection	• Safety, health, and/or environmental protection involving specific regulatory requirements or providing protection against loss of process or energy containment.
	L2P	• Loss Prevention For: ---Major Equipment Damage ---Significant Business Interruption ---Significant Environmental Risk	• Financial risk or asset protection that is unacceptable to the responsible business area.
L3	L3S	• Critical Safety Protection --- Protection that is designed to prevent an immediate, life-threatening process safety incident when a demand occurs.	• Considered a safety instrumented function (SIF) per IEC 61511
	L3P	• Critical Product Protection	• Critical product protection as defined by internal company work process.

At Air Products, the importance level helps classify the level of security, administrative controls and mechanical integrity requirements. The importance level also provides guidance and requirements for the level of segregation and independence required between the safety instrumented system (SIS) and basic process control system (BPCS).

Process hazard analysis information that was prepared during the design phase was reviewed to identify instrumented safeguards. Air Products typically employs the HAZOP methodology for process hazard identification analysis. An example of a typical PHA HAZOP worksheet is listed in Figure 2. The PHA report details the possible hazards, the consequence of the hazard, and any safeguards that may exist. Note the safeguards may include equipment and process design, mechanical protection systems, instrumented protection systems and administrative controls.

DEVIATION	CAUSES	CONSEQUENCES	SAFEGUARDS	Class	RECOMMENDATIONS
High pressure	Column steam reboiler pressure control failure, causing excessive heat input.	Column overpressure and potential mechanical failure of the vessel and release of its contents.	Mechanical design of vessel. High pressure alarms with operator intervention Pressure relief valve	L1 L1 L3	Install SIF within the SIS to stop reboiler steam flow upon high column pressure.
High Pressure	Steam reboiler tube leak causing high pressure steam to enter vessel.	Column Overpressure and potential mechanical failure of the vessel and release of its contents.	High pressure alarms with operator intervention. Pressure relief valve.	L1 L3	See previous item.

Figure 2: Typical Process Hazards Analysis HAZOP Worksheet

Air Products' developed layer of protection analysis (APLoPA) and fault tree analysis (APTree) software programs were used during the SIL Assessment and Verification work process. APTree was used to determine initiating cause frequencies for the layer of protection analysis when needed and the PFDavg for identified SIF's. Layer of protection analysis accounts for multiple layers of independent protection that are typically part of a process design.

Typical layers of protection may include:

1. Basic Process Control System (BPCS) – Regulatory control
2. Alarms that provide enough time for knowledgeable operators to recover from initial upsets prior to the hazardous event

3. Safety Instrumented Functions
4. Inherent pressure containing strength
5. Pressure Relief Devices
6. Secondary containment
7. Barricades
8. Administrative controls
9. Exposure probability

The LOPA results are reported in a table format from the APLoPA software. Figures 3 and 4 illustrate an example (developed for training purposes) that demonstrates the execution and methodology used during the SIL Assessment portion of the case study.

Please note that it is not appropriate to reference the data in this paper as it was developed for illustration purposes only. Actual analyses require data references and risk targets that are both auditable and defensible by the company performing the analysis.

Figure 3 shows a representation of a typical control and safety instrument system used for overpressure protection of a pressure vessel. Figure 4 details the APLoPA report for the example in Figure 3.

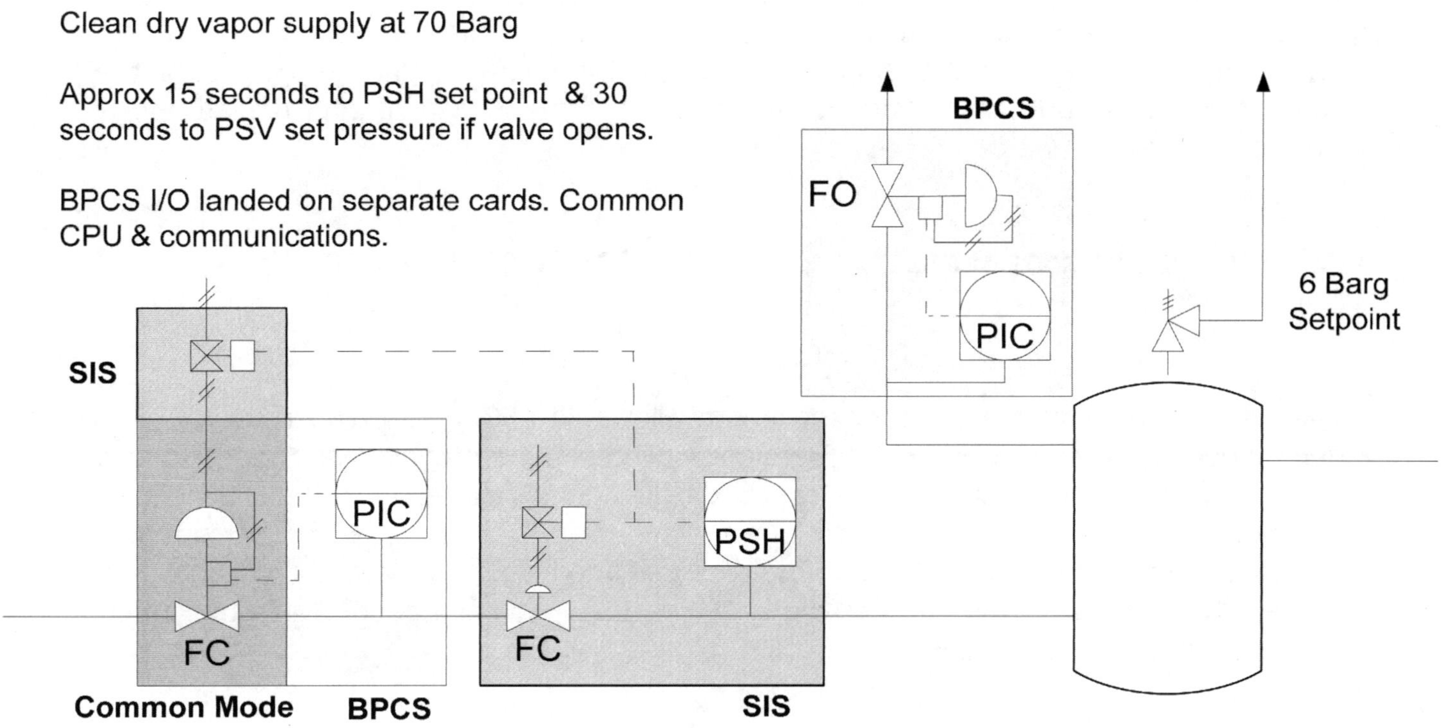

Figure 3: Overpressure Protection Diagram

Layer of Protection Analysis Report - Generated 09-Feb-2007 15:04:58										
Scenario:	pressure vessel rupture due to overpressure (No SIF Credit)									
Calculation Complete	References Complete									
Target Frequency: 5.0E-06 Calculated Frequency: 1.8077E-04 Required Performance: 2.766E-02										
Intermediate Event Frequency	Initiating Cause	Initiation Frequency (1/yr)	Alarm Protection Layer Fails to Protect	PIC vent control fail to vent	1oo2 HP SIF fail to function	1oo1 HP SIF fail to function	Spring Operated Relief Valve Fails to Open	Probability Vessel Ruptures	Prob Person Present	Prob of Fatality if Person Present
1.63E-04	Pressure letdown PIC (loop) fails dangerous	0.1	1	0.1	1	1	0.0163	1	1	1
8.15E-06	Common Mode BPCS dangerous output	5.00E-04	1	1	1	1	0.0163	1	1	1
9.62E-06	Control valve spuriosusly fails open	5.90E-03	1	0.1	1	1	0.0163	1	1	1
Date: 09-Feb-2007 15:04:43 Analyst: DAYJD Version: 2.1.0 - 2.0.0										

Figure 4: LOPA Report Example (Without SIS Credit)

Figure 4 Notes:

1. This analysis assumes that the pressure letdown control and the vent control utilize totally independent input and output cards.
2. No credit for alarm layer of protection due to lack of operator response time.
3. No credit for SIS protection taken during this stage of analysis

The LOPA results detail the initiating causes and frequency of each event that was evaluated. All layers of protection receiving credit are listed with its respective probability of failure. In the example in Figure 4, the overall event likelihood that the vessel will rupture due to overpressure is the sum of the individual intermediate event likelihoods. This can be calculated as follows:

Overall Event Likelihood = $1.63\text{x}10^{-4} + 8.15\text{x}10^{-6} + 9.62\text{x}10^{-6} = 1.81\text{x}10^{-4}$ (yr^{-1})
(Without SIS Credit)

The required performance level or maximum allowable PFD_{AVG} required for the SIS is determined by dividing the Target Frequency by the Overall Event Likelihood. In this example, an additional $2.77\text{x}10^{-2}$ PFD_{AVG} risk reduction is required from the SIS to meet a target frequency of $5\text{x}10^{-6}$. This value falls within the SIL 1 range, but we perform our SIL verification to ensure that the actual "as designed and maintained SIF" does not exceed this value, i.e. the SIF PFDavg must be less than or equal to this maximum allowable PFDavg.

The assessment results were summarized in a SIL Assessment Report. This report included the following information:

1. Executive summary of the SIL ratings (maximum allowable PFD_{AVG}) required to meet overall company targets
2. System description, hazards and safeguards
3. Reference design documents
4. APLoPA Results
5. APTree Results (if applicable)
6. Failure rate data with references used in analysis

This report is being stored in a common work area that is shared within the company's Process Safety community for reuse on future projects and evaluations.

SIL Verification

Once a safety instrumented function was identified during the assessment phase, a detailed review and quantification of the safety instrumented function risk reduction capability was performed. This included review and collection of the following information:

1. Equipment Operating and Maintenance Manuals

2. SIF electrical schematics
3. SIF vendor model number and catalog data
4. Component failure rate/performance data
5. Safety manuals that might exist.

If applicable failure rate data was not available from the manufacturer or was insufficient, an internal failure rate database was used. The PFD_{avg} of each individual SIF was calculated as a function of proof test frequency using APTree software. An example of a 1 out of 1 High Pressure SIF evaluated using APTree is shown in Figure 5.

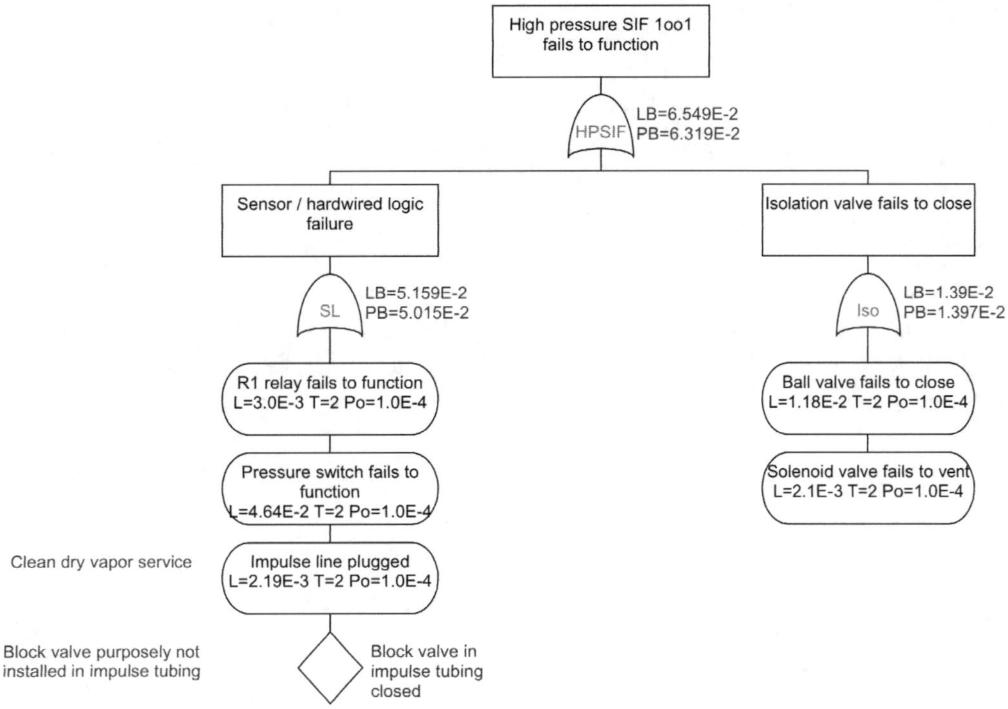

Figure 5: APTree Example

Figure 5 Terminology

<u>Fault Tree Component Inputs</u>

1. L = Failure Rate/Yr
2. P_0 = Probability of being failed at time zero
3. T = Proof test interval (Yr)

<u>Fault Tree Gate Outputs</u>

1. LB = Average Failure Rate (Yr^{-1})
2. PB = PFD_{AVG} (probability)

The PFD_{avg} was determined for multiple proof test frequencies. An example of the summary of PFD_{avg} for both of the 1oo1 and 1oo2 High Pressure SIF's from Figure 3 are given in Table 2.

Table 2: PFDavg Summary Results

Proof Test Interval (years)	PFDavg 1 oo 1 (Yr^{-1})	PFDavg 1 oo 2 (Yr^{-1})
0.5	0.0169	0.0131
1	0.0325	0.0257
2	0.0632	0.0504
3	0.0925	0.0743

Following calculation of the PFD_{AVG} for each SIF, these layers of protection were inserted back into the original LOPA analysis to confirm that our corporate risk targets (Target frequency) were being met. Figure 6 shows the final LOPA analysis using the PFD_{AVG} calculated during the SIL Verification phase of the case study. In this example, a SIF proof test frequency of 1 year is required to meet the target frequency.

Layer of Protection Analysis Report - Generated 09-Mar-2007 16:44:30

Scenario: Pressure vessel rupture due to overpressure (SIF Credit per Verification)

Calculation Complete: References Incomplete

Target Frequency: 5.0E-06 Calculated Frequency: 4.7111E-06 Required Performance: Target Achieved

Intermediate Event Frequency	Initiating Cause	Initiation Frequency (1/yr)	Alarm Protection Layer Fails to Protect	PIC vent control fail to vent	1oo2 HP SIF fail to function	1oo1 HP SIF fail to function	Spring Operated Relief Valve Fails to Open	Probability Vessel Ruptures	Prob Person Present	Prob of Fatality if Person Present
4.19E-06	Pressure letdown PIC (loop) fails dangerous	0.1	1	0.1	0.0257	1	0.0163	1	1	1
2.09E-07	Common Mode BPCS dangerous output	5.00E-04	1	1	0.0257	1	0.0163	1	1	1
3.13E-07	Control valve spuriosusly fails open	5.90E-03	1	0.1	1	0.0325	0.0163	1	1	1

Date: 09-Mar-2007 16:44:07 Analyst: DAYJD Version: 2.1.1 - 2.0.0

Figure 6: Final LOPA Analysis Report with SIS Credit

Overall Event Likelihood = $4.19 \times 10^{-6} + 2.09 \times 10^{-7} + 3.13 \times 10^{-7} = 4.71 \times 10^{-6}$ (yr^{-1})
(With SIS Credit)

The final results for each SIF were summarized in a SIL Verification Report. This report included the following information:

1. Detailed SIF description
2. PFD_{avg} results
3. Reference design documents
4. Design, Configuration and Mechanical Integrity assumptions
5. APTree reports

This report is being stored in a common work area that is shared within the Air Products Engineering community for reuse on future projects and evaluations.

CONCLUSION

Air Products and Chemicals has begun implementation of a standardized approach to perform SIL Assessments and Verifications when necessary. This risk-based approach included development of training materials and software tools customized to our modified internal work process. An application library of commonly used instrumented protection schemes allows for pre-evaluated and consistent designs across multiple business areas. This approach allows for a cost effective and timely design of safety instrumented protections for our business.

References

1. IEC 61511 "Functional safety - Safety instrumented systems for the process industry sector".

2. ANSI/ISA-84.00.01-2004 Parts 1-3 (IEC 61511 Mod) "Functional Safety: Safety Instrumented Systems for the Process Industry Sector".

3. ANSI/ISA-84.01-1996: Application of Safety Instrumented Systems for the Process Industry.

4. CCPS Guideline Book, *Layer of Protection Analysis – Simplified Process Risk Assessment*, New York, New York, 2001.

5. ISA TR84.00.04 Part 1, *Guidelines for the Implementation of ANSI/ISA-84.00.01-2004 (IEC 61511 Mod)*, 2005.

The Evaluation of Risks of Ethoxylation Processes

Ernesto Salzano
Istituto di Ricerche sulla Combustione, Consiglio Nazionale delle Ricerche
Via Diocleziano 328, 80124 Napoli (I). Email: salzano@irc.na.cnr.it

Ciro Tortora
Chelab srl, Via Fratta 25, 31023 Resana, Treviso (I)

Martino Di Serio, Elio Santacesaria
Dipartimento di Chimica, Università di Napoli »Federico II«, Via Cintia 80126, Napoli (I).

Abstract

Alkoxylation is commonly performed in industry for the production of surfactants and polyglycols by means of simple batch liquid-stirred reactor. More recent processes use semi-batch Venturi Loop reactors or Spray Tower Loop Reactors, where liquid phase is dispersed into the gaseous through recirculation loops, thus eliminating hazardous rotating parts. In this paper, the analysis of consequences related to accidental scenarios which are added by the presence of recirculation loop has been performed, aiming at evaluating the risk of different alkoxylation reactors by varying the operating conditions. More specifically, the analysis of consequences for the ethoxylation of dodecanol at different partial pressure of ethylene oxide and process temperature are shown in terms of safety distances, by varying atmospheric wind velocity, for a range of release sections on recirculation loop or from small releases from flange and valves along the same pipeline. Results show that dramatic accidental scenarios may occur only for large leakage sections, which however are comparable with accidental scenarios derived from gaseous or liquid ethylene oxide.

INTRODUCTION

Ethoxylation of fatty alcohols, alkyl phenols, or fatty acids is commonly performed in industry for the production of surfactants and polyglycols. To this aim, more than two hundred patented reactors are on the market. Simple liquid-stirred reactors are classically used, and sometimes external circulation loops are installed in order to increase the reactor heat exchange capacity. Hence, gaseous Ethylene Oxide (EO) is continuously fed to the liquid substrate, followed by post-treatment operations such as catalyst removal or bleaching operations, in order to achieve the required quality features of the product. As cited above, the main problems for this process is the difficulty in eliminating mass-transfer and heat-transfer limitations, which are however generally associated with conventional stirred tank alkoxylators, and to the presence of rotating mechanical parts in contact with the gaseous phase. Indeed, EO decomposes in gaseous phases with strong exothermic reaction, followed by runaway reactions and explosive conditions.

Other than classical batch processes, two alternative types of reactors are Venturi Loop Reactors (VLR) (e.g. Buss Loop Reactor Technology, now licensed by Davy Process Technology AG (Switzerland)) and Spray Tower Loop Reactors (SLR) (e.g. Pressindustria, now licensed by Scientific Design Company, Inc.). Quite obviously, as cited above, several different designs and combinations are available commercially, e.g. for the number and the designs of venturi or spray jets or for the possibility of simultaneous feeding with EO and PO, but the categorization of reactors in VLR and SLR is helpful for further discussion. In VLR the gas phase, as for stirred reactors, is dispersed into the liquid phase whereas in SLR the liquid phase is dispersed into the gaseous one. For both reactors, high productivity and efficiency with heat transfer is ensured by external thermal exchanger. Quite clearly, process reactors based on Venturi or on Spray Loop do not use rotating metallic devices as in the stirred tank reactor. This aspect has been considered as a main point for safety: Avoiding mechanical agitation means to reduce or eliminate the ignition hazard, hence an inherently safer practice, i.e. a permanent choice for safety [1,2]. However, further considerations are needed for sound risk assessment of the three processes and over-simplification for very hazardous processes as those involving EO should be avoided.

The starting point of the analysis reported in this paper is the analysis of Gustin [3], which has recently reviewed the hazards of ethoxylation reactions and more generally the hazards of processing ethylene oxide (EO), including the hazard related to the flammability of EO in the presence of air or oxygen, the gas phase explosion hazard due to EO vapour instability, i.e. decomposition flames with pressure effects in vessels even in the absence of oxygen, the liquid EO phase decomposition hazard, since the decomposition of liquid may be initiated by high temperatures, hot spots or catalysts, the runaway reaction hazard, due to improper operating conditions, or self-heat due to accumulation, hot spot for catalyst and if the temperature is not controlled. These points have been extensively developed and analyzed in the literature, due to the high risks of industrial processes with EO (see www.ethyleneoxide.com). Details are not reported here for the sake of brevity [4-7].

In the following, the risk of recirculation loop on main reactor is discussed in terms of maximum hazardous distances for people and equipment (domino effects), either for toxicity or flammability or explosibility. Hence, no discussion will be reported for the risks of leakage of pure EO or EO/N2 mixture from the feed line or from the reactor shell, which are common associated with all the three reactor types.

The Ethoxylation process

In this paper, the liquid charge considered is 1-dodecanol as it is a common substrate for ethoxylation. After the charge of dodecanol in the reactor the gaseous phase is purged with nitrogen to eliminate oxygen. Hence, after the purging, a positive pressure of nitrogen is set in the reactor (1-2 bar) to dilute the EO in gaseous phase and to assure a gaseous composition within safety limits (y_{EO} < 0.4) (see [2]; and the reference therein).For the three processes, the reactor temperature about 120-180 °C) is kept constant by an external regulation and the EO is fed in the reactor with a flow rate of P_{EO} which is able to assure a gas composition within safety limits, and an internal pressure within fixed limits (4-6 bar). Hence, after then the fixed amount of EO has been fed the reaction is continued until all EO is reacted and the gaseous phase contains only nitrogen.

Table 1 reports a typical reactor geometry and operating conditions. Quite clearly, this is only an example of ethoxylation equipment, but the parameters given can be usefully considered for the sake of comparison of the different processes and for the analysis of risks related to recirculation loop..

Table 1. Chemical and geometrical characteristics of alkoxylation reactors analysed and the typical operation conditions for ethoxylation processes.

Reactor Volume (m^3)	10
Height (m)	2.5
Dodecanol initially charged (Kg)	2000
Time of EO feeding (min)	about 120
Initial Pressure of N_2 (bar)	1.5
Total reactor pressure (bar)	4.0 – 5.0
Partial EO pressure (bar)	1.0 – 3.0
Temperature (°C)	125.0 – 180.0
Pressure of EO storage tank (bar)	4.0 – 5.0
Catalyst (KOH) (kg)	6.5
Kinetic constant ($cm^3\ mol^{-1}\ s^{-1}$)	6.18 exp(-6640.0/T)

For the reactor and operating conditions reported, Figure 1 shows the total pressure, the density of the liquid mixture ρ_{mix}, the partial pressure of EO and the ratio of EO moles fed to the VLR reactor with respect to the initial moles of dodecanol (the ethoxylation degree of dodecanol $\xi = n_{EO}/n^{\circ}_{dodecanol}$) during the entire time range of operation, as calculated from the data of current authors in previous work [2].

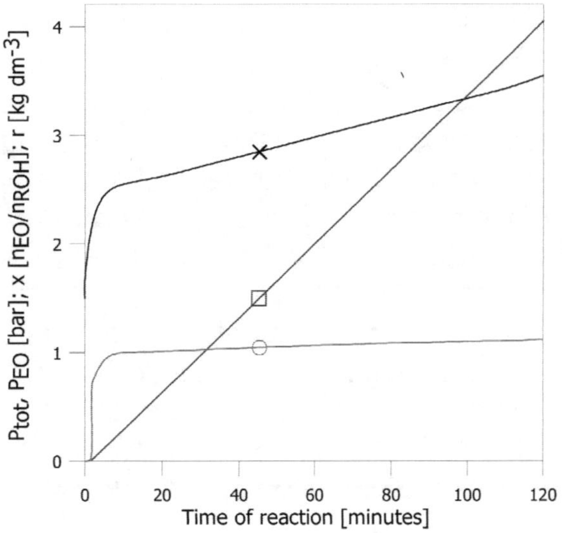

Figure 1: Total pressure P_{TOT} (×), EO partial pressure P_{EO} (○), and ethoxylation degree $\xi = n_{EO}/n°_{dodecanol}$ (□), as function of reaction time

The partial pressure of EO is quite constant along the run until EO feeding is stopped. Finally, it's worth noting that the diameter of recirculation loop pipeline ranges typically between 6 and 10 inches: i.e., any damage to the loop can raise high concern for the accidental scenarios which may derive from EO evaporation.

RESULTS AND DISCUSSION

Following classic definition, the risks associated with recirculation loop in ethoxylation processes are dependent on the magnitude of accidental scenarios which may derive from the release of EO/dodecanol from the recirculation system (analysis of consequences) and on the probability of failure of the pipelines, or leakages from valve, flanges and pumps. With respect to probability of failure or leakage probability, no analysis has been presented in the literature for the specific ethoxylation process. Furthermore, this paper is mainly addressed to the analysis of consequences, as the type and the intensity of accidental scenarios which may derive from the release or leakage are not known. However, general information on the probability of failure can be applied for the pressurized pipelines and for the pumps operating along the recirculation loop. To this aim, values of $2 \cdot 10^{-6}$ y^{-1} and $3 \cdot 10^{-6}$ y^{-1} are found in the literature respectively for pressure vessel failure and for sudden catastrophic failure of vessels [8,9]. With respect to pipework, frequency of failures as obtained for butane as defined in Canvey report [10] can be used in first approximation. More specifically $5 \cdot 10^{-4}$ y^{-1} for pumps and $3 \cdot 10^{-4}$ km^{-1} y^{-1} for large section butane pipeline. These values can be used in conjunction with the results of analysis of consequences.

Accidental release of EO/dodecanol/condensate mixture from leakages on flanges, valves, pumping systems, i.e. small release sections, are characterized by rapid flash of EO from jet/spray release of pressurized liquid dodecanol, whereas more dramatic large ruptures on the recirculation pipeline are characterized by large pools for which EO evaporation rate has to be evaluated. In both cases, EO may produce fires, explosions or toxic dispersion after EO is dispersed into atmosphere. Furthermore, both scenarios depend strongly on EO concentration in the liquid dodecanol, which in turns depend on EO solubility at process condition. In the case of VLR, it can be assumed that, in the operation conditions, the mass transfer is highly efficient and that the liquid phase is well mixed, so in this case the EO liquid concentration is equal to the EO solubility [2]. Quite clearly, this is a worst scenario as it corresponds to the higher possible concentration. So far, in the following we will refer to VLR reactor when recirculation loop hazards are of concern. Results are however on the safe side with respect to SLR.

The Evaluation of Solubility of EO at Operating Conditions

The solubility of EO (S_{EO}) as function of temperature and EO partial pressure in dodecanol and in ethoxylated products was determined previously by the current authors [11]. As S_{EO} is not affected strongly by ethoxylation degree ξ [2], we can use the following empirical relation which fits experimental data for pure dodecanol (Fig. 2):

$$S_{EO} = 0.01 \cdot P_{EO} \cdot e^{\left[-3544\left(\frac{1}{T}-\frac{1}{293.15}\right)\right]} \tag{1}$$

where T is the process temperature (K), P_{EO} is expressed in bar and S_{EO} is the solubility expressed in moles of EO for cm^3 of mixture.

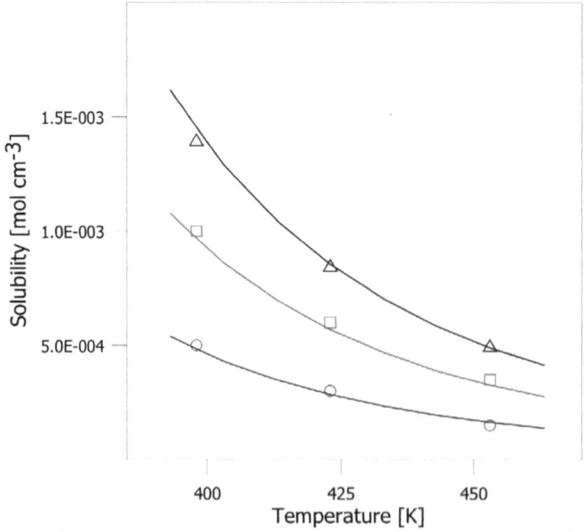

Figure 2. Solubility of EO in dodecanol; points: experimental data; lines: calculated data (Eq. 2). P_{EO}: ○ = 1 bar; □ = 2 bar; Δ = 3 bar.

By simple analysis, it can be then assessed that Henry's constant dependence of EO in dodecanol with temperature, in the range of temperature here analysed, follows Clausius-Clapeyron equation:

$$h_{EO}(T) = h_{R,EO} \cdot e^{\left[-\frac{\Delta H_{V,EO}}{R}\left(\frac{1}{T}-\frac{1}{T_R}\right)\right]} \tag{2}$$

where R is the universal gas constant, $h_{R,EO}$ = 0.01 mol bar^{-1} cm^{-3} is Henry's constant at the given reference temperature T_R (=293.15 K) as obtained experimentally by the current authors and $\Delta_{HV,EO}$ is latent heat of evaporation of EO (= 2.5 · 104 J/mol). The solubility of EO is essential for the knowledge of the total amount available for the flammable and toxic dispersion of gas, if a leakage occurs.

The Evaluation of Evaporation Rate of EO

The specific rate of release (rate of evaporation per square meter of pool) of EO from dodecanol (Na) towards atmosphere ($P_{EO,air}$ = 0) when recirculation loop fails and an accidental release occurs is important for the implementation of proper mitigation procedures. To this aim, neglecting the variation of composition of EO in the liquid phase during the evaporation (well mixing assumption), the equation of Kawamura and Mackay [12] can be applied to vapour phase, i.e. between the liquid-air interface and the atmospheric bulk:

$$N_a = k_g \frac{(P_{EO,i} - P_{EO,air})}{RT_p} \tag{3}$$

where $P_{i,EO}$ and $P_{EO,air}$ are respectively the partial pressure of EO respectively at the interface and in the gas phase bulk, which is null in the free atmosphere, k_g is mass transfer coefficient, R is the gas constant, and Tp is the liquid pool temperature.
Following US-EPA and US-NOAA suggestion [13], k_g (m s^{-1}) can be assessed by the equation of Kawamura and Mackay for the mass transfer coefficient in an atmospheric turbulent flow field, which is:

$$k_g = 0.004786\, V_{air,10}^{0.78}\, Z^{-0.11}\, Sc^{-0.67} \tag{4}$$

where $V_{air,10}$ is the wind speed at a height h_{10} = 10 m, calculated by means of equation:

$$V_{air,10} = V_{air}\left(\frac{h_{ground}}{h_{10}}\right)^p \tag{5}$$

where V_{air} is the wind velocity at ground height h_{ground}, the coefficient p is 0.2 of the ground wind velocity for the D class of stability and 0.4 for the F class of stability [13], Z is the pool diameter in the along-wind direction, and Sc is the laminar Schmidt number (= $\mu/\rho D_{EO}$), where the diffusivity constant D_{EO} can been assessed by the classical laws of Graham [14] or by Fuller, Schettler and Giddings equation [15], which has been used in the following. All thermodynamic and fluid dynamic variables used throughout the paper for the vapour have been calculated averaging the values between pool interface condition (the temperature has been considered equivalent to the bulk pool temperature) and atmosphere condition (T = 300 K; P = 1 bar). Alternatively, k_g may be calculated by considering a plate in laminar flow, by means of the following equation for forced flow [16]:

$$Sh = 0.332\, Re^{0.5}\, Sc^{0.33} = \frac{k_g Z}{D_{EO}} \tag{6}$$

where Sh, Re and Sc are respectively Sherwood, Reynolds and Schmidt numbers.

For the evaluation of Re number, the characteristic length has been considered as the diameter of the pool interacting with the air flow. Schmidt number has been calculated by using the diffusivity constant as evaluated in the previous method. Finally, by assuming that $P_{i,EO}$ is equivalent to P_{EO} (which however is a conservative option) we can calculated, for a range of operating conditions, the specific rate of evaporation.

A comparison of the laminar and turbulent approach is given in Fig. 3, where Na is given for different process temperatures and wind velocities, for P_{EO} = 3 bar (we have reported only the results for a single driving force, though conservative on the safe side, for the sake of brevity), and for the temperature of 453 K (similar results are obtained however for the other reference process temperatures), for the F class of stability only (see below for further details).

The evaporation rate falls within the two limits for laminar and turbulent phenomena. It's clear the negligible effects of temperature on both laminar and turbulent approach, at least in the range of temperature here analyzed. Indeed, in the first case, the pool temperature affects only the temperature term of Na in Eq. 3) and the influence is negligible. Besides, in the laminar case, the values of ρ and μ (and of course D_{EO}) affect also the mass transfer coefficient k_g, through Re and Sh. The great effect of wind velocity when high partial pressure of EO are considered and turbulent evaporation models are considered is also observable.

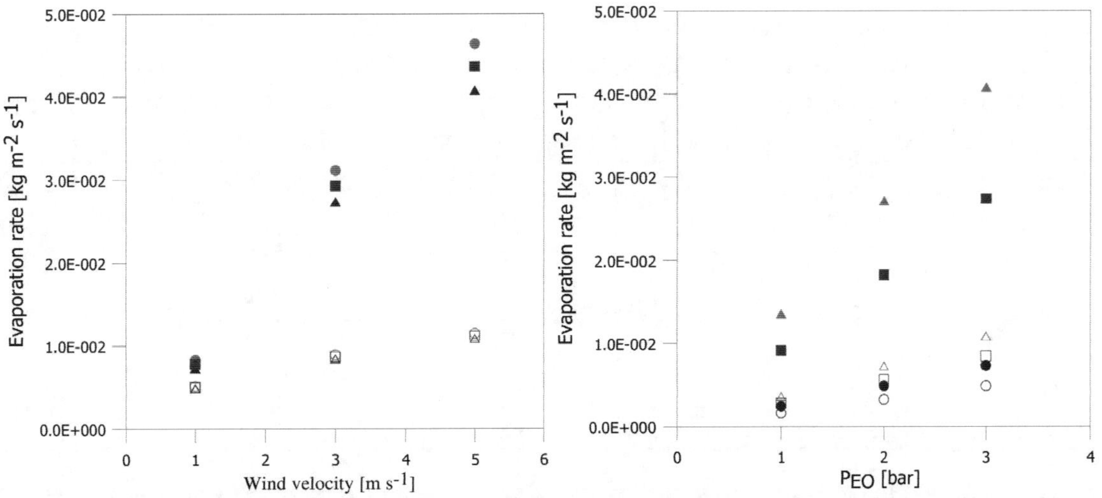

Figure 3. Left: Calculated mass flow rate of EO vapour from dodecanol mixture to atmosphere vs.: Left: Wind velocity, P_{EO} = 3 bar; T = 398 K (○); 423 K (□); and 453 K (Δ). Right: EO pressure (P_{EO}), T = 453 K, V_{air} = 1 m s^{-1} (○); 3 m s^{-1} (□); 5m/s (Δ). Filled symbols: evaporation rates calculated by turbulent correlation Empty symbols: laminar correlation.

The values of Na and the solubility data (Eq. 2) allow the calculation of the total amount of EO dissolved in the entire mixture contained in the process reactor, and the total time required for a complete transfer of EO from the released liquid mixture to the atmosphere, provided the total liquid amount released and the pool dimension formed after the release are known. To this regard, it's worth saying that due to the EO addition, an increase of mass of liquid starting from the initial dodecanol charged should be considered. This aspect is indeed important for safety consideration because the total amount of EO dissolved in dodecanol is higher at the final stages of process (up to the double) if P_{EO} and T are kept constant. In the following, the mass released for any accidental scenario will be considered as correspondent to the initial phases after charge of liquid. For the evaluation of pool dimension, liquid density of mixture ρ_{mix} is needed. As showed in Fig. 1, ρ_{mix} may vary with ethoxylation degree ξ (slightly) and process temperature. Eventually, the following empirical polynomial correlation [8] can been used:

$$\rho_{mix} = 0.86 + 0.025\xi - 4.76 \cdot 10^{-4}\xi^2 - 2.69 \cdot 10^{-5}\xi^3 - 7.7 \cdot 10^{-4}(T - 273.15) \qquad (7)$$

Tables 2-3 report the results for EO evaporation from accidental pool for any process condition and wind velocity, assuming pool depth = 0.01 m for the sake of comparison.

Table 2. Total amount of EO (in kg) dissolved in the dodecanol mixture contained in the reactor, for process conditions reported in Table 1, at the initial stage (dodecanol mass = 2000 kg; ξ = 0), and at the final stage of reaction (ethoxylated dodecanol mass = 4000 kg; ξ = 4).

Dodecanol mass [kg]	T [K]	P_{EO} =1 bar	P_{EO} =2 bar	P_{EO} =3 bar
2000	398	58	115	161
	423	36	71	101
	453	18	43	61
4000	398	103	206	289
	423	63	127	180
	453	33	76	108

Table 3. Time [s] for the release of dissolved amount of EO per square meter of pool, by laminar and turbulent approach for evaporation rate.

	Laminar *Wind velocity [m s⁻¹]*			**Turbulent** *Wind velocity [m s⁻¹]*		
P_{EO} = 1 bar	1	3	5	1	3	5
T [K]	time [s]	time [s]	time [s]	time [s]	time [s]	time [s]
398	112	64	50	60	16	11
423	129	74	58	80	21	14
453	79	46	35	51	14	9
P_{EO} = 3 bar	1	3	5	1	3	5
T [K]	time [s]	time [s]	time [s]	time [s]	time [s]	time [s]
398	104	60	47	56	15	10
423	120	69	54	74	20	13
453	75	43	34	48	13	9

Results show that the worst case condition refers to P_{EO} = 3 bar, T = 398 K, and wind velocity 5 m s^{-1}. In this case, the possibility of prevention measure or mitigation systems is strongly affected by the large amount of EO in dodecanol and by dramatically high rates of evaporation from the dodecanol mixture, which is of few seconds.

CONSEQUENCES ANALYSIS FOR THE RECIRCULATION LOOP

For the consequence-based comparison of the reactors cited above, we have assumed a release of EO+dodecanol mixture from the recycle pipeline by varying the flowing section from leakages section with diameter lower than 1" to the almost complete rupture (10" diameter) of the larger recirculation tube typically adopted in ethoxylation processes. To this end, the following accidental scenarios due to the release of ethylene oxide should be considered: a) jet fire of EO flashing from jet release of dodecanol mixture; b) pool fire of EO from liquid pool, which affects equipment and human being for the heat radiation, c) toxic dispersion of EO from either pool or from jet release, d) flash fire of flammable cloud; and e) vapour cloud explosion, the latest if particular conditions of congestion and confinement apply to the scenario environment.

Jet Fire

Jet fire of EO flashing rapidly from dodecanol mixture may occur if ignition occurs after the failure or leakage through small section, valves, or flanges and jet release of liquid mixture. In this case, however, ignition should occur before the intervention of safety isolation valve. Furthermore, safety distances are small if compared with other scenarios. Toxic effects and flash fire deriving from EO flashing from jet release of dodecanol are discussed in the following as dispersion of EO from semi-continuous release from small leakages.

Pool Fire

Mass flow from any accidental section A_h on the recirculation loop can be calculated by classical equation for the conservation of mechanical energy. Due to the reactor pressure condition, the time for emptying is dramatically short for release rupture greater than 2″ diameter: for these values, a large pool is almost instantaneously formed and its extension depends only on the type of ground (e.g. slope) or on the presence of dikes or bunds. For release rate lower than 2″ diameter, pool radius have be calculated by assuming a pool depth of 0.01 m and the time dependence of mass flow pool radius. The results have been also evaluated by PHAST code [17]. Again, large pools are formed shortly after the leakage, i.e. in a time at least comparable to the typical time of intervention of safety system, due to the pressure head in the reactor. Results of simplified correlation and PHAST code are close, thus confirming the validity of preliminary assumption on pool depth.

Leakage diameter smaller or close to 0.5″ diameter are essentially hazardous only for the operators located in the proximity of leakage, Indeed, it can be assumed that after few tenths of seconds the intervention of safety systems (automatic isolation valves, safety interlock system) can isolate effectively the recirculation lines. Consequently, the entire amount of dodecanol cannot be released through the leakage. It should be noted that flash point of 1-dodecanol is reported in the open literature as 127°C (closed cup), whereas dodecanol-EO condensate (polyethoxylated ethoxylated dodecanol or polyoxyethylene lauryl ether or ethoxylated lauryl alcohol) shows a flash point of 113 °C [15]. Hence, ignition can trigger pool fire for any process temperature condition unless cooling of mixture occurs after the release. Furthermore, ignition is always likely for EO due to the decomposition reaction or for the flash fire of EO vapour.

When large leakage or pipeline failure occurs, the total liquid amount is released almost instantaneously. Hence, assuming a pool depth of 0.01 m, the radius of the pool can be calculated as about 10 m. When smaller leakages occur, if 3 minutes for intervention are required and this time is considered also as the time for closing isolation valves installed on the main reactor, a conservative maximum diameter of respectively 2 m and 4 m for 0.5″ and 1″ diameter releases have been considered. Quite clearly, if isolation valves or the fire system fail, a maximum pool diameter of about 10 m (as for larger failures) should be considered for further analysis.

Small effects for pool fire are due to the dissolved EO in the mixture, as the low amount, whereas ignition can be facilitated by the presence of EO vapour, which are either highly inflammable or instable. To this regard, it's worth mentioning that the total evaporation of EO is completed in few seconds and hence EO does not contribute consistently to heat radiation due to pool fire.

Typical heat radiation level that may possibly cause damage are 5.0 kW/m^2 and 12.5 kW/m^2, respectively for human (second degree burns from 40-second exposure) and for equipment (domino effects). Here it's worth noting that surrounding equipment can be affected by fires only if long duration fires are of concern [19], hence only ineffectiveness of fire systems allow for domino effects. The distance from pool fires of flammable liquids with boiling points above ambient temperature can be determined by TNO guidelines [20], and the simplifying assumption of EPA Risk management program guidance for offsite consequence analysis [21]. Given any radiation per unit area received by the receptor (W/m^2), the distance x from pool fire to source where the radiation is experienced is given by the following equation:

$$x = H_c \sqrt{\frac{3.18 \cdot 10^{-5} A}{H_v + C_p (T_b - T_{air})} \frac{1}{q}} \tag{8}$$

where H_C (about 38000 J/kg, for complete combustion to CO_2 and H_2O) and H_V (0.49 = J/kg) are the heat of combustion and the latent heat of evaporation of dodecanol [22] , A is the pool area (= πZ^2), C_p is the liquid heat capacity (NIST database). T_b (= 534 K) and T_{air} are respectively the boiling temperature of the liquid and the ambient temperature.

As the very high latent heat of evaporation of dodecanol or its condensate, very low mass burning rates are calculated and very short distances ($x \approx 3$ m) from pool border are obtained even for the low heat radiation intensity q = 5 kW/m^2. Few hazards from pool fire can be then assumed either for the low probability of ignition or for the low intensity of consequences.

The Dispersion of EO to Atmosphere as Toxic and Flammable Substance

Dispersion analysis should be aimed to the definition of flammable mass, i.e. the mass with concentration included in the range of upper and lower flammability limits (UFL, LFL), which is essential for the hazard of explosion (see next paragraph), and aimed to the definition of maximum distance run by EO, for the hazard assessment of flash fire and toxic dispersion.

Quite clearly, dispersion analysis depends on the atmospheric condition. To this regard, it has been showed by Woodward [23] that the distances reached by the heavy vapors of EO are relatively insensitive to atmospheric stability class. On the contrary, the flammable mass is very high with stable atmospheres (E and F stability). Hence, in the following, the stability class F has been only considered for simplicity. Ambient temperature is 300 K. So far, Richardson number Ri (or equivalently, Froud number) for the pool formed accidentally should be first calculated for the assessment of behaviour of EO vapour. The following equation as reported by Havens and Spicer [24,25] for dispersion models of flammable cloud analyses has been then considered:

$$Ri = \frac{g(\rho_{EO,v} - \rho_{air})}{\rho_{air}} \frac{V_{i0}}{Z \rho_{EO,v} V_{air} u_*^2} \qquad (9)$$

where g is the acceleration due to gravity, ρ_a and $\rho_{EO,v}$ are respectively the density of air and of EO vapour (Na Z), V_{io} is the release rate of EO from pool which refers to the data reported previously, and u_* is the friction velocity, which has been considered as one tenth of the wind velocity at 10 m height. A threshold value for dense gases (heavier than air) is Ri > 1 is given by Spicer and Havens [24,25]. Further references [26] gives values between 1 and 50 as the transition values for Ri between pure heavier and lighter than air gas/vapors.

The calculated values of Ri for any weather or process condition are not reported here for the sake of brevity. As a matter of fact, Ri is smaller than one whatever the weather and process (T,P) condition, with both laminar or turbulent evaporation condition, unless low wind velocity (< 1 m s^{-1}) and very large pools (large release section) are considered, for which an average value of 10 (hence however within the boundaries of 50 for pure passive behavior) is obtained. As a conclusion, short duration or instantaneous Gaussian models can be always considered for $V_{air} \geq 1$ m s^{-1}, whereas heavy gas dispersion analysis should be used with very low wind velocity. When very small release sections are of concern (<<1" diameter), Ri is lower than one even for low wind velocity but semi-continuous source rather than short duration or instantaneous release should be taken into account.

Regardless the dispersion models, the threshold concentrations used in the following refer to half of the lower flammability limit (½ LFL = 1.3% v/v EO in air or about 13000 ppm), IDLH (Immediately Dangerous for Life and Health concentration value for EO = 800 ppm, or 0.08% v/v). The distances are measured starting from the center point of the EO+dodecanol pool formed after the release from the pipeline constituting the recycle line for the VLR and SLR reactor.

Passive Dispersion For Near-Instantaneous Source

In the case of near-instantaneous release of passive EO vapors, the ratio of flammable mass mf to the total available mass of EO (m_{EO}) can be calculated by means of the following equation in the case of Gaussian point-source near-instantaneous release, as in the case of large failure of recirculation loop (release section >1" diameter) and wind velocity greater than 1 m s^{-1} [23]:

$$\frac{m_f}{m_{EO}} = \mathrm{erf}\left[\sqrt{\ln\left(\frac{c_o}{LFL}\right)}\right] - \mathrm{erf}\left[\sqrt{\ln\left(\frac{c_o}{LFL}\right)}\right] + \frac{2}{c_o\sqrt{\pi}}\left[UFL\sqrt{\ln\left(\frac{c_o}{UFL}\right)} - LFL\sqrt{\ln\left(\frac{c_o}{LFL}\right)}\right] \quad (10)$$

where c_o is the initial concentration of the cloud (essentially pure EO, i.e. c_o = 100% v/v), LFL (= 2.6% v/v) and UFL are respectively the lower and the higher flammability limits of EO and *erf* is the error function as given by Press et al. [27].

When evaluating the flammable mass, the upper flammability limit (UFL) of EO in air is generally considered equal to 100 %v/v as EO can decompose also in the absence of oxygen. This value keeps into account mainly the hazards of ignition or EO explosion behavior in closed reactors, for the decomposition reaction. When open environment and blast wave produced by explosions are of concern, as it occurs after accidental release, the effects of this phenomenon are negligible, as the decomposition burning velocity is very low (0.05 m s^{-1}). Besides, for explosion issues, combustion reaction is characterized by very high fundamental laminar burning velocity (= 1.07 m/s) at stoichiometric concentration and luminous (non-decomposition) flames are seen up to about 67% v/v [23]. As a conclusion, assuming conservatively c_o = 0.95 at the interface, the maximum value of flammable to total mass ratio is about 0.83 and 0.94 respectively for UFL = 67% and 100%. These high values are confirmed by the analysis of Van Buijtenen [28], which shows very high flammable mass ratio when high UFL/LFL ratios. The flammable mass obtained for UFL = 67% will be then used for the assessment of explosion behavior of EO clouds, as shown in the next sections, whereas the value obtained for UFL = 100% will be used within flash fire analysis, due to the high decomposition temperature of EO, which still harmful to human.

At this point, the distance run by EO vapors in the downwind direction, considering the threshold values for concentration c* (= ½ LFL, IDLH), is essential for toxic and flash fire risk assessment. To this aim, for the very low Richardson number obtained for high wind velocity and rapid evaporation, we have adopted the classical, general equation for passive dispersion in the atmosphere of Pasquill-Gifford for an instantaneous point-source (a very short duration source release or puff), taking into account downwind x direction only, at the height level z comprised between ground level and 2 m as flash fire and toxic effects are only dangerous with respect to human, because negligible hazard to equipment may occur from these scenarios [19].

The maximum distances x* where EO concentration is higher than toxic (x^*_{tox}) and flammable (x^*_{ff}) concentration threshold c*, for the class stability F, are given in the following tables for any process T,P condition and wind velocity of 3 m s^{-1}. A value of roughness z_0 = 0.1 m has been considered in our analysis. Due to the high toxicity and high range of flammability of EO, the effects on person are likely at great distances from release point, in dependence of the process condition. It's important noting that higher process temperatures are less hazardous than lower process temperature, as lower amount of EO is dissolved in dodecanol. Regarding flash fire, it should be remembered that early ignition of vapors by hot surface and/or exothermic EO decomposition reaction are very likely in open industrial environment when catastrophic accidental scenarios occur. Furthermore, effective prevention and mitigation systems (as sprinkler systems or water curtain or fast isolation valves) are possibly installed on the recirculation loop connection to the ethoxylation process. Eventually, the distances calculated are conservative as they do not take into account these latest safety issues and assume the instantaneous emptying of the entire reactor content forming large hot pools and the dispersion over flat unobstructed field.

Table 4: Total mass available for explosion ($m_{EO,expl}$) and maximum distance reached by EO for flash fire (1/2 LFL, x^*_{ff}) and toxic dispersion (IDLH, x^*_{tox}) for instantaneous passive dispersion at the end of alkoxylation process (total mass of dodecanol = 4000 kg). Wind velocity = 3 m s^{-1}.

	P_{EO} = 1 bar			***P_{EO} = 2 bar***			***P_{EO} = 3 bar***		
T [K]	$m_{EO,expl}$ [kg]	x^*_{ff} [m]	x^*_{tox} [m]	$m_{EO,expl}$ [kg]	x^*_{ff} [m]	x^*_{tox} [m]	$m_{EO,expl}$ [kg]	x^*_{ff} [m]	x^*_{tox} [m]
398	86	44	167	171	62	223	240	72	257
423	53	35	136	105	49	182	143	57	211
453	27	24	102	63	38	147	87	45	171

The analyses for wind velocities higher (or even lower) than 3 m s^{-1} are not reported as very slightly different values for safety distances are obtained. On the other hand, roughness z_0 influences the safety distance more effectively. Results have been obtained by considering z_0 = 1 m, which is typical for industrial environment, but the safety distances increase by 20% by halving the value of roughness to 0.5 m. Smaller distances are obtained for atmospheric class E.

Passive Dispersion For Semi-Continuous Release (Plume) Of Short Duration

When small leakages occur (< 0.5″ diameter), regardless the wind velocity, a passive behavior and semi-continuous release of EO flashing from the liquid flow should be taken into account. In this case, unlike the catastrophic scenarios described in the previous section, the intervention of safety systems is much more effective and does not allow long evaporation time, in particular for hazardous process as ethoxylation of dodecanol. We then assume a total time of 180 seconds (often considered very conservatively for the full effectiveness of automatic isolation and safety systems as sprinklers) for the release into the atmosphere. For the calculation of EO dispersion, the classical Pasquill-Gifford model has been used. Results for the safety distances for toxic effects and flash fire are close to those evaluated in the previous section when release section are close or slightly smaller than 0.5″ diameter, whereas the evaluated threshold distances for flash fire are negligible for release sections smaller than 0.5″ diameter, whatever the process and weather conditions, i.e. small leakages are not hazardous but on the localized area of release in terms of flash fire. Flammable mass within concentration limits for flammability, even by considering conservative choice for UFL is always negligible for release section lower than 0.5″ diameter.

Heavy Gas Dispersion For Steady Source Of Short Duration

For low wind speed (< 1 ms^{-1}) and large release sections, the short duration (near-instantaneous) release of EO has to be analyzed following dense gas behavior models for the evaluation of maximum distance reached by EO, with specific reference to flash fire (½ LFL), toxicity (IDLH) and for the flammable mass available for vapour cloud explosion.

A number of codes have been presented in the literature for the analysis of heavy gas dispersion (see [26] for review and comparison). However, a good estimation for the complex case of dispersion of EO evaporating from dodecanol pool at temperature close to boiling point can be obtained by the methodology of Britter and McQuaid [29], as reported by Lees [30]. To this regard, it is important to note that the temperature of the pool (or the temperature T_o of evaporating EO, which is equivalent to process temperature for near-instantaneous source) and the ambient temperature T_{amb} are different. Hence, the corrected concentration c* for non-isothermal concentration is:

$$\underline{c}^* = \frac{c^*}{c^* + (1 - c^*)\left(\frac{T_{amb}}{T_o}\right)} \qquad (12)$$

Results are reported in Table 5 for the characteristic wind velocity of 1.0 m s^{-1}. For the total flammable mass needed for explosion evaluation, due to heavy gas dispersion behavior and to the large flammability range of EO, 95% of the entire EO amount dissolved (see Table 2) has to be considered, as defined previously for Gaussian behavior.

Table 5: Maximum distance for the flammable cloud (flash fire, x^*_{ff}) and for the toxic dispersion (IDLH, x^*_{tox}), for released mass of dodecanol of 4000 kg.

T [K]	P_{EO} = 1 bar		P_{EO} = 2 bar		P_{EO} = 3 bar	
V_{air} = 0.5 m s^{-1}	x^*_{ff} (m)	x^*_{tox} (m)	x^*_{ff} (m)	x^*_{tox} (m)	x^*_{ff} (m)	x^*_{tox} (m)
398	58	234	69	278	75	302
423	50	201	59	239	65	260
453	41	164	51	203	55	222
V_{air} = 1.0 m s^{-1}	x^*_{ff} (m)	x^*_{tox} (m)	x^*_{ff} (m)	x^*_{tox} (m)	x^*_{ff} (m)	x^*_{tox} (m)
398	82	330	98	393	106	427
423	71	284	84	337	92	368
453	58	232	72	287	78	314

Quite clearly, the distances reached by EO in the case of heavy gas dispersion are dramatically higher than values obtained when Gaussian dispersion analysis are considered, in particular for toxic effects. Strong effects of wind velocity are obtained by varying wind velocity up to 1 m s^{-1}. However, data are less reliable for this wind velocity as passive-heavy gas transition behavior can be considered by checking Ri number, which has however low values (between 5 and 12 for this wind velocity), only for turbulent dispersion.

Explosion Of EO Cloud

Blast wave produced by explosion can produce catastrophic damages even at great distances from the release point. If EO flammable vapour cloud explodes, the decay of peak overpressure with respect to the distance from the explosion source can be calculated knowing the total available energy from the heat of combustion of EO (=$2.76 \cdot 10^7$ J/kg) and the total amount of flammable mass formed accidentally after the leakage from recirculation loop and EO evaporation, as calculated previously. To this regard, if large release section are considered, negligible differences on the total flammable mass are calculated for heavier or lighter than air EO vapors. Hence, 95% of available EO mass will be considered in the following conservatively. The obtained values of explosion energy are then used with energy-scaled plot, typically adopted for the evaluation of peak pressure vs. distance in case of gas explosion. Main models are Multi-Energy Method and Baker and Sthrelow Model. Details on physics of explosions and methodologies can be found in CCPS guidelines [31], as their description and limits

are out of the scope of this work. For any blast propagation model, the main issue is the knowledge of initial peak overpressure, which depends not only on the available energy but also on the geometry (level of congestion and confinement).
Assuming conservatively a fast-deflagration behavior of EO and stoichiometric concentration (i.e. regardless of non-homogeneous concentration profile), by using the energy-scaled plot of MEM, safety distances can be assessed by using the threshold pressure for the lethality of human and for strong damage to buildings, typically 0.30 bar, and for the structural integrity of equipment (0.16 bar, in the case of low-strength atmospheric equipment, see [32] for further details). Results are reported in Table 6, for the process conditions reported in Table 1.

Table 6: Safety distances x_{exp} for EO vapour explosion at the end of process (total mass = 4000 kg). Overpressure $P_{th,h}$ (=0.30 bar) and $P_{th,e}$ (=0.16 bar) are the threshold values respectively for human lethality or building collapse and structural integrity of atmospheric equipment (domino effects).

	$P_{EO} = 1$ bar		$P_{EO} = 2$ bar		$P_{EO} = 3$ bar	
T [K]	x_{exp} ($P_{th,e}$) (m)	x_{exp} ($P_{th,h}$) (m)	x_{exp} ($P_{th,e}$) (m)	x_{exp} ($P_{th,h}$) (m)	x_{exp} ($P_{th,e}$) (m)	x_{exp} ($P_{th,h}$) (m)
398	66	13	95	16	107	18
423	64	11	81	14	91	15
453	51	9	68	11	77	13

These values should be evaluated starting from the center of the pool. However, even if distances are considered from the pool border, it should be noted that EO explosions are dangerous for human only in the close surrounding of pool, where the toxic effects, heat radiation, pool temperature effects are surely superimposed. Regarding equipment integrity, larger distances are still found. However, structural integrity only affects economic losses. Besides, domino effects (i.e. the escalation of secondary scenario) need higher pressure and many details on the surrounding equipment [32]. For instance, a value of 0.22 bar can be assumed as threshold value for atmospheric storage tank. This pressure is found at distances of about 50 m from the release point of EO, the pool center, in the worst case condition reported in Table 6.

CONCLUSIONS

Results of consequence analysis have show that dramatic accidental scenarios should be considered only if large leakages (section > 2" diameter) occur on recycling loop/pipelines, in particular for very low wind velocity (i.e. for heavy gas behaviour). It has however to be assumed that the automatic protection systems and mitigation devices (extinguishing systems, foaming agent, water deluge, either on source or on target equipment) are very likely installed in high-risk chemical process as those here analyzed. Finally, explosion can be sufficiently energetic and give high overpressure only in the case of near-instantaneous release of reactor content (large release section).

References

[1] van Dierendonck L.L., Zahradnik J., and Linek V. (1998), Loop Venturi Reactors. A Feasible Alternative to Stirred Tank Reactors?, Ind. Eng. Chem. Res., 37, 734-738.

[2] Di Serio M., Tesser R., and Santacesaria E. (2005), Comparison of Different Reactor Types Used in the Manufacture of Ethoxylated, Propoxylated Products, Industrial & Engineering Chemistry Research, Ind. Eng. Chem. Res., 44, 9482 - 9489.

[3] Gustin J.L. (2000), Safety of ethoxylation reactions, Hazards XV—the. Process, its Safety and the Environment, Symposium Series no. 147. (IChemE, Rugby).Lees F.P., Loss Prevention in the process industries, II Ed., Butterworth-Heinemann, Oxford (UK), 1996.

[4] June R.K., and Dye R.F. (1990), Explosive decomposition of ethylene oxide, Plant/Operations Progress, 9, 67-74.

[5] Hostache G., and Ebner W. (1991), Spontaneous decomposition of ethylene oxide. Influence of inerting and vessel size on the final pressure, energies of ignition, Chem. Ing. Tech. 63, 1262-1263.

[6] Thibault P., Britton L.G., and Zhang F. (2000) Deflagration and detonation of ethylene oxide vapor in pipelines, Process Safety Prog. 19, 125-139.

[7] Pekalski A.A., Zevenbergen J.F., Braithwaite M., Lemkowitz S.M., and Pasman H.J. (2005), Explosive decomposition of ethylene oxide at elevated condition: effect of ignition energy, nitrogen dilution, and turbulence. J. Hazard. Mater., 118, 19–34.

[8] Marshall J., Mundt A., Hult M., McKealvy T., Meyers P., and Sawyer J. (1995), The relative risk of pressurized and refrigerated storage for six chemicals, Process Safety Progress, 14, 3, 200–211.

[9] Hovey D.J. and Framer E.J. (1993), Pipeline accident, failure probability determined from historical data', Oil and Gas Journal, 104-107.

[10] Health and Safety Executive (1981), Canvey: a Second Report. A Review of the Potential Hazards from Operations in the Canvey Island/Thurrock Area Three Years after Publication of the Canvey Report (London: HM Stationery Office)

[11] Di Serio M., Tesser R., Felippone F., and Santacesaria E. (1995), Ethylene Oxide Solubility and Ethoxylation Kinetics in the Synthesis of Nonionic Surfactants, Ind. Eng. Chem. Res., 34, 4092-4098.

[12] Kawamura P. I., and Mackay D. (1987), The evaporation of volatile liquids, J. Haz. Mat., 15, 343-364.

[13] Reynolds R.M. (1992), ALOHA™ (Areal Locations of Hazardous Atmospheres) 5.0 theoretical description. NOAA Tech. Memo. NOS ORCA-65. National Oceanic and Atmospheric Administration/Hazardous Materials Response and Assessment Division. Seattle, WA.

[14] Rohsenow W.M. (1961), Heat, Mass, and Momentum transfer, Prentice-Hall, New Jersey.
[15] Fuller N., Schettler P.D., and Giddings J.C. (1966), A new method for prediction of binary gasphase diffusion coefficients, Ind. Eng. Chemistry, 58, 5, 18-27.
[16] Skelland A.H.P. (1974), Diffusional mass transfer, Wiley.
[17] DNV (2006), Det Norske Veritas, PHAST code, vers. 6.5.
[18] Sigma-Aldrich (2006), Material Safety Data Sheet.
[19] Cozzani V., Gubinelli G., and Salzano E. (2005), Criteria for the escalation of fires and explosions, American Institute of Chemical Engineers, Spring National Meeting, Process Plant Safety Symposium, Atlanta.
[20] TNO (1992), Methods for the Determination of Possible Damage to People and Objects Resulting from Releases of Hazardous Materials, Director-General for Social Affairs and Employment, The Hague, The Netherlands.
[21] EPA, General guidance on risk management programs for chemical accident prevention, 40 CFR PART 68, RMP SERIES 2004.
[22] Svensson CH. (1979), Enthalpies of vaporization of 1-decanol and 1-dodecanol and their influence on the CH2-increment for the enthalpies of formation, J. Chem. Thermodyn., 11, 593.
[23] Woodward J.L. (1998), Estimating the flammable mass from vapour cloud, Center for Chemical Process Safety, AIChE, New York.
[24] Spicer T., and Havens, J. (1996). Application of dispersion models to flammable cloud analyses, Journal of Hazardous Materials, 49, 115-124.
[25] Havens, J. A., and Spicer T. O. (1985), Development of an atmospheric dispersion model for heavier-than-air gas mixtures. Report No. CG-D-22-85. Fayetteville, Arkansas: University of Arkansas Chemical Engineering Department.
[26] Hanna S.R., Drivas P.J., and Chang J.J. (1996), Guidelines for Use of Vapor Cloud Dispersion Models, 2nd Edition, CCPS, Center for Chemical Process Safety, American Institute of Chemical Engineering, New York.
[27] Press W. H., Flannery B.P., Teulolsky S. A., and Vetterling W.T. (1986), Numerical Recipes, Cambridge Univ. Press, Cambridge, U.K.
[28] Van Buijtenen C.J.P. (1980), Calculation of the amount of gas in the explosive region of a vapor cloud released in the atmosphere, J. Hazard. Mater., 3, 201–220.
[29] Britter R. E., and McQuaid J. (1988), Workbook on the Dispersion of Dense Gasses, HSE Contract Research Report No. 17/1988, Health & Safety Executive, Sheffield, UK.
[30] Lees F.P. (1996), Loss Prevention in the Process Industries: Hazard Identification, Assessment and Control, 2 Ed., Butterworth-Heinemann.
[31] CCPS (1994), Guidelines for evaluating Vapour Cloud Explosion, Flash Fires and BLEVE, Center for Chemical Process Safety, American Institute of Chemical Engineering, NY.
[32] Cozzani V., and Salzano E.(2004), Threshold values for domino effects caused by blast wave interaction with process equipment, J. Loss Prev. Proc. Ind., 17, 437-447.

AUTHOR INDEX

AUTHOR INDEX